Image – Action – Space

IMAGE – ACTION – SPACE
SITUATING THE SCREEN IN VISUAL PRACTICE

Luisa Feiersinger, Kathrin Friedrich, Moritz Queisner (Eds.)

DE GRUYTER

This publication was made possible by the *Image Knowledge Gestaltung. An Interdisciplinary Laboratory* Cluster of Excellence at the Humboldt-Universität zu Berlin (EXC 1027/1) with financial support from the German Research Foundation as part of the Excellence Initiative.

Gefördert durch die

The editors like to thank Sarah Scheidmantel, Paul Schulmeister, Lisa Weber as well as Jacob Watson, Roisin Cronin and Stefan Ernsting (Translabor GbR) for their help in editing and proofreading the texts.

ISBN 978-3-11-046366-8
e-ISBN (PDF) 978-3-11-046497-9
e-ISBN (EPUB) 978-3-11-046377-4

Library of Congress Control Number: 2018956404

Bibliographic information published by the Deutsche Nationalbibliothek
The Deutsche Nationalbibliothek lists this publication in the Deutsche Nationalbibliographie; detailed bibliographic data are available on the internet at http://dnb.dnb.de.

The book is published with open access at www.degruyter.com, https://www.doabooks.org and https://www.oapen.org.

Cover illustration: Malte Euler
Typesetting and design: Andreas Eberlein, aromaBerlin
Printing and binding: Beltz Bad Langensalza GmbH, Bad Langensalza
Printed in Germany

www.degruyter.com

Inhalt

7 **Editorial**
Image – Action – Space. Situating the Screen in Visual Practice

11 **Aud Sissel Hoel**
Operative Images. Inroads to a New Paradigm of Media Theory

29 **Erkki Huhtamo**
The Spell of the Catoptric Television. Media Archaeology, Topos Study, and the Traces of Attention

41 **Tristan Thielmann**
Early Digital Images. A Praxeology of the Display

55 **Jan Distelmeyer**
Carrying Computerization. Interfaces, Operations, Depresentations

69 **Luisa Feiersinger**
Spatial Narration. Film Scenography Using Stereoscopic Technology

79 **Lasse Scherffig**
From Action Capture to Interaction Gestalt

93 **Kathrin Friedrich**
Screening Bodies. Radiological Screens and Diagnostic Operations

103 **Luci Eldridge**
Working on Mars. An Immersive Encounter through the Screen

115 **Nina Franz and Moritz Queisner**
The Actors Are Leaving the Control Station. The Crisis of Cooperation in Image-guided Drone Warfare

133 **Shannon Benna**
Systems and Practices to Produce Stereoscopic Space on Screen

147 **Jens Schröter**
Viewing Zone. The Volumetric Image, Spatial Knowledge and Collaborative Practice

159 **Carolin Höfler**
Sense of Being Here. Feedback Spaces Between Vision and Haptics

177 **Timo Kaerlein and Christian Köhler**
Around a Table, around the World. Facebook Spaces, Hybrid Image Space and Virtual Surrealism

191 **Sarah Atkinson**
Synchronic Simulacinematics. The Live Performance of Film Production

203 **Ina Neddermeyer**
I Want to See How You See. Curatorial Practices of Exhibiting Virtual Reality

Editorial

Image – Action – Space

Situating the Screen in Visual Practice

With the improved capability of imaging, sensor and display technology, screens have become mobile or touchable and, most recently, transparent. While a visualization on-screen is not necessarily related to the spatial context beyond the screen, transparent displays allow users to see simultaneously the physical space behind the display and a visualization on-screen. The two observers on the cover of this volume look at a future construction site through the transparent display of a head-mounted device that superimposes an architectural model on their view (fig. 1). Its rendering coincides with the scale of the actual building, and the visualization adapts to the observer's point-of-view in real time. While a juxtaposition of building and model on a separate screen would require continuous comparison between image and object, the head-mounted display combines them in a joint perceptual space. The missing offset between image and object puts forward a new practice of interacting with spatially related information: users can navigate through space by superimposing a transparent mobile interface onto their field of view.

This example shows how screen-based media transform the way we see and act: transparent displays constitute a form of images that only *work* when they are situated. They shift focus onto the *situation* rather than to the result of an imaging process. Of course, screens are always embedded in the context of a situation, particularly those visual practices that require the linking between screen-based visualization and physical space. A smartphone mapping app that indicates one's current location and orientation, for instance, requires that users situate themselves in space based on a two dimensional map. During a surgical intervention, to give another example, surgeons must cope with the limiting architecture of image display in the operating room, in which information on screen may not align with their perspective on the patient's body or with the scale and orientation of relevant anatomy. Accordingly, surgeons must ascribe an image of the patient, for instance a computer tomography, to the patient's body cognitively in order to *act* appropriately in a particular situation. Even a regular television screen on a living room shelf creates a specific viewing situation. But this situation is not adaptive to the images on-screen – it does not correlate image and space.

An example for the way in which screen-based visual practices dissolve the distinction between image and space, creating a hybrid and adaptive form of visuality, is the location-based augmented reality game *Pokémon Go*. The application encourages users to explore physical space in order to catch virtual figures displayed within the camera stream of a mobile phone (fig. 2). By aligning camera image and physical space, *Pokémon Go* players perform operations both within and beyond the boundaries of the screen. What seems to be a simple moment in a game is actually an intricately structured visual practice: The in-game view layers photographic and animated elements depending on the player's location and within the camera's field of view. The mechanism of merging image, action and space in *Pokémon Go* transforms *viewing* into *using* and emphasizes an active role of the image in guiding a user's action and perception.

The visual practice of *Pokémon Go* points to the methodological issue of how to analyze and theoretically frame the situation and situatedness of screens. By shifting the focus towards their "screenness", we intend to examine visual practices by asking what a screen *does* rather than asking what a screen *is*.[1] From this perspective, the question is less about what becomes visible, or what can be seen, but rather about how the interaction with and through screens structures action and perception. The terms *image, action* and space serve as analytical reference points for investigating how screens engender a situated and dynamic relation between them.

This volume draws on the evolving debate about the screen as "a concept in progress", which has started to inquire its defining status.[2] While the screen "has become convenient catch-all used to describe the research and study of what we access through screens, perpetuating the idea of the screen as passive conduit", is only until recently that screen studies have started to investigate the impact and application of screens in particular situations and with regard to its actionability and material affordances.[3] Concepts from media theory and visual studies, such as dispositif, spatial images, operative images or mobile screens help to scrutinize *screenness*.[4]

The authors analyze how screens are situated in visual practices by scrutinizing the dynamic, transformational and performative characteristics of screen-based media with regard to image, action and space. By focusing on their dynamic yet deterministic relations the volume presents an approach to *screenness* that focuses on the *actionability* of screen-based media in all their different hardware and software configurations. Speaking of actionability emphasizes the way in which multiple forms and configurations

1 Lucy Suchman proposes looking at the socio-material practices surrounding the application of screens by using the term "screenness". Brit Ross Winthereik, Peter A. Lutz, Lucy Suchman, Helen Verran, Attending to Screens and Screenness. Guest Editorial for special issue of Encounters, in: *STS Encounters* 4.2 (2011), pp. 1–6. In a different vein, Lucas Introna and Fernando Ilharco develop the notion of screenness for their endeavour of introducing a "Heideggerian phenomenological analysis of screens". Lucas D. Introna, Fernando M. Ilharco, On the Meaning of Screens. Towards a Phenomenological Account of Screenness, in: *Human Studies* 29 (January 2006), pp. 57–76, DOI: 10.1007/s10746-005-9009-y.

2 Dominique Chateau, José Moure, Introduction. Screen, a Concept in Progress, in: Dominique Chateau, José Moure (eds.), *Screens. From Materiality to Spectatorship – A Historical and Theoretical Reassessment*, Amsterdam: Amsterdam University Press, 2016, pp. 13–22. See further Stephen Monteiro (ed.), *The Screen Media Reader. Culture, Theory, Practice*, London: Bloomsbury, 2017; Workshop "Touching the Screen", University of Oslo, 2015; PhD course "Framing Screens: Knowledge, Interaction and Practice", IT University Copenhagen, 2010; Workshop "Screen Operations. Conditions of Screen-based Interaction", Humboldt University Berlin, 2016; Conference "Situation Space. How Spatial Images Define the User's Disposition", Humboldt University Berlin, 2017.

3 Monteiro 2017 (as in fn. 2), p. 3.

4 Giuliana Bruno, *Atlas of Emotion. Journeys in Art, Architecture and Film*, New York: Vers, 2002; Frank Kessler, Dominique Chateau, José Moure, The Screen and the Concept of Dispositif – A Dialogue, in: Dominique Chateau, José Moure (eds.), *Screens. From Materiality to Spectatorship – A Historical and Theoretical Reassessment*, Amsterdam: Amsterdam University Press, 2016, pp. 264–271; Miriam Ross, Stereoscopic visuality. Where is the screen, where is the film?, in: *Convergence. The International Journal of Research into New Media Technologies* 19.4 (2013), pp. 406–414; Jens Schröter, *3D. History, Theory and Aesthetics of the Transplane Image*, London: Bloomsbury, 2014; Aud Sissel Hoel in this volume, pp. 11–27; Erkki Huhtamo, Elements of Screenology. Toward an Archaeology of the Screen, in: *ICONICS. International Studies of the Modern Image* 7 (2004), pp. 31–82; Nanna Verhoeff, *Mobile Screens. The Visual Regime of Navigation*, Amsterdam: Amsterdam University Press, 2012; Ramón Reichert, Annika Richterich, Pablo Abend, Mathias Fuchs, Karin Wenz (eds.), *Mobile Digital Practices, Digital Culture & Society (DCS)* 3.2 (2017).

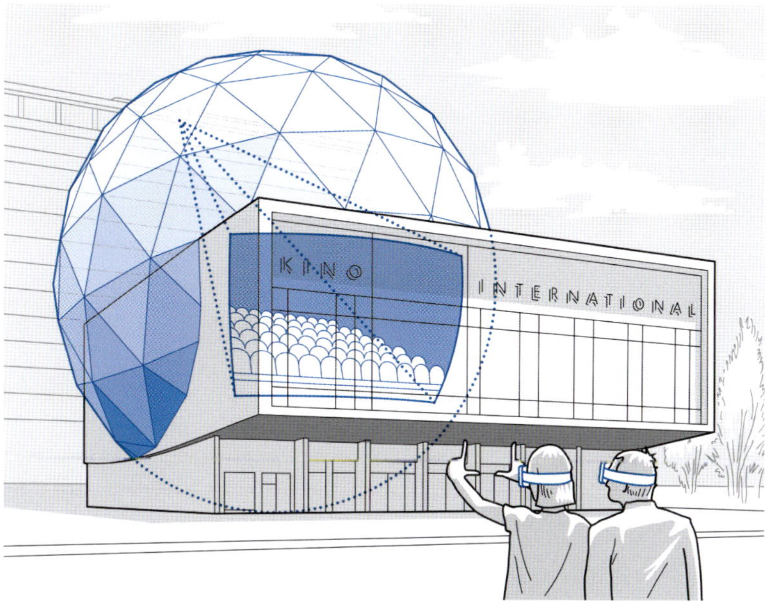

1 Mixed reality in architectural planning superimposes building and model.

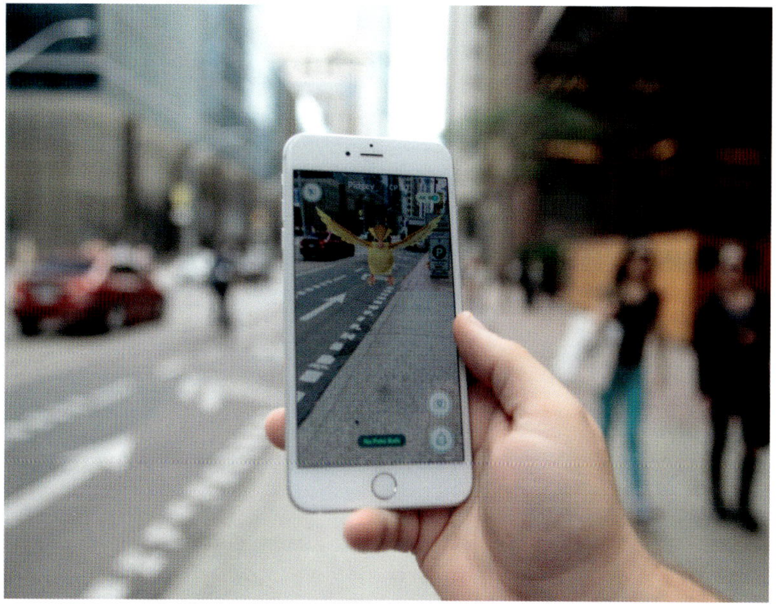

2 Augmented reality game merging image and place according to user sight-line.

of screens challenge users to integrate imaging techniques and visual information into their action routines, viewing habits and working processes.

Foci of investigation are: first, the analysis of codes, data and software that form the technological basis for both acquiring and displaying visual information that already define and structure action as well as decision-making. Second, the exploration of interface design that constitutes both conceptual and epistemic considerations that render interfaces and their affordances as a screen-based space of its own. And third, the perceptual level and the investigation of how the screen intertwines human senses, cognitive capacities and physical actions. These analytical perspectives on screens and screenness stem from the idea of a pragmatic and theoretical triangulation of image, action and space, which stresses the adaptive and situated alignment of visualization, operationality and spatiality.

Figures

1 Malte Euler, Creative Commons, Attribution-NonCommercial-NoDerivatives 4.0 International (CC BY-NC-ND 4.0), 2018.

2 Picture alliance, REUTERS, Chris Helgren, 2016.

Aud Sissel Hoel

Operative Images

Inroads to a New Paradigm of Media Theory

There is much talk these days about images being somehow *operative*. This notion, which foregrounds the active doing of images, is often invoked to make sense of the disruption in the image economy brought about by computerization. In today's digital media environments, human dealings with the world increasingly take place via various kinds of images and screens that do more than just display visual information. As cameras become ubiquitous, images networked, image data geotagged and databases navigable in real time, the status of images seems to be rapidly changing. Among scholars of the image, there is a growing realization of the shortcomings of existing theories and concepts when it comes to explicating key features of today's digital image applications. The current focus on the operational aspects of images, therefore, is frequently accompanied by a call for conceptual revisions.

This article contributes to the ongoing attempts to develop an operational basis for understanding images. To this end, it considers a selection of contemporary approaches that, each in their own way, grant centrality to the operational aspects of images. In the literature under consideration, there is a great deal of focus on machine vision and automation, as well as on the roles of new media in warfare and political conflict. These topics, of course, are not at all new. They were also at the forefront of 1980s media theory, with Jean Baudrillard and Paul Virilio as notable figures. While these thinkers continue to be influential, the central argument of this article is that there is something new about how today's scholars of operative images approach the topic of mediation. What is new is that there seems to be a shift in underlying assumptions about the nature and roles of media. The thinkers considered in this text, therefore, are treated as transitional figures standing on the verge of a new paradigm of media theory. The emerging, operational paradigm of media theory is characterized by its deeper recognition of the active dimension of images and media.

The new line of research into the agency and efficacy of images is highly promising, breaking new ground by putting image theory on an altogether new track. More work needs to be done, however, when it comes to articulating what is meant by the term *operation* in this context. Addressing this need, the article probes the literature on operative images, discussing and comparing different approaches to operative images along four lines: from the perspective of art (section 1), from the perspective of new media production and use (section 2), from the perspective of media archaeology aspiring to become exact science (section 3), and from the perspective of visual studies (section 4).[1] In

1 These four approaches are certainly not exhaustive of how operative images are conceived in the current literature, but they suffice to unearth systematic differences in how the notion of operation is currently understood.

all these sections, I seek to lay bare how the emphasis on the operational aspects of images puts pressure on established notions of images.

The rough overview of the literature undertaken in sections 1–4 shows that there is a tension regarding the boundaries and scope of operativity. While some approaches conceive operative images as a new kind of images that supplements the larger category of traditional images, other approaches aim for a deeper revision that challenges the very idea of what an image is. The overview also shows that the notion of operation is under-theorized as a media-theoretical concept, since in many cases it is simply imported from other research fields, such as computer science.

One note before I proceed: The reader may have noticed that the question relating to the active doing of images is addressed here in the wider context of media theory. While this may cause some initial confusion, it is certainly no coincidence. As we shall see, the slippage into media theory happens continually in the literature on operative images, and it happens for a reason – indeed, as a consequence of the operational approach: If we are to follow through with the ideas suggested by the thinkers considered in this article – that images are instruments, interfaces, measuring media, manipulable diagrams – the boundaries between *image* and *medium* start to become porous, leaving both terms transformed.

Representation versus Operation

Harun Farocki's three-part installation *Eye/Machine* (2001–2003) is a key reference point in the literature on operative images. The installation, which explores the relation between humans, machines and modern warfare, announces the advent of a new visual regime, and simultaneously of a new stage in the history of machine vision where the machines have started to *see* for themselves. The catalyzing event for the *Eye/Machine* trilogy was the outrage and sensation of the 1990–1991 Gulf War, where point-of-view footage from laser-guided bombs (popularly known as smart bombs) was widely broadcasted to TV audiences. The military deployment of *eye machines* prepared the way for a new type of warfare – a "war at a distance"[2] facilitated by a new kind of images that Farocki terms *operative images* (*operative Bilder*).

Farocki's work addresses the changing status of images in the context of intelligent machines. Commenting on Farocki's work, Trevor Paglen notes: "Instead of simply representing things in the world, the machines and their images were starting to 'do' things in the world".[3] Volker Pantenburg adds that Farocki was "one of the first to examine in depth the various uses of images as instruments".[4] Both aspects, the interventional and the instrumental, are reflected in Farocki's much-cited definition, which holds that operative images "do not represent an object, but rather are part of an operation".[5] Thus conceived, operative images are utility images – working images that typically serve practical purposes tied to specialized tasks, such as, in this case, guiding remote-controlled missiles. Similarly referring to Farocki's work, Thomas Elsaesser goes further by characterizing operative images as "instructions for action" – and not only that, in the digital media environment, Elsaesser

2 Which is also the title of the English single-track film based on the *Eye/Machine* installation. Harun Farocki, *War at a Distance*, Germany 2003.

3 Trevor Paglen, Operational Images, in: *e-flux* 59 (2014), http://e-flux.com/journal/59/61130/operational-images/ (accessed May 27, 2018).

4 Volker Pantenburg, Working Images. Harun Farocki and the Operational Image, in: J. Eder, C. Klonk (eds.), *Image Operations. Visual Media and Political Conflict*, Manchester: Manchester University Press, 2017, p. 49.

5 Harun Farocki, Phantom Images, in: *Public* 29 (2004), p. 17.

maintains, the instructive function seems to have become "the new default value of all image-making".[6]

The example of the smart bomb accentuates another feature of operative images. In the words of Farocki, they provide phantom perspectives on things. The term *phantom* here alludes to the use of phantom shots in early cinema, that is, of recordings taken from positions not normally occupied by humans (Farocki gives the example of a camera hung under a train).[7] The phantom perspective relates, in other words, to the capacity of machine-made images to leap beyond the human scale, reporting on events outside the scope of human sensibility. In this respect, Farocki's *Eye/Machine* series resonates with the exploration of machine vision in art works such as *Man with a Movie Camera* (1929) by the modernist, avant-garde filmmaker Dziga Vertov.[8] Nevertheless, in the current context of intelligent machines, the leap beyond the human scale seems to be of a more radical nature. Hal Foster puts it thus:

> *[The images treated by Farocki] are not authored, and, as they mostly survey the predetermined, they appear to be more automatically monitored than humanly viewed. In this way Farocki intimates that a new 'robo eye' is in place, one that, unlike the 'kino eye' celebrated by modernists like Dziga Vertov, does not extend the human prosthetically so much as it replaces the human robotically.*[9]

Farocki himself also alludes to the idea of replacement, characterizing today's picture-processing apparatuses as "sensory automatons" destined to replace and outperform the work of the human eye.[10] The main novelty of operative images, then, seems to be that they, in the words of Martin Blumenthal-Barby, "require neither human creators nor human spectators".[11] What sets operative images apart from other images is that they are "not originally intended to be seen by humans"; instead they are "supposed to function as an interface in the context of algorithmically controlled guidance processes".[12]

The last remark, that operative images function as interfaces, is a key observation to which I will return. For now, I will focus on the "posthuman"[13] aspects of operative images, which have lead scholars to question their very status as images. Pantenburg, for example, comments:

> *[T]he operational image emulates the look and feel of traditional images, but on closer inspection, this turns out to be a secondary function, almost a gesture of courtesy extended by the machines: The computer does not need the image.*[14]

6 Thomas Elsaesser, Alexander Alberro, Farocki: A Frame for the No Longer Visible. Thomas Elsaesser in Conversation with Alexander Alberro, in: *e-flux* 59 (2014), http://e-flux.com/journal/59/61111/farocki-a-frame-for-the-no-longer-visible-thomas-elsaesser-in-conversation-with-alexander-alberro/ (accessed May 27, 2018).

7 Farocki 2004 (as fn. 5), p. 13, p. 20.

8 The continuity between these works has been explored in the literature. As pointed out by Volker Pantenburg, the connection is made explicit by Farocki in his installation *Counter-Music* (2004). Pantenburg 2017 (as fn. 4), p. 59, fn. 3; For a detailed exploration of the connection, see David Tomas, *Vertov, Snow, Farocki. Machine Vision and the Posthuman*, New York: Bloomsbury Academic, 2013.

9 Hal Foster, The Cinema of Harun Farocki, in: *Artforum* (November 2004), p. 160.

10 Farocki 2004 (as fn. 5), p. 17.

11 Martin Blumenthal-Barby, 'Cinematography of Devices'. Harun Farocki's *Eye/Machine* Trilogy, in: *German Studies Review* 38.2 (May 2015), p. 329.

12 Ibid.

13 The term *posthuman* is sometimes invoked in the discussion of machine vision. See for example Tomas 2013 (as fn. 8).

14 Pantenburg 2017 (as fn. 4), p. 49.

In the strictest sense, therefore, operative images "would have to be characterized as visualisations of data that could also take on other, different guises".[15] Fortunately, Pantenburg does not leave it at that. He goes on to call attention to how Farocki's work is deeply influenced by the philosopher Vilém Flusser and his ideas about technical images. According to Flusser, "technical images" (such as photographs and television images) differ from "traditional images" (Flusser gives the example of cave painting) in that they "owe their existence to technical apparatuses".[16] Consequently, technical images and traditional images *mean* in completely different ways: While technical images are "computations of concepts" that arise "through a peculiar hallucinatory power that has lost its faith in rules"; traditional images are "observations of objects" that arise through "depiction".[17]

While Flusser's category of technical images comprises pre-digital images such as photographs and television images, contemporary scholars typically draw the line in a different place. William Uricchio, for example, in his attempt to conceptualize the distinguishing features of digital images, emphasizes the "algorithmic construction of the image", which is understood to disrupt "the long regime of three-point perspective".[18] In applications such as Microsoft Photosynth and augmented reality systems, the interventions of algorithms between the viewing subject and the object viewed introduce "cracks in the façade of the subject-object relationship characteristic of the modern era".[19] A similar idea is exposed by Ingrid Hoelzl and Rémi Marie, who conceptualize the digital transformation of the image in terms of a shift from geometry to algorithm, and from projection to processing. Because of this shift, the image is "no longer a passive and fixed representational form, but is active and multiplatform, endowed with a signaletic temporality that is not only the result of digital screening (or compression), but also a transfer across digital networks".[20] This implies that the image is "no longer a stable representation of the world, but a programmable view of a database that is updated in real-time", and hence, that it "no longer functions as a (political and iconic) representation, but plays a vital role in synchronic data-to-data relationships".[21]

The answer to Pantenburg's question, whether operative images are images at all, depends, of course, on how one chooses to define the term *image*. Thanks to the frequent use of a contrasting rhetoric by scholars of the image, we get a rough sense of what the default notion of images might be: depictions based on an observation of objects, passive and fixed representations based on stable subject-object relationships. Farocki, too, provides clues to such a default notion of images through his numerous negative definitions of operative images. In the intertitles of the *Eye/Machine* series, for example, we learn that operative images are devoid of social intent, that they are not meant for edification, and nor for contemplation. We learn further that operative images are not really intended for human eyes, and that they exceed the human scale. Elsewhere, he adds that operative images are made "neither to entertain nor to inform", building toward

15 Ibid., pp. 49–50.

16 Vilém Flusser, *Into the Universe of Technical Images*, translated by Nancy Ann Roth, introduction by Mark Poster, Minneapolis/London: University of Minnesota Press, 2011, p. 7.

17 Ibid., p. 10.

18 William Uricchio, The Algorithmic Turn. Photosynth, Augmented Reality and the Changing Implications of the Image, in: *Visual Studies* 26.1 (March 2011), p. 26.

19 Ibid., pp. 25–26.

20 Ingrid Hoelzl, Rémi Marie, *Softimage. Towards a New Theory of the Digital Image*, Bristol, UK: Intellect, 2015, p. 3.

21 Ibid., p. 4.

what seems to be the main negative characteristic given in his most-cited definition: operative images "do not represent an object".[22] A traditional image, then, seems to be an image that represent an object in a way that, somehow, conforms to the human scale.

The characteristics of operative images that have been unearthed so far are already starting to show signs of inconsistency. Operative images are utility images, and as such they belong to a wider family of instruments and tools, which are constructed by humans to serve practical human purposes. Yet, in the literature under consideration, image-instruments are typically identified with intelligent machines and automated systems, and as such they are conceived as images that no longer cater to human eyes, that disrupt the human scale and that roam about freely as if in defiance of petty, human intentions. Humans constructed them, but we no longer know whose purposes they serve. Even if he sometimes alludes to the imminent replacement of humans by machines, Farocki himself, however, is reluctant to take the human completely out of the loop. The ambiguity is marked already in the title of Farocki's installation, namely by the slash separating *Eye* and *Machine*. As noted by Foster, the slash raises the question of relation: "Does the slash signify a split between eye and machine [...] or a new elision of the two, or somehow both – a split that has produced an elision?"[23] As we shall see, the ambiguity prevails in the literature on operative images.

22 Farocki 2004 (as fn. 5), p. 17.
23 Foster 2004 (as fn. 9), p. 160.

Cultural Operations

The next approach to be considered, that of Lev Manovich, differs from Farocki's in that it no longer revolves around the idea of automation. The relevance of Manovich's work is confirmed by Werner Kogge, who, in an article discussing Manovich's *The Language of New Media* (2001), proposes operative images as a "paradigm of new media".[24] While Manovich himself does not use the term *operative image*, Kogge's suggestion is not unwarranted. According to Manovich, new media "calls for a new stage in media theory": If we want to understand the logic of new media, we need to turn to computer science, borrowing terms such as "interface", "database" and "operation".[25]

As Kogge points out, Manovich's approach is refreshing in that it avoids sweeping generalizations about media, which seemed to be the trademark of 1980s media theory, exemplified, say, by Baudrillard's ruminations about total simulation.[26] Manovich also avoids overemphasizing the newness of new media, sketching "archaeologies" that connect computer screens with classical screens, or computer-based techniques of media creation with previous techniques of representation and simulation.[27] Still, as Manovich makes clear, there are aspects of new media that lack historical precedents. The newness of new media relates to their "programmability",[28] which results from the merging into one of two separate historical trajectories, that of modern

24 Werner Kogge, Lev Manovich. Society of the Screen, in: A. Lagaay, D. Lauer (eds.), *Medientheorien. Eine philosophische Einführung*, Frankfurt/New York: Campus Verlag, 2004, pp. 297–315, p. 302.
25 Lev Manovich, *The Language of New Media*. Cambridge, MA: The MIT Press, 2001, pp. 11–12, p. 48.
26 Kogge 2004 (as fn. 24), p. 303.
27 Manovich 2001 (as fn. 25), pp. 95–103, pp. 145–160.
28 Ibid., p. 47.

media and that of the computer. This meeting changes the identity of both, giving rise to a "universal media machine".[29]

The programmability of new media relates to how all new media objects are numerical representations.[30] Composed of digital code, all new media objects can be described formally in mathematical terms, making them susceptible to algorithmic manipulations. This means that if the contents of old media are to be stored, displayed or distributed via computers, they must be converted into numerical representations through a process of digitization. This requirement relates to what, in Manovich's view, is the most consequential effect of computerization: the transformation of media into computer data.[31] As a result, new media objects in general can be said to consist of two distinct layers: a "cultural layer" whose structural organization "makes sense to its human users" and a "computer layer" whose structural organization instead "follows the established conventions of the computer's organization of data".[32] He gives the example of a digital image, which on one level is a "representation"[33] that "belongs on the side of human culture", and which on another level, is a "computer file" that belongs, rather, to the "computer's own cosmogony".[34] Since today, media are for the most part created and accessed via computers, we can expect the computer to influence the traditional cultural logic of media by imposing its own distinct computer logic. Importantly, however, as Manovich sees it, this influence is not a one-way street. Just like traditional artists before them, new media designers and users perceive the world and approach media through various cultural filters and representational schemes.[35] Moreover, like all media, the computer works by "remediating"[36] older media.[37] Beyond that, Manovich reminds us that the computer level is not fixed and finished once and for all but continues to evolve as the computer is set to perform new tasks. The influence between the levels, therefore, goes both ways, which means that the "new computer culture" is best conceived as a "blend of human and computer meanings, of traditional ways in which human culture modeled the world and the computer's own means of representing it".[38]

Manovich's idea about the two layers may seem reminiscent of the split between the human and the machinic as discussed in the previous section. Still, it is interesting to note that, in Manovich's case, the computer is not really outside the human as such. When he talks about the cultural layer, the term *culture* is taken in a narrow sense, reflecting his focus on "cultural software" – software that supports cultural actions such as "creating cultural artifacts and interactive services which contains representations, ideas, beliefs and aesthetic values".[39] Furthermore, when he says that a digital image on the level of representation "belongs on the side of human culture", he means that it belongs to the historical trajectory of visual representation with its characteristic cultural forms, languages and conventions.[40] Thus, when the cultural layer is contrasted with the computer layer, the implication is not that the latter exists beyond

29 Ibid., p. 4, pp. 25–26, p. 69.
30 Ibid., p. 27.
31 Ibid., p. 45.
32 Ibid., p. 45.
33 It is a representation in the sense of featuring recognizable objects. Ibid., p. 45.
34 Ibid., pp. 45–46.
35 Ibid., pp. 117–118.
36 Jay D. Bolter, Richard Grusin, *Remediation. Understanding New Media*, Cambridge, MA: The MIT Press, 1999.
37 Manovich 2001 (as fn. 25), p. 89.
38 Ibid., p. 46.
39 For a longer list, see Lev Manovich, *Software Takes Command*, New York: Bloomsbury, 2013, p. 23.
40 Manovich 2001 (as fn. 25), p. 45.

human culture in a wider sense. The point is, rather, that the computer layer belongs to a separate historical trajectory with its own distinct conventions – which is why the "language of cultural interfaces", as we encounter it on most of today's computer screens, is often an "awkward mix between the conventions of traditional cultural forms and the conventions of HCI – between an immersive environment and a set of controls".[41]

Manovich's refusal to draw a sharp line between the two layers – the human and the machinic – also informs his approach to the notion of operation. While he acknowledges that operations behind computer programs can be automated, and hence that "human intentionality can be removed from the creative process, at least in part",[42] he refrains from identifying the notion of operation with the machinic. Instead, "operations" are defined more widely as "typical techniques of working with computer media".[43] As Manovich sees it, in the computer age, typical operations such as copy, cut, paste, search and filter are also used outside the computer, as "general cognitive strategies" employed in the culture at large.[44] Operations, in other words, are conceived as "technologically-based cultural practices" that, despite being embedded in software, are not tied to it.[45] Thus, when he sets out to analyze operations, Manovich focuses on general techniques (or "commands") that are common to many different software programs, such as selection, which relates to how in computer culture authentic creation tends to be replaced by selection from predefined menus, and compositing, which relates to the fitting together of heterogeneous elements into a single, seamless object.[46] The operations of selection and compositing both center on media production and use. It is only when he turns to teleaction that Manovich addresses the kind of operations that are topical in the literature on operative images. Manovich admits that teleaction is "qualitatively different" from selection and compositing in that it no longer concerns the "traditional cultural domain of representation".[47] Teleaction results from another meeting of historical trajectories, this time between media, computers and telecommunication.[48] In Manovich's view, "teleaction" is a more precise term of what is commonly referred to as "telepresence", which he defines "as one example of *representational technologies used to enable action, that is, to allow the viewer to manipulate reality through representations*".[49]

True to his habit of questioning the newness of new media, Manovich emphasizes that today's action-enabling images also have a prehistory. The common focus on "the history of visual representation in the West in terms of illusion", makes us prone to overlook the separate history of image-instruments.[50] To support his case, Manovich draws on the work of Bruno Latour,[51] who, interestingly, uses perspectival images as well as photographs as examples of image-instruments. Paraphrasing Latour, Manovich maintains that image-instruments are characterized by their "precise and reciprocal relationship between objects

41 Manovich 2001 (as fn. 25), p. 91.
42 Ibid., p. 32.
43 Ibid., p. 118.
44 Ibid., p. 118.
45 Ibid., p. 118, p. 121.
46 Ibid., pp. 123–35, pp. 136–60.
47 Ibid., p. 161.
48 Ibid., p. 162.
49 Ibid., p. 165 [original emphasis].
50 Ibid., p. 167.
51 More precisely on Bruno Latour, Visualization and Cognition. Thinking with Eyes and Hands, in: *Knowledge and Society. Studies in the Sociology of Culture Past and Present* 6 (1986), pp. 1–40.

and their signs".[52] By systematically capturing features of reality, a perspectival image, for example, is "more than just a sign system that reflects reality – it makes possible the manipulation of reality through the manipulation of signs".[53] Yet, in the history of image-instruments, the convergence with the trajectory of telecommunication makes a difference, since the electronic transmission of video images and the instantaneous construction of representations enable real-time remote control – something that provides a unique kind of power: "I can drive a toy vehicle, repair a space station, do an underwater excavation, operate on a patient, or kill – all from a distance".[54] This is why, seen from the history of action-enabling images, teleaction is a more radical technology than, say, virtual reality, because it "allows the subject to control not just the simulation but reality itself".[55]

It is worth noting that, the way Manovich defines image-instruments (as representations that systematically capture features of reality), it is not a requirement that the representations in question be produced mechanically. Leaning on Latour, Manovich seems rather to assume a continuity between perspectival images and photographs (characterizing the latter as perspectival images *par excellence*).[56] By emphasizing such a continuity, Manovich differs from thinkers like Flusser as well as from thinkers like Friedrich Kittler and Wolfgang Ernst (to be considered in the next section), for whom the introduction of technical images involves a momentous, cultural rupture. Manovich, on his side, instead of identifying the operational and the instrumental with the machinic, concentrates on the establishment of a systematic and reciprocal relation between objects and signs, which is what enables humans to use images to manipulate reality. Yet there are tensions in Manovich's approach to image-instruments. While he continues to refer to them as "representations" and "signs", his explorations of image-instruments lead to the realization that an image-instrument is "more than just a sign system that reflects reality".[57] Manovich, however, stops there and does not take the further step of considering *why* the traditional notions of representation and sign seem unable to properly account for the reciprocity between instrument and reality – not to speak of their shortcomings when it comes to elucidating the interventional and instructional aspects of instrumental mediation.

There is also a second way that Manovich's considerations about image-instruments puts pressure on the traditional notion of images. While telepresence is typically associated with live video images, Manovich shows that teleaction does not depend on video. Instead, he observes that "different kinds of teleaction require different temporal and spatial resolutions".[58] In the case of radar-images, for example, "the image is so minimal that it hardly can be called an image at all".[59] Lacking information about shape, texture and color, radar-images record nothing but the position of an object – which, however, suffices to destroy it.[60] It seems, then, that for image-instruments to perform their operational roles, the visual aspect is not really needed. If they do not reflect reality, and if they are no longer visual, in what sense are image-instruments still images? Manovich does not answer this question. Overall, his investigation of

52 Manovich 2001 (as fn. 25), p. 167.
53 Ibid., p. 168.
54 Ibid., p. 169.
55 Ibid., p. 166.
56 Ibid., p. 167.
57 Ibid., p. 168.
58 Ibid., p. 170.
59 Ibid.
60 Ibid.

image-instruments remains an excursion, the bulk of his work being geared towards new media production. Thus, while his investigation of image-instruments certainly puts pressure on received notions of images, Manovich himself never explicitly questions their status as representations. Consequently, in Manovich's account of teleaction, the *action* is conceived as human action: Teleaction is the manipulation of reality *by* humans (the viewer, the subject, the teleoperator) *through* representations. Images themselves are not considered actors.[61]

Technical Operations

While in *The Language of New Media* Manovich experiments with terms borrowed from computer science (including *interface* and *operation*), he later comes to criticize this work for its tendency to regard computer science "as a kind of absolute truth".[62] Emphasizing even more strongly than before that "computer science is itself part of culture", he now aligns himself with the emerging field of software studies that approaches software as something more than a matter of engineering: "computers and software are not just 'technology' but rather the new *medium* in which we can think and imagine differently".[63] A very different approach to the notion of operation is found in the work of Wolfgang Ernst, who seems to go in the opposite direction: Emphasizing the technical and engineering aspects of images and media, Ernst dissociates the notion of operation from the human-computer interface, seeking instead to explicate the operational processes that play out below the "surface" of software.[64]

Ernst's approach belongs to a line of research that is commonly referred to as "German media theory"[65] and that was opened by Friedrich Kittler, another notable figure of 1980s media theory. Kittler's work stands out due to its strong focus on the materiality and technicality of media. Taking inspiration from Michel Foucault's 1969 treatise *Archaeology of Knowledge*, Kittler concerns himself with epistemic ruptures in systems of knowledge, which in Kittler's view are related to media shifts. Hence, in Kittler's work, the "historical apriori" of Foucault turns into a "technical apriori".[66] As Kittler sees it, "media determine our situation"[67] by providing the material conditions under which something may become knowledge. Focusing less on discourses and more on the material substrates of media, he conceives media as inscription systems. According to Kittler, the introduction of the first technological media ("phonographs and cinematographs") marks a major epistemic rupture in that they, in contrast to previous media ("texts and scores"), were able to store time.[68] The introduction of technological media marks a rupture, more precisely, in that

61 Which they might have been, say, if Manovich had engaged more closely with Latour's work and adopted the broader notion of agency advanced by actor-network theory. Bruno Latour, *Reassembling the Social. An Introduction to Actor-Network-Theory*, Oxford: Oxford University Press, 2005.

62 Manovich 2013 (as fn. 39), p. 10.

63 Ibid., p. 13 [original emphasis].

64 Wolfgang Ernst, *Digital Memory and the Archive*, edited and with an introduction by Jussi Parikka, Minneapolis/London: University of Minnesota Press, 2013, p. 71.

65 Anthony Enns, Foreword. Media History versus Media Archaeology, in: Wolfgang Ernst, *Chronopoetics. The Temporal Being and Operativity of Technological Media*, translated with a foreword by Anthony Enns, London/New York: Rowman & Littlefield, 2016, p. xiv.

66 Bernhard Siegert, Cultural Techniques. Or the End of the Intellectual Postwar Era in German Media Theory, in: *Theory, Culture & Society* 30.6 (2013), p. 50.

67 Friedrich A. Kittler, *Gramophone, Film, Typewriter*, translated with an introduction by Geoffrey Winthrop-Young and Michael Wutz, Stanford, CA: Stanford University Press, 1999, p. xxxix.

68 Ibid., p. 3.

they involve a shift from inscription systems whose time is "(in Lacan's term) symbolic" to systems whose time runs "on a physical or (again in Lacan's terms) real level".[69] Furthermore, the shift from the *symbolic* to the *real* implies that humans are no longer the ones doing the inscribing. The introduction of technological media, in other words, occasions a displacement of "so-called man" (as Kittler tends to put it): machines, he maintains, and especially the intelligent machines introduced by digital technology, "are not there for us humans".[70] This idea, that the machines are not there for us, ties into Kittler's anti-humanist take on history, whose ultimate subject is not humans but technology. With the advent of intelligent machines, the self-processing of nature[71] no longer needs human intermediaries: "Instead of wiring people and technologies, absolute knowledge will run as an endless loop.[72] Again according to Kittler, the introduction of digital technology also has the effect of erasing the differences among individual media, due to the way that it reduces sound and image, voice and text to "surface effects, known to consumers as interface".[73] This, clearly, undermines the role of the human senses just as much as it undermines meaning: "Sense and the senses turn into eyewash".[74]

Since the heyday of Kittler's anti-humanist theory, many of his followers have gradually moved away from the exclusive focus on the material properties of media technologies, centering instead on the notion of "cultural techniques".[75] Other followers seem intent, rather, to "out-Kittler Kittler"[76] by affirming even more strongly the anti-humanist tendencies in Kittler's work. This is the case with Ernst, whose resolute focus on machine agency is what makes his approach particularly relevant.

Ernst's approach to media has been characterized as an "operative media archaeology".[77] In his own efforts to explicate his approach, Ernst positions himself against media archaeology as cultural history on the one hand, and against media phenomenology on the other. As Ernst sees it, historical discourse and human perception are both prone to interpretation and ridded with subjectivity, which is why he seeks instead a "technoascetic" approach that "takes the point of view of the machine itself".[78]

This implies that, in the work of Ernst, *archaeology* does not mean *genealogy*. Drawing on Foucault's notions of archive and archaeology,[79] media archaeology is defined, rather, as "a kind of epistemological reverse engineering, and an awareness of moments when media themselves, not exclusively humans anymore, become active 'archaeologists'

69 Ibid., p. 4.
70 Kittler cited in Enns 2016 (as fn. 65), pp. xiv-xv.
71 Kittler's idiosyncratic take on history (including the role of technology) is succinctly summarized by Geoffrey Winthrop-Young as follows: "the ultimate subject of history is technology, understood in a very broad sense as the processing of nature that for an extended period of time was dependent on human intermediaries, but that now, with the arrival of digital technology, is closer to a self-processing of nature that leaves humans behind". Geoffrey Winthrop-Young, *Kittler and the Media*, Cambridge, UK: Polity, 2011, p. 80.
72 Kittler 1999 (as fn. 67), pp. 1–2.
73 Ibid., p. 1.
74 Ibid.
75 Enns 2016 (as fn. 65), p. xvi; For an overview of approaches centering on the notion of *cultural techniques*, see Bernhard Siegert, Cultural Techniques. Or the End of the Intellectual Postwar Era in German Media Theory, in: *Theory, Culture & Society* 30.6 (2013), pp. 48–65; and Geoffrey Winthrop-Young, Cultural Techniques. Preliminary Remarks, in: *Theory, Culture & Society* 30.6 (2013), pp. 3–19.
76 Winthrop-Young 2013 (as fn. 75), p. 15.
77 Jussi Parikka, Operative Media Archaeology. Wolfgang Ernst's Materialist Media Diagrammatics, in: *Theory, Culture & Society* 28.5 (2011), pp. 52–74.
78 Ernst 2013 (as fn. 64), p. 24, p. 72.
79 For definitions, see ibid., p. 211, note 4.

of knowledge".[80] In contrast with Manovich, who, according to Ernst, remains on the surface by investigating "monitors and interfaces" and what they "offer to the human user", Ernst is concerned with "technoepistemological configurations underlying the discursive surface".[81] Thus conceived, the archaeology of media "is not simply an alternative form of reconstructing beginnings of media on the macrohistorical scale", it describes, rather, "technological 'beginnings' (*archai*) of operativity on the microtechnological level".[82] These technological beginnings relate to the very essence of technical media, which Ernst conceives in operational terms: "It belongs to the specificity of technical media that they reveal their essence only in their operation".[83] The essence of technical media relates to "microtemporal processes" that are critical for the operations of technical media, that is, for their performance as "processual hardware".[84] This means that, with a view to their operational essence, technical media are not arbitrary or subject to discursive cultural relativization; they have an "epistemological existence" of their own, due to the way they produce their own machine-specific time – what Ernst refers to as their "*Eigenzeit*".[85] Thus, the primary focus of Ernst's kind of media archaeology is "time-criticality", the time-giving and time-differentiating aspects of technical media – the way technical media "do not simply exist *in time* but result in *timing* agencies".[86]

80 Ibid., p. 55.
81 Ibid.
82 Ibid., p. 57.
83 Ibid.
84 Ibid. p. 50, p. 177.
85 Ibid. p. 57.
86 Wolfgang Ernst, *Chronopoetics. The Temporal Being and Operativity of Technological Media*, translated with a foreword by Anthony Enns, London/New York: Rowman & Littlefield, 2016, p. vii [original emphasis].

This implies that the operational lifespan of technical media objects is not identical to their cultural lifespan. He gives the example of an old radio found in a museum, whose outer world has vanished. If such a radio, a historical museum object, is reactivated so as to broadcast today's radio programs, it undergoes a change in status from "*historical* to *processual* hardware".[87] Operationally speaking, therefore, the radio is still present, since "[t]here is no 'historical' difference in the functioning of the apparatus now compared to then".[88] Thus, when the radio is reactivated, it truly becomes a medium again, which means that "there is a media-archaeological short circuit between otherwise historically clearly separated times".[89] This then is why, for Ernst, traditional historical approaches will not do: By subjecting media processes to a literary narrative, they misread and misrepresent the *Eigenzeit* of technical media.

According to Ernst, time-critical media provide a different (and better) kind of evidence of the past than the evidence provided by historical-discursive accounts. As Ernst sees it, machines have the power to "temporarily liberate" us from the limitations of literary narrative and human perception.[90] The unique evidential power of technical media is directly connected with their time-giving agencies, which, according to Ernst, induce "disruptions in human temporal perception" due to their "asynchronous being in what is known as 'historical' time".[91] Technical media (including computers) differ from the "traditional symbolic tools of cultural engineering (like writing the alphabet)" in that they register and process "not just semiotic signs but physically

87 Ernst 2013 (as fn. 64), p. 177 [original emphasis].
88 Ibid., p. 57.
89 Ibid.
90 Ibid., p. 56.
91 Ernst 2016 (as fn. 86), p. vii.

real signals".[92] Like Kittler before him, he articulates this opposition in terms of the *symbolic* versus the *real*: Technical media "emancipate" the object from "an exclusive subjection to textual analysis", and in so doing, they remind us about "the insistence and resistance of material worlds".[93]

Ernst further develops the idea of the unique evidential power of technical media by invoking the Peircean notion of "index": Media archaeology is "on the side of the indexical",[94] which is seen as opposed to the side of the iconic and the symbolic.[95] Hence, when it comes to photography, he agrees with Roland Barthes, who "emphasizes photography as a decisive mutation in informational economies".[96] According to Ernst, photography is an example of a "true media-archaeological tool" due to its "automatic registration and self-inscription of light".[97] A similar rupture is found in gramophonic recording, "which can record as well the accompanying noise (i.e., the index) of the physically real within and outside the recorded voice".[98] Technical media such as these provide a unique kind of evidence due to the way that they "immediately couples human perception with the signal flow [...], with or without their translation into the iconological regime of cognition".[99] The immediate coupling occasioned by technical media is then contrasted to the "indirect, arbitrary evidence symbolically expressed in literature and musical notation".[100] Hence, as Ernst sees it, media archaeology is "media studies as exact science": an approach that investigates "media-induced phenomena on the level of their actual appearance", that is, as "physically real (in the sense of indexical) traces of past articulation".[101]

Ernst's kind of media-archaeology, then, as pointed out by Jussi Parikka, is conceived as a "a way of stepping outside a human perspective to the media-epistemologically objective mode of registering the world outside human-centered sensory perception".[102] Technical media (including computers) are conceived by Ernst as "measuring media" – media that, in contrast to mass media, are "able to decipher physically real signals technoanalogically".[103] In Ernst's view, measuring media are closer to reality because they "behave 'analogously' to physics itself".[104] More precisely, they are assumed to be closer to reality because they operate on the level of numbers and not on the "phenomenological multimedia level" of text, image and sound.[105] Media archaeology as conceived by Ernst is "close to mathematics", which in turn is seen as close to nature.[106] Hence, when human senses are coupled with technological settings, "man is taken out of the man-made cultural world".[107] In this way, Ernst aspires toward a "cool" media-archaeological gaze, which can be performed by algorithmic machines better than by

92 Ernst 2013 (as fn. 64), p. 58.
93 Ibid., p. 43.
94 Ibid., p. 45.
95 In Ernst's treatment, the iconic and symbolic tend to be lumped together, since they are both associated with culturally variant human perception and history.
96 Ernst 2013 (as fn. 64), p. 38; see also Barthes Roland, Rhetoric of the Image, in: Roland Barthes, *Image Music Text*, essays selected and translated by Stephen Heath, London: Fontana Press, 1977, p. 45.
97 Ernst 2013 (as fn. 64), p. 47.
98 Ibid., p. 64.
99 Ibid., p. 67.
100 Ibid., p. 173.
101 Ibid.
102 Jussi Parikka, Archival Media Theory. An Introduction to Wolfgang Ernst's Media Archaeology, in: W. Ernst, *Digital Memory and the Archive*, edited and with an introduction by Jussi Parikka, Minneapolis/London: University of Minnesota Press, 2013, p. 9.
103 Ernst 2013 (as fn. 64), p. 178.
104 Ibid., p. 62.
105 Ibid., p. 71.
106 See ibid., pp. 71–73 for more details about how this (problematic) argument goes.
107 Ibid., p. 177.

human perception, since it is no longer dominated by "semiotically iconic, musically semantic, literally hermeneutic ways of seeing, hearing, and reading".[108] Thus, in contrast to Manovich, who emphasizes how the cultural layer and the computer layer mutually influence each other, resulting in a "blend of human and computer meanings", Ernst pursues a firm anti-humanist approach that seeks instead to rid the analytical gaze of everything human.[109] What is gained by this approach is that the non-human and time-critical agencies of technical media come into view. However, again as noted by Parikka, by pursuing a "happy positivism", Ernst comes close to "mythologizing the machine as completely outside other temporalities, including the human".[110] Moreover, by defining the operational in a strictly technical sense, he seems to bracket out the very mediating aspects of media: their roles as interfaces to the world and other people, their status as meaningful forms of expression (images, texts, sounds).

Efficacious Images

While Harun Farocki's artistic explorations of operative images opens a complex array of questions relating to pressing social, political and ethical issues, the approaches of Lev Manovich and Wolfgang Ernst both stay "close to the machine"[111] – focusing on software and hardware, respectively. In this section, I consider operative images from the perspective of visual studies,[112] where the discussion revolves around the efficacy of images, which, as we shall see, need not necessarily be identified with new media or the machinic.

When it comes to visual studies approaches to operative images, an interesting case in point is the international conference *Image Operations*,[113] which, together with other recent academic events,[114] have contributed to the establishment of the field of "image operations studies".[115] In an edited volume following the conference, image operations are discussed with a special emphasis on their roles in warfare, insurgency/counterinsurgency and political activism. In the introduction, Jens Eder and Charlotte Klonk consider three cases where imagery has been directly involved in highly charged political situations: Kevin Carter's Pulitzer Prize winning photograph showing a starving and collapsed Sudanese child with a vulture in the background; a classified US military video released by WikiLeaks showing gunsight footage from an attacking helicopter that opens fire against a group of men including two Reuters news staff; and a YouTube video showing the beheading of the American journalist James Foley by a member of the militant jihadist group ISIS. In what sense are these cases to be considered as image operations? Eder and Klonk provide some indications: they are image operations, first, in that they all provoked "a whole series of largely uncontrollable events" that went "beyond

108 Ibid., p. 27.
109 Manovich 2001 (as fn. 25), 46.
110 Parikka 2013 (as fn. 102), p. 7, p. 10.
111 Interestingly, both thinkers use this exact phrase, see Manovich 2001 (as fn. 25), p. 117 and Ernst 2013 (as fn. 64), p. 59.
112 In the German-speaking parts of the world, more frequently referred to as *Bildwissenschaft*.

113 The conference took place at the Institute for Cultural Inquiry in Berlin on April 10–12, 2014.
114 These events include the conference *Media Acts* (Trondheim 2011), the conference *What Images Do* (Copenhagen 2014), a series of three conferences *Dynamis of the Image: An Archaeology of Potentialities* (Düsseldorf 2014, Basel 2014 and Paris 2015), the workshop *Screen operations: Conditions of Screen-based Interaction* (Berlin 2016) and the PhD course *Operative Images* (Berlin 2017).
115 Zoya Brumberg, Book Review. Jens Eder and Charlotte Klonk (eds), Image Operations: Visual Media and Political Conflict, in: *Journal of Visual Culture* 16.3 (2017), p. 391.

the original intentions of their producers".[116] In these cases, the series of events lead to, among other things, Carter's suicide, the imprisonment of the soldier who was charged for disclosing the military video, and a rigorous ban on the footage showing Foley's beheading. Furthermore, they are image operations in that the production and circulation of images "led directly or indirectly to the physical death of real people".[117] Even if the images "operated within the seemingly disembodied digital sphere of the Internet", they all had "serious consequences", affecting bodies in "vital ways".[118] Finally, they are image operations in that, in all three cases, the images were "crucial factors in the dynamics" of the conflicts in question, and as such, "the *agens et movens* in the unfolding of events".[119] Thus, as conceived by Eder and Klonk, image operations are primarily defined in terms of their consequences, which in turn seem to be based primarily on the representational function of the imagery. Due to their disturbing contents, the images incite a series of uncontrollable events that have serious, real-world effects. At the same time, Eder and Klonk repeatedly emphasize that, in all these cases, images do more than "just reflect or represent conflicts"; rather, they "play performative and constitutive roles within them".[120] They also call attention to how, in the digital media environment, the performative and constitutive roles of images grow stronger, amplifying "the volume, speed, reach and level of conflictual involvement".[121]

After having proposed these characteristics, Eder and Klonk proceed to ask the pertinent question: "So who or what is operating in image operations?".[122] They answer by pointing to a "complex network of agencies" in terms of actor-network theory.[123] Certainly, people and organizations use images as tools, but there is an important sense that "images themselves also act".[124] This idea, that images have a "dynamic of their own", is key to a highly influential line of research in contemporary visual studies.[125] It is somewhat surprising, therefore, that, when they go on to clarify the notion of images, they choose to focus their book on the rather traditional idea of "visual pictures" understood as "anything that visually represents or expresses something else without being written language".[126] In fairness, Eder and Klonk present a range of very different conceptions of images, including "image games" (invoking Wittgenstein's notion of language games) and "image acts" (invoking Searle's notion of speech acts) – the overall impression being that the introduction wavers between established approaches to images in terms of representation and revisionist approaches centering on the idea of image agency. The implication of this all-embracing approach is that the operational comes across as a mere supplement to the more established approaches. This becomes clear, for example, when Eder and Klonk set out to clarify the specific powers of images, listing the operational – which is now, rather unexpectedly, defined in terms of the interactive use of images in digital media – as a fourth potential of images following their

116 Jens Eder, Charlotte Klonk, *Image Operations. Visual Media and Political Conflict*, Manchester: Manchester University Press, 2017, p. 1, p. 4.
117 Ibid., p. 3.
118 Ibid., pp. 3–4.
119 Ibid., p. 3 [original emphasis].
120 Ibid., p. 4.
121 Ibid., p. 4.
122 Ibid., p. 6.
123 Latour 2005 (as fn. 61).
124 Eder, Klonk 2017 (as fn. 116), p. 6.
125 W. J. T. Mitchell, *What Do Pictures Want? The Lives and Loves of Images*, Chicago/London: The University of Chicago Press, 2005; Gottfried Boehm, Ikonische Differenz, in: *Rheinsprung 11. Zeitschrift für Bildkritik* 1 (2011), pp. 170–176, http://rheinsprung11.unibas.ch/archiv/ausgabe-01/glossar/ikonische-differenz.html (accessed May 27, 2018); Horst Bredekamp, *Der Bildakt*, Berlin: Klaus Wagenbach, 2015.
126 Eder, Klonk 2017 (as fn. 116), p. 9.

mimetic, symbolic and aesthetic (including sensual and affective) potentials.[127]

Like Farocki and Ernst, Eder and Klonk accentuate that images have an agency of their own, which implies that images cannot be fully understood by reconstructing the intentions of their producers.[128] In Eder and Klonk's view, this is because, when images start to circulate, they have unforeseen effects that may even go against the original intentions of their producers. Hence, in contrast with Farocki and Ernst, the operational is identified with the real-world performative effects of images as they circulate in society, and not so much with their machinic element. The advantage of this approach is that it brings prominence to the ethical dilemmas that arise on the level of images, pointing to the need for a renewed focus on image ethics. The notion of image operation, however, is rather vaguely defined and remains, as noted by Zoya Brumberg, a "nebulous concept".[129]

While Eder and Klonk for the most part approach the operational as a supplement to more established approaches to images, they also at times seem to push in the direction of a deeper revision of the image category. If followed through, the idea that images have an agency of their own profoundly challenges received notions of images in terms of representation. Some thinkers, therefore, such as Sybille Krämer, regard the current focus on the operational as an occasion for a much-needed rethinking of the very idea of images. While in line with contemporary research advocating the agential powers of images, Krämer's approach stands out in its explicit focus on "operational iconicity" (*operative Bildlichkeit*).[130] Krämer's take on operative images emphasizes two interrelated points: the necessity of going beyond the text-image dichotomy, and the promise of the diagrammatic approach.[131] The background here is that traditional notions of images, including classical ways of distinguishing between semiotic modalities, are intimately bound up with more fundamental divisions – a highly influential example being Immanuel Kant's opposition between the two stems of human knowledge: sensibility and understanding. While the classical ways of conceptualizing the boundaries between images, texts and numbers typically conform to such long-established, fundamental oppositions, recent attempts to rethink images are *deep revisions* in that they no longer assume the dualist worldview at the basis of the old distinctions – challenging received notions of images, therefore, at their very root. Krämer contributes to the ongoing revisionist endeavors, showing how the old philosophers themselves provide resources to overcome unproductive dualisms, such as Kant with his notion of schema[132] and Charles S. Peirce with his notion of diagram.[133]

So why, then, this renewed interest in Peirce and diagrammatics? Late in his career, Peirce developed a broadened notion of diagrams that is highly relevant to the current attempts to conceptualize operative images, for two reasons: First, because it provides a dynamic and operational notion

127 Ibid., pp. 9–10.
128 Ibid., p. 1.
129 Brumberg 2017 (as fn. 115), p. 389.

130 Sybille Krämer, *Operative Bildlichkeit. Von der 'Grammatologie' zu einer 'Diagrammatologie'? Reflexionen über erkennendes 'Sehen,'* 2009, http://userpage.fu-berlin.de/~sybkram/media/downloads/Operative_Bildlichkeit.pdf (accessed May 27, 2018).
131 Ibid., pp. 1–3, pp. 10–12.
132 Ibid., pp. 12–15.
133 Ibid., pp. 10–12; In fact, Peirce's notion of diagram takes its inspiration from Kant's notion of schema. Charles S. Peirce, *(PAP) [Prolegomena for an Apology to Pragmatism]*, in: Charles S. Peirce, *The New Elements of Mathematics*, Vol. IV: Mathematical Philosophy, The Hague: Mouton, 1976, p. 318; Ibid., pp. 10–12.

of iconicity[134] that pushes beyond static ideas of images in terms of similarity (including Peirce's own previous definitions of iconicity); and second, because it provides a new notion of evidence that overcomes mechanistic accounts (including those based on indexicality). The diagrammatic approach, in other words, provides a fresh take on images that scrambles the icon-index-symbol trichotomy as we know it from textbooks in semiotics. Beyond that, the true merit of a Peircean diagram is that it has the unique power to generate new and surprising information when manipulated in systematic ways.[135] Thus conceived, a diagram is not necessarily visual. It is not a "visual picture" in the terms of Eder and Klonk because the iconic element of diagrams has more to do with their demonstrative powers.[136] The diagrammatic structure is not exclusive to images and does not serve to distinguish them from texts or numbers, since, as Peirce sees it, there is iconicity at the heart of linguistic propositions and mathematical formulas, just as there are rules at the heart of images.[137] The diagrammatic approach, in other words, redraws the boundaries between images, texts and numbers as we have come to know them, emphasizing interconnections rather than oppositions. In the same vein, it is the diagrammatic structure that connects images to the wider family of instruments. For these reasons, Peirce's dynamic and operational notion of iconicity promises to throw new light on the nature and workings of image-instruments, whether we are interested in how and why perspectival images or photographs are "more than just sign systems that reflect reality" (to paraphrase Manovich), or more concerned with the evidential and instructional powers of digital image applications.[138]

Concluding Remarks

By maintaining that operative images are not representations but rather instruments that form part of operations, Harun Farocki sets the stage for the ensuing discussions presented above. Identifying the notion of operation with automation, he frames the human-machine relationship as antagonistic. Farocki's *Eye/Machine* installation and its commenters also introduce the idea of the imminent replacement of humans by machines: Disrupting the human scale, sensory automations outperform the human eye. Having no need for human spectators, operative images serve, rather, as interfaces in algorithmically controlled processes. Issues relating to the human-machine antagonism continue to resonate in the subsequent two sections. While Lev Manovich seeks to resolve the antagonism by domesticating the machine, Wolfgang Ernst instead chooses the opposite strategy of bracketing everything human to secure the purity of machinic operations. Jens Eder and Charlotte Klonk, on their side, identify the notion of operation with the performative effects of images as they circulate in society, articulating the active dimension in terms of distributed networks of agencies.

134 A key source for Peirce's operational notion of iconicity is an unpublished manuscript that is referred to as "PAP". See ibid. For a more detailed discussion, see Frederik Stjernfelt, *Diagrammatology. An Investigation on the Borderlines of Phenomenology, Ontology, and Semiotics*, Dordrecht: Springer, 2007, pp. 89–116; Aud S. Hoel, Lines of Sight. Peirce on Diagrammatic Abstraction, in: F. Engel, M. Queisner and T. Viola (eds.), *Das bildnerische Denken. Charles S. Peirce*, Berlin: Akademie Verlag, 2012, pp. 253–271; Aud S. Hoel, Measuring the Heavens. Charles S. Peirce and Astronomical Photography, in: *History of Photography* 40.1 (2016), pp. 49–66.

135 Stjernfelt 2007 (as fn. 134), p. 90.

136 Eder, Klonk 2017 (as fn. 116), p. 9.

137 These rules are generative rules, and not the arbitrary rules of semiological structuralism.

138 Manovich 2001 (as fn. 25), p. 168.

While all four approaches, each in their own way, strongly confirm the idea of images having an active dimension, none of them provides a developed account of the operational as a media-theoretical concept. In this respect, the accounts considered in sections 1–4 remain too ambiguous, too cultural, too technical and too wide, respectively. An indication that the notion of operation remains under-theorized as a media-theoretical concept can be seen in that none of the approaches in question gives a satisfactory account of the new role of operative images as interfaces – as interfaces, that is, not only in the HCI sense discussed by Manovich, but in the epistemological and ontological sense as intermediaries to the world and other people. In their operational role as intermediaries, images cannot be reduced to Kittlerian "surface effects".[139] Moreover, a developed account of the notion of operation as a media-theoretical concept would also have to include a more satisfactory take on the relation between technology and the human senses, not relegating the latter to the "phenomenological multimedia level" (as Ernst does).[140] The tendency in the literature to distinguish between images and media that supposedly conform to the human senses and those that induce a disruption in the familiar patterns of perception is yet another instantiation of an unproductive opposition between human and machines. Observing that the boundary between the two can be drawn in several ways, Manovich raises a pertinent question: "But what is human nature, and what is technology?"[141]

The guiding idea of this article is that there is something new in the way that the scholars of operative images approach the topic of mediation, which has to do with a deeper recognition of the active dimension of images and media. Moreover, as already hinted in the final paragraphs of the previous section, if the idea that images have a dynamic of their own is followed through, we may come to question the classical ideas of images at their very root – including their underlying assumptions. This, then, is why, to the extent that we are currently standing on the verge of an emerging, operational paradigm of media theory, this paradigm will have to be a comprehensive one, not restricted to technical images, digital images or new media.

139 Kittler 1999 (as fn. 67), p. 1.
140 Ernst 2013 (as fn. 64), p. 71.
141 Manovich 2001 (as fn. 25), p. 171.

Thanks to Kathrin Friedrich and Moritz Queisner for their thoughtful comments on a previous version of this article. Thanks also to Annamaria Carusi for her helpful input.

Erkki Huhtamo

The Spell of the Catoptric Television

Media Archaeology, Topos Study, and the Traces of Attention

We are always turning in the same circle,
always rolling the same stone.

Gustave Flaubert in a letter to Louise Colet,
April 22, 1854

The Power of Topoi

How can we research media attention as a historical phenomenon? This essay suggests one way from a media archaeological perspective. Media archaeology is a *traveling discipline*, a bunch of approaches loosely tied together by some common threads, but also with significant differences among its practitioners.[1] Because of this, it is necessary to define in what sense the concept *media archaeology* is being used here. I call the variant I have been developing for the past twenty-five years *media archaeology as topos study* or simply *topos archaeology*. The idea of *topos*, which refers to recurring elements that travel within and across cultural traditions for hundreds and even thousands of years came from the German literary scholar Ernst Robert Curtius.[2] Topoi can be compared to empty vessels or molds that are filled with new content as they reappear in changing cultural contexts. They represent continuities, but can also point to ruptures and transitions. The interpretations and meanings of topoi change in the course of their migrations, as I explained in condensed form in my article *Dismantling the Fairy Engine. Media Archaeology as Topos Study* and will demonstrate in greater detail in a forthcoming book.[3]

Unlike Curtius, I do not think topoi only exist within literary traditions. Visual imagery functions in similar ways, as Aby Warburg demonstrated with his notion *Pathosformel* (pathos formula) that likely influenced Curtius' topos. Warburg's unfinished *Mnemosyne Atlas* also anticipated media archaeology.[4] It explored visual culture by means of thematic tableaux, collages that challenged the linear style-based accounts typical for art history. Warburg pointed out migrations and transformations of visual motifs in terms of fields instead of surface vectors of cause and effect. He broke down artificial disciplinary barriers by linking academic art with non-canonical forms like magazine illustrations and advertisements, questioning the orthodoxy of academic art history. The *lives* of topoi do not respect institutional and

1 Erkki Huhtamo, Jussi Parikka, An Archaeology of Media Archaeology, in: E. Huhtamo, J. Parikka (eds.), Media *Archaeology. Approaches, Applications, and Implications*, Berkeley: University of California Press, 2011, pp. 1–21, p. 3.

2 Ernst Robert Curtius, *Europäische Literatur und lateinisches Mittelalter*, Bern, 1948, translated into English in 1953 as *European Literature and the Latin Middle Ages*, new Version trans. by Willard R. Trask, London: Routledge & Kegan Paul, 1979.

3 See Erkki Huhtamo, Dismantling the Fairy Engine. Media Archaeology as Topos Study, in: Huhtamo, Parikka 2011 (as fn. 1), pp. 27–47.

4 The best edition is Aby Warburg, *L'Atlas Mnémosyne*, translated by Sacha Zilberfarb, Paris: L'écarquillé – INHA, 2012.

cultural boundaries, least of all those separating high from popular culture. Neither do those who research them. The nature of topos archaeology is interdisciplinary.

Cultural contextualization is more important for me than it was for Curtius. He operated *within* literary traditions, demonstrating how topoi such as *the world upside down* migrated from one text to another.[5] That is not sufficient for media archaeology. Topoi must be treated as both transhistorical and as symptomatic manifestations of the times and places where they appear. As I understand it, the task of media archaeology is therefore Janus-faced: it traces topoi *between* contexts and analyzes them *within* contexts. In media culture topoi serve at least three roles: as connectors to older and broader cultural traditions; as discursive commentaries on media cultural forms, themes and fantasies; and as motifs exploited by the culture industry. These days they are modified and disseminated by potentially anyone on the Internet.[6] *Topos transmissions* can be detected in all areas of media culture. They are not limited to the distant past; the Internet is both a topos disseminator and a generator, recycling age-old topoi and giving birth to new ones.

Let us look at Mark Ulriksen's satirical painting *Capturing the Memories*, which was used in the cover of *The New Yorker* in 2012.[7] A family of four has been lined up for a vacation snapshot in a tropical paradise, their backs turned to the stereotypical lagoon with palm trees (a topos). Each family member is fingering a mobile phone, utterly lost in its spell. The invisible person who is taking the photo is using one as well, as we can tell from the shadow cast on the sand. Similar scenes have become a token of everyday life. As merely sitting in a crowded restaurant can often demonstrate, direct face-to-face contacts have been challenged by gazes into handheld screens.[8] Ulriksen deliberate exaggerates, but only a little. Normally people still glance into the lens when they are being photographed (or taking selfies themselves). The example seems to confirm Guy Debord's insight from *The Society of the Spectacle*: "[T]he real world becomes real images, mere images are transformed into real beings – tangible figments which are the efficient motor of trancelike behavior."[9]

Ulriksen's cover illustration is worth comparing with a television commercial created to promote Microsoft's already forgotten Windows Phone 7 (2010). It too depicts a society mesmerized by mobiles. The commercial, code-named *Really*, recalls, perhaps unintentionally, René Clair's 1927 silent film *The Crazy Ray* (*Paris qui dort*), where Paris is collectively mesmerized by a mad scientist operating a diabolic machine.[10] All Parisians, except those who had been up high enough (in the Eiffel Tower or in an airplane), have been turned into zombie-like sleepwalkers without a will of their own. In *Really* there are also characters who are still *awake*, but for a different reason. They are free from the spell of the mobile phone, craving direct contact: a woman frustrated with a phoning massage therapist; a man expressing

5 For an exploration of this topos, see Frédérick Tristan, *Le monde à l'envers*, Paris: Atelier Hachette & Massin, 1980.

6 Erkki Huhtamo, Obscured by the Cloud. Media Archaeology, Topos Study, and the Internet, in: Thorsten Lomker (ed.), *ISEA 2014 Dubai. Location. Proceedings of the 20th International Symposium of Electronic Art*, Dubai: Zayed University Books, 2015, pp. 22–35.

7 *The New Yorker*, July 23, 2012.

8 As an indication of how fads change, the 2015 version might show all five people posing together for the smartphone, attached to the end of a telescoping *selfie stick*. Taking a selfie may still command a momentary eye contact, albeit mediated.

9 Guy Debord, *Society of the Spectacle*, New York: Zone Books, 2004, paragrap. 18.

10 Rene Clair, *The Crazy Ray*, France 1924. *The Crazy Ray* is also a meta-cinema, an investigation of a medium to manipulate space and time. Clair was the director of the dadaistic film classic *Entr'Acte*, France 1924, where similar tricks were already used.

condolences over a mobile dropped in a toilet; a sexy wife losing her temper because of her husband's disinterest; and a boy hitting his absent-minded father with a baseball.[11] The commercial ends with an ambiguous slogan: "It is time for a phone to save us from our phones [...]. Designed to get you in and out and back to life."[12] It is hard to say if the irony is deliberate or not: the only way to solve the current ills of social life caused by a gadget is another – essentially similar – gadget.

On surface level both the magazine cover and the television commercial are comments on the same timely problem. Topos archaeology can demonstrate that they are also re-activated versions of a pre-existing topos, which has been encountered numerous times in the formative development of media culture, commenting on the consequences of excessive attention to new gadgets. Before rushing to conclusions about the unique impact of smartphones, it is worth studying the *media manias* that are claimed to have broken out ever since the Kaleidoscope became a fad in the 1810s.[13] Media maniacs lose their loved ones to invisible suitors or end up in all kinds of accidents, over and over again. Some people manage to stay unaffected, ridiculing or pitying the addicted.[14] The recurrence of the topos demonstrates that the current discourse on sleepwalkers staring into mobile screens, oblivious of their surroundings, is unprecedented only to a degree, more because of the scale of the phenomenon than because of the motif itself. The excavations could probably be extended even further back in time to include collective manias not related to technology.[15]

Media archaeology should not limit itself to textual or visual analysis, which would run the risk of ignoring or misrepresenting the tangible – the material and performative – aspects of media culture. That is why I analyzed the moving panorama in my book *Illusions in Motion. Media Archaeology of the Moving Panorama and Related Spectacles* on three levels, which I called the painted, the performed, and the discursive panorama.[16] I tried to demonstrate why it is important to relate media archaeological discourse analysis to investigations of material and technological factors. I also pointed out the necessity of discussing media dispositives and their actualizations within media practice as part of the topos archaeological endeavor. My research has led me to the conviction that topoi can manifest themselves in hardware design, technological features as documented in patent documents, feedback mechanisms, and user interfaces. However, they cannot be independent of contexts of conception and use.

11 Created by the ad agency Crispin, Porter and Bogusky. See Windows7Phone, Really: New Windows Phone Ad, https://youtube.com/watch?v= 55kOph D64r8 (accessed January 8, 2018). In the campaign's pilot commercial, *Season of the Witch*, time has slowed down or stopped completely in a crowded street – a bicyclist has fallen, a car has hit a pole, and a load of fruit have dropped to the ground. A few people are moving like sleepwalkers, bumping into each other, but most are entirely frozen; everyone is staring at a portable device. See windowsphonepro, Season of the Witch: New Windows Phone 7 Ad, https://youtube.com/watch?v=Lg1gbVGk19k (accessed January 8, 2018).

12 Except for this voiceover message and a few short remarks, the commercials are only accompanied by music, Donovan's *Season of the Witch* and Edvard Grieg's *I Dovregubbens hall* (*In the Hall of the Mountain King*) from the *Peer Gynt Suite*.

13 I will analyze this issues in detail in a forthcoming book.

14 For examples, see Erkki Huhtamo, *Illusions in Motion. Media Archeology of the Moving Panorama and Related Spectacles*, Cambridge, MA/London: MIT Press, 2013, pp. 77–79 and passim.

15 Charles Mackay pointed a possible way to proceed in his Victorian classic *Memoirs of Extraordinary Popular Delusions and The Madness of the Crowds*, Ware, Hertfordshire: Wordsworth Editions Ltd, 1995 [1841, 1852]. Remarkably, the book is still in print.

16 Huhtamo 2013 (as fn. 14).

The Magic Mirror as a Magickal Transmitter and Receiver

Applying this media archaeological approach to the screen in visual practice can link today's ubiquitous personal screens and their usages, dispositives, and discourses to an example known by many names: *magic(k) mirror, enchanted mirror, sorcerer's mirror, scrying mirror,* etc. It is interesting not only because of its ambiguous relationship with media history, but also because of its intertwined material and discursive identities. The magic mirror does not only belong to the past, as an Internet search can easily demonstrate. A website promoting New Age energy therapy defines it as a "divination tool for looking into past lives, and past, present, and future events". The writer provides the following operating instructions: "Sit before your mirror and begin to imagine objects on its surface, one after another. You should try to see these images clearly in the mirror with your eyes open, just as if they were there in reality." Most people, the text continues, "do not see the images appear with the physical eyes on the mirror's surface but see within the mirror and in the mind's eye. The mirror acts as a focal point, a gateway within."[17] The website sells round and oval scrying mirrors made of glass or acrylic plastic, but the Internet also offers *how to* instructions for making one's own by covering a clear glass plate with black matte spray paint.[18]

There are people who consider the magic mirror as an operational device that actually works when used in scrying practice.[19] The web page "How to use a Scrying Mirror / Magick Mirror" describes it as a "powerful psychic tool" for "seeing into darkness", stating that it "allows communication with higher realms and the subconscious".[20] The scrying mirror is claimed to have numerous uses: "To contact spirit guides, To access knowledge, For healing and self improvement, As a magickal [sic] transmitter and receiver, For divining the past, present and future, As a Portal to the Astral Plane, For shamanic journeying, For ritual invocation and evocation, To improve visualization skills." The comments left on the website discuss the making, purchase, and consecration rituals of the scrying mirror. Someone has recorded concerns: "1/isn't it risky to go almost [sic] into another world? 2/what if we wont be able to return?" etc.[21] For the commentators the scrying mirror seems a perfectly practical gadget, and in that sense little different from the mobile phone. One of the comments could almost have been left by a media archaeologist: "Basically a scrying mirror is the first television and phone thing that ancient people used back then?"[22]

Except for their paranormal concerns, these descriptions evoke issues that are often associated with the *interface logic* of media machines, that are subsumed by the term screen. Concentration is required to turn the magic mirror into a *gateway* – to make it, as a Wiccan puts it, to "appear as

17 Scrying Mirrors – Natures Energies, https://naturesenergieshealth.com/metaphysics/divining/scrying-mirros/ (accessed June 22, 2017).
18 See in particular Donald Tyson, *How to Make and Use a Magic Mirror. Psychic Windows Into New Worlds*, Custer, Washington: Phoenix Publishing, 1995.
19 The name "magic mirror" is also used for Asian metal mirrors with illusionistic properties. Janet Leigh Foster and Stephen Herbert have authored the definitive book on them, *Magic Mirrors. A 2000-Year History of Asia's Most Sensational Secret*, forthcoming. Magic mirrors could also be related to the Claude Mirror, black glass mirrors used by landscape artists as sketching aids.
20 Somenthingwiccan, How to use a Scrying Mirror / Magick Mirror, https://ebay.com/gds/How-to-use-a-Scrying-Mirror-Magick-Mirror-/10000000009366585/g.html (accessed June 22, 2017).
21 Ibid., comment by Adarsh Barman, posted March 5, 2017.
22 Ibid., comment by Rylan Sasaki, posted June 6, 2017.

a dark tunnel or window".[23] The surface is dynamic; it "will begin to change and fade; a dark mist will appear".[24] The Wiccan emphasizes that the magic mirror is a personal medium, warning: "Do not let others look into its surface, except in ritual context."[25] Anyone using a mobile phone in a public environment certainly shares this idea, although for different reasons. For another scrying practitioner, "the images are interactive".[26] Such comments resonate with issues that occupy scholars interested in the media user's sharing habits. Is mental absorption into the screen more effective, safer (especially when the user is in motion) or more dangerous than distracted multitasking? What is at stake in interactive media experiences? Does physically manipulating the user interface keep the user from getting totally immersed into the audiovisual realm or contrariwise lead to a tighter bond with what is represented on the screen? How are social and individual media experiences related with each other?

Investigating the bonds with seemingly superfluous devices like the magic mirror may have lessons in store for researchers of media reception and interaction. Media archaeology does so by searching the past for excavation sites where *magickal transmitters and receivers* may lie buried. Most often they are found as fragments of discourse rather than as archives or material remains.[27] From time immemorial, reflecting surfaces have been used for observing one's own likeness, but also "for looking into past lives, and past, present, and future events".[28] In *Le miroir* (1978), a media archaeological magnum opus *avant la lettre*, Jurgis Baltrušaitis excavated a wealth of material and discursive magic mirrors, covering huge stretches of space-time.[29] Although he did not use that particular word, as a window or gateway to an *elsewhere* the magic mirror can be interpreted as a *topos*, a recurring figure or formula. Together with countless other topoi, it contributed to the slow formation of media culture. Baltrušaitis used the anachronistic concept "téléviseur catoptrique" ("catoptric television") about the visions said to have appeared on magic mirrors, implying that magic lore and technological screen-based media can be related with each other.[30]

The best known of Baltrušaitis's examples is a prophesy the court sorcerer Michel de Nostredame (Nostradamus, 1503–1566) is said to have presented to Catherine de' Medici (1519–1589), the queen consort of France, by means of a magic mirror.[31] According to the story, the highly superstitious Catherine was anxious to know the future of her children, and especially to find out who would succeed her on the throne. Nostradamus made each of her sons appear in turn in a magic mirror, passing in front of Catherine's credulous

23 Ibid.
24 Ibid.
25 Ibid.
26 Honor Seed, comment at "Black Mirror Scrying – Has anyone tried it?", See http://abovetopsecret.com/forum/thread190611/pg1 (accessed June 22, 2017). The same writer used the expression: "These interactive visions I saw." Both messages posted January 12, 2006.
27 An Aztec magic mirror, made of obsidian and thought to have been used by the Elizabethan alchemist John Dee in his scrying practice, is kept at The British Museum. It is almost circular but with a protruding part serving as a handle. In *Hudinbras* (1663), Samuel Butler is thought to have referred to this mirror, used by Dee's associate Edward Kell(e)y, as "The Devil's Looking Glass." For further information search the British Museum, Collections Online, for "Dr Dee's Magical Mirror / Dr Dee's Magical Speculum" http://britishmuseum.org (accessed June 22, 2017).
28 Sabine Melchior-Bonnet, *The Mirror. A History*, trans. Katharine H. Jewett, New York, London: Routledge, 2002, pp. 108–110, pp. 195–196, pp. 262–264.
29 Jurgis Baltrušaitis, *Le miroir. Révélations, science-fiction et fallacies*, Paris: Elmayan/Le Seuil, 1978. It is regrettable that this major work was never translated into English.
30 Ibid., p. 208. Chapter VIII gives many examples about the tradition of magic mirrors.
31 Ibid., pp. 184–187, pp. 206–208; Melchior-Bonnet 2002 (as fn. 28), pp. 108–110, p. 195.

eyes as many times as he would have years to reign. The story is still often evoked in paranormal circles and treated as fact. An extreme case is a book of conversations with Nostradamus by Dolores Cannon, a "past-life regressionist and hypnotherapist, who specializes in the recovery and cataloging of 'Lost Knowledge'".[32] The conversations were conducted via a spirit medium.[33] At the beginning of each session, after the medium had fallen into a state of trance, she established a contact with the famous magus through a magic mirror she saw in Nostradamus's study. Nostradamus agreed to meet the medium "in some other dimension" which was "gray and formless with no more substance than drifting clouds" (obviously this is what the medium saw in the mirror). Nostradamus confirmed that the magic mirror was the very same one he had used to concoct the vision for Catherine.

Was Catherine surprised to see the figures move in the mirror, the medium asked? Not really, Nostradamus answered, because she was accustomed to court magicians, but "felt uncomfortable, because she saw that all of her sons wouldn't survive." Had Cannon studied Baltrušaitis's work, she would have found out the apocryphal nature of the story.[34] In *Trésor des histoires admirables* (1614), Simon Goulart recounted a version said to have been told by Albert de Gondi (1522–1602), duke of Retz, who was Catherine's grandchild, but it may have been oral folklore.[35] Variants were produced over the following centuries. In Honoré de Balzac's version, a woman (or a man, say others), gifted with second-sight and brought to the Castle of Chaumont by Nostradamus, placed the Queen "in front of a magical mirror in which a spinning wheel was reflected, each child's face appearing at the end of a spoke, the soothsayer made the wheel turn, and the Queen counted the number of turns. Each turn was a year of a reign".[36] Balzac's imaginary device brings to mind a lottery wheel spinning on a television screen. In the tradition of *magia naturalis*, writers from the phantasmagoric Étienne-Gaspard Robertson (*Mémoires*, 1831) to Fulgence Marion (*L'Optique*, 1867, trans. *The Wonders of Optics*) suggested that Nostradamus had performed a trick of technological sleight-of-hand, which needed a rational explanation.[37] In other variants there is no magic mirror at all, just spirits moving around within a magic circle drawn on the floor.[38]

32 From Cannon's Amazon.com author's page, https://amazon.com/Between-Death-Life-Conversations-Spirit/dp/1940265002/ref=la_B001K818HK_1_4?s=books&ie=UTF8&qid=1525703563&sr=1-4 (accessed June 22, 2017).

33 Nostradamus, Dolores Cannon, *Conversations with Nostradamus*, Vol. 2, Revised Edition, Huntsville, AR: Ozark Mountain Publishing Inc., 1992.

34 It is ironical that the magic mirror became used in the title of collections of Nostradamus's writings. See Michel de Nostredame, *The Complete Fortune-teller: being the Magic Mirror of Michael Nostradamus; Also, the Infallible Divination by means of Figures, or Arithmancy of Count Cagliostro*, London: Lawrence & Bullen, 1899; Nostradamus, *The Magic Mirror; being the complete fortune-teller of Michael Nostradamus*, New York: Zend Avesta Pub. Co., 1931.

35 Simon Goulart, *Le troisiesme et quatriesme volume du Thresor des Histoires Admirables et Memorables de nostre temps* [...], Cologny: Samuel Crespin, 1614, pp. 438–439. Goulart gives the name of the source as "Marechalle de Raiz." His version mentions that the scene in the mirror took place in a "hall". The trick would therefore have featured some kind of a stage set. If it ever took place, it would most likely have happened in the summer of 1556 in Paris.

36 Honoré de Balzac, *About Catherine de' Medici and Other Stories*, trans. Clara Bell, Philadelphia: The Gebbie Publishing Co., Ltd, 1900, p. 254.

37 Robertson, *Mémoirs* récréatifs scientifiques et anecdotiques, I, Paris: Chez l'Auteur et a la Librairie de Wurtz, 1831, p. 344; Fulgence Marion, *L'optique*, Paris: Hachette, 1867; *The Wonders of Optics*, trans. and ed. Charles W. Quin, New York: Charles Schribner & Co., 1871, pp. 199–200. For an illustration of one possible arrangement of the dispositive, see fig. 56 (between pages 200 and 201). In his *letter IV* David Brewster gave a scientific explanation of such devices (without referring to Nostradamus) in *Letters on Natural Magic*, London: John Murray, 1833, pp. 59–62.

38 Nicolas Pasquier's version, see Baltrušaitis 1978 (as fn. 29), p. 187. In an engraving published by Baltrušaitis both versions have been combined. Nostradamus is shown doing his tricks within a magic circle, while the vision appears in a large horizontal mirror placed above a fireplace (ibid.). The source of the illustration is unknown. It can be traced back to Émile-

The appearances of views into a distance via an optical device, or *catoptric televisions*, to adopt Baltrušaitis's term, form an extensive topos tradition. Beside the many forgotten texts, magic mirrors can be found in classics like Shakespeare's *Macbeth* (Act IV, Scene I) and Sir Walter Scott's *My Aunt Margaret's Mirror* (1828), where two noble ladies consult the advice of a mysterious savant from Padua, who has a "very tall and broad mirror". It happens that "as if it had self-contained scenery of its own, objects began to appear within it".[39] The scene the ladies saw was "as if represented in a picture, save that the figures were movable instead of being stationary".[40] Francis H. Underwood's story *The Exile of von Adelstein's Soul* (1872), set in fifteenth-century Vienna, also featured a magic mirror.[41] Baron von Adelstein drives his coach over a young man, who happens to be the son of a witch, Frau Eldzeit. The young man dies, and the witch casts a spell, forcing the soul of the baron to leave his body at nights and to settle into the embalmed body of the dead man. The baron's chaplain and confessor, Father Wilhelm, begins to investigate why the baron becomes lifeless in the night and only recovers in the morning. He consults Albrecht Werner or Albertus Nyktalops, a mystical "philosopher" who has "a wonderful speculum or mirror of steel [...] by means of which distant objects and even spirits of the dead are brought within view at pleasure".[42]

Jules Grillot de Givry, *Le Musée des sorciers, mages et alchimistes*, Compiègne: Impr. de Compiègne; Paris: Libr. de France, 1929, but not further.

39 Sir Walter Scott, My Aunt Margaret's Mirror, in: *Waverley Novels, Vol. 40, Chronicles of Canongate. First Series. The Surgeon's Daughter. In Two Volumes, II*, Boston: Samuel H. Parker, 1833, pp. 198–235, pp. 224–225.

40 Ibid.

41 Francis Henry Underwood, The Exile of von Adelstein's Soul, in: *Cloud-Pictures*, Boston: Lee and Shepard; New York: Lee, Shepard & Dillingham, 1872, pp. 1–78. *The Exile of von Adelstein's Soul* was written in 1858, but published for the first time in this volume.

42 Ibid., p. 35.

The device is kept "in a darker room without angles or resting-places for the eyes, – all its lines, as in a perspective, tending to one point, in which was placed the speculum".[43] Father Wilhelm felt how "[t]he floating nebulous light that hung over its surface struck him with apprehension, for no lamp or other means of illumination was visible".[44] The brightness was so intense that "after looking at the wonderful light in the speculum, all other objects around were invisible in the gloom".[45] The mirror could only be consulted at night as "with the coming of dawn its brightness is dimmed, and at sunrise its power of reflection is gone."[46] Whether this disk of polished steel, which Albrecht had "obtained, while travelling in the East, from an Arabian philosopher", was a product of black or natural magic, "always within the limits set by the great First Cause", is left open.[47] Father Wilhelm's experience of looking at "the present occupation of some of the Baron's friends"[48] is worth quoting:

> *The priest fixed his eyes steadfastly upon the mirror, and thought of the Baron. Slowly the mirror seemed to become a window, expanding every moment like the opening of an iris, and growing more transparent, until at last there was before his vision the family group in the palace; the Baron rising from his chair, his mother shedding tears as she was about to accompany him to his chamber, and his sister hanging pensively upon his arm.*[49]

43 Ibid., pp. 36–37.

44 Ibid.

45 Ibid., p. 38.

46 Ibid., pp. 41–42.

47 Ibid.

48 Ibid., p. 36.

49 Ibid., p. 38.

It is tempting to interpret Underwood's imaginary device as a media dispositive. The positioning of the mirror as the luminous focal point for observation in a darkened space evokes cinema rather than television; the latter is often watched in lighted or semi-illuminated spaces. But there is a difference: unlike cinema, the system is interactive. Instead of buttons or switches, it uses – anachronistically speaking – a mind-machine interface (MMI): *channels* are *switched* by thinking about the person one wishes to see. Further, as Father Wilhelm discovered when he wanted to consult the scene of the accident where the young man had been killed a month earlier, the magic mirror was capable of presenting *reruns* of past events. He saw "microscopic figures passing through a distant square" as the incident unfolded.[50] The scenes were silent, which may well have been influenced by a contemporary media form: the room camera obscura, a popular attraction at seaside resorts and elsewhere.[51] The luminous real time image of events unfolding outside was projected on a circular table in the center of the darkened chamber. These comparisons make huge leaps between times and places far apart from each other, but they are not necessarily arbitrary. Material forms of media culture have been anticipated by discursive ones, technological solutions by imaginary or even *magic* ones.

A perfect example of the ways in which a topos tradition can migrate between cultural and mediatic contexts is the story *La Belle et La Bête* (1751), which became a carrier of topoi. Jeanne-Marie Leprince de Beaumont (1711–1780) adapted it in 1756 from a fairy tale published in 1740 by Suzanne Barbot de Villeneuve, which in turn had been inspired by earlier sources. Leprince de Beaumont's version became the inspiration for Marmontel and Grétry's successful opéra comique *Zémire et Azor* (1771). Jean Cocteau's classic feature film *La Belle et la Bête* (1946) inspired Philip Glass's opera (1994), while Disney's animation film (1991) is only one of the many re-tellings of the topos within popular culture.[52] The magic mirror as a way of observing people from a distance was already featured in Villeneuve's tale. Its role in Cocteau's *La Belle et la Bête* is particularly interesting because the film was released at the moment when television was just beginning its triumphal march.[53] Although its adoption in France was slow, television was already an established idea. It is likely that some of the spectators, who saw Cocteau's film, associated its magic mirrors with the television screen.

The television pioneer Alan A. Campbell Swinton stated in 1912: "[I]f there could be added to each telephone instrument what would indeed be a magic mirror, in which we could see even only in monochrome the faces of those with whom we were communicating, the material advantages would be great. In addition, there would be much senti-

50 Ibid., p. 48.

51 Underwood acknowledged he was "aware that the device of employing the magical speculum [...] is not a new one." Ibid., p. vi.

52 The magic mirror tradition lives on, beside fairytales, in Halloween postcards. The magic mirror is depicted as a device for seeing one's future husband or wife. This idea was associated with Nostradamus and Catherine de' Medici almost entirely erroneously in a masonic journal: "The famous Nostradamus conjured up in a magic mirror the phantasmal form of her future husband for Marie [sic] de Medicis." Mysticus, A Corner of the Library: The Magic Mirror, in: *The New Age*, XXIX.6 (1921), p. 274.

53 Since the 1920s, Cocteau's work often referred to contemporary media machines, often in relation to myths from classical antiquity. Magic mirrors play important roles in his poetic films *Le Sang d'un Poête* (1930) and *Orphée* (1950) as concretized metaphors and entry points into alternate realities. Lewis Carroll's *Through the Looking Glass* (1871) can be related to the same historical trajectory. It was also evoked in television promotion. See the double-page ad "... and through this looking glass. the Wonderland of NBC Big Color TV!" (c. 1956–1958), in: Steve Kosareff, *Window to the Future. The Golden Age of Television Marketing and Advertising*, San Francisco: Chronicle Books, 2005, pp. 116–117.

mental and other value."[54] Such associations became commonplace. Advertisers began comparing television sets with crystal balls and magic mirrors early on. They were a form of modern wizardry, "man's strangest dream come true in your home".[55] In 1939 a British publication characterized television as "the magic mirror of the living room".[56] In 1944, on the eve of the television era, a children's book looked back into the past while anticipating the future: "If you have a magic mirror you can see a play at a theater without leaving home, you can be with friends who live in another city. The magic mirror makes everything near. What is the magic mirror? It is a *television* machine."[57] The Admiral Corporation grasped the connection, branding its line of TV-sets *Magic Mirror Television*.[58] In 1940, the *Archery News* asked the fundamental question: "May not the magic mirror be nothing but an early conception of the possibilities of Television?"[59] The media archaeologist's answer is affirmative, but excludes the words *nothing but*, for the magic mirror has been other things as well.

54 Alan A. Campbell Swinton, Presidential Address. November 7th, 1911, in: *The Journal of the Röntgen Society* VIII.30 (1912), p. 7. Swinton's ideas about electronic scanning contributed to the development of television.

55 Wulf Herzogenrath, Thomas W. Gaehtgens, Sven Thomas und Peter Hoenisch (eds.), *TV Kultur. Fernsehen in der Bildenden Kunst seit 1879*, Amsterdam, Dresden: Verlag der Kunst, 1997, p. 146, p. 147, p. 157.

56 Reference to "magic mirror of the living room" is from *The Nation's Business* 27 (1939), p. 97.

57 Michail Il'in, Elena Segal, *A Ring and a Riddle*, trans. Beatrice Kinkead, Philadelphia: Lippincott, 1944, p. 72 [original emphasis].

58 *Questions and Answers about Admiral Magic Mirror Television*, booklet, II printing, Chicago: Admiral Corporation, March 1948. Author's collection. An advertisement in *Life* explained that the "magic mirror" was an aluminum sheet inside the tube to make the picture "*twice* as bright as ordinary TV." *Life* 37.11, (1954), p. 25. The expression "the magic mirror of television" remained in generic use *Television Magazine* 13 (1956), p. 62.

59 *Archery News* 19 (1940), p. 36. Only a snippet view can be seen on Google Books.

Conclusion: the Topos Reified

Returning to the issue of attention, it should be clear by now that normally media archaeology can say little about actual experiences. On the contrary, it casts doubt on the truthfulness and accessibility of *direct* observations recorded in discourses. What seems authentic and personal often turns out to be mere topoi in disguise. As members of societies, humans inhabit vast *topos spaces* brimming with received ideas and motifs that are used as molds for expressing contemporary issues and concerns. The topos transmissions happening in these spaces may not be fully acknowledged as such by those living within their reach and even taking part in their dissemination. However, there are those who are fully aware of the power of the topoi and use them to foster ideological and commercial ends. This applies to advertisers, image and identity makers and other professionals of the culture industry. Their businesses are centered on discovering recognizable formulas and revising them just enough to maintain the customers' interest. More often than not, the formulas are found from the past. Topoi are effective as *tools* because their stereotypical components appeal to mainstream taste while they can also be dressed up and presented as *the Coolest Thing on the Planet*. Camouflaging the Old as the New suggests both a moment of soothing recognition and an awe-inspiring encounter with the *unprecedented*.

The magic mirror is a perfect example because its presence is so widespread in today's popular culture. It is less due to Nostradamus than to Walt Disney, whose classic animation feature *Snow White and the Seven Dwarfs* (1937) has been seen by millions and has inspired countless products, both industrial and artisanal.[60] The iconic still image of the Evil Queen staring into her magic mirror has turned into an Internet meme, provided with captions and photo-shopped into other personalities, including (predictably) Donald Trump.[61] Compared with the examples discussed so far, something else is at stake: instead of displaying far-away scenes, Disney's mirror provides the Queen a *facetime session* with an incarcerated spirit via both sound and image. The talk is all about her. The theme is obsessive narcissism and the compulsion to constantly verify one's own beauty. Disney's magic mirror is therefore closer to the mundane mirror uses we all are familiar with. The popularity of this variant today must be associated with the selfie obsession, which has turned into a global phenomenon thanks to the ubiquity of the smartphone. Millions of people use their phone screens as interactive mirrors, capturing, posting and transmitting their likenesses to others, even naked. The captive spirit questioned by the Evil Queen has been replaced by peers posting replies online. Anything but a *Like* can cause lurid or even tragic reactions.

The idea, presented in the introduction, that a topos can materialize as a technological device has been confirmed by recent developments. As I am writing this, *magic mirror* has become a popular buzzword among both developers and fans of new technology. Applications vary, but the basic system configuration is relatively uniform: the magic mirror of 2017 is a screen overlaid with a two-way mirror (often a thin sheet of plastic) and connected to a computer. It allows the user to see one's own likeness while consulting information from the computer, the web or social media. The information floating around the user's mirror image is called forth by touches, gestures or voice commands. The screen is normally positioned vertically in the *portrait mode*, but taking the mirror comparison literally, Samsung introduced in 2015 a perfectly round wall-mounted interactive display.[62] Commercial *Magic Mirror Photo Booths* and *Magic Selfie Mirrors* enhance the posers' likenesses with digital features (much like Sega's Purikura arcade machines did years ago). Companies have also begun promoting magic mirrors to fashion stores, persuading customers to try design clothes and make-up virtually on their screen doubles.[63] The magic of the digital mirror is that of commodity fetishism; the awe it produces pure appearance.

The magic mirror has also turned into a do-it-yourself phenomenon. Anything from old PC monitors and tablet computers to flat panel LED screens have been converted for the purpose. Encouragement and instructions can be

60 There are online discussions of whether Disney's Evil Queen used the expression "Mirror, Mirror on the Wall" or "Magic Mirror on the Wall" when invoking the spirit. The latter is correct, although most people seem to think it is the former. Such mistaken collective memories are sometimes called *The Mandela Effect*. Fiona Broome, posted on December 15, 2015, at http://mandelaeffect.com/mirror-mirror-research/ (accessed July 13, 2017). The question about the Mandela Effect is interesting, but controversial. On her website Fiona Broome defines herself as "author, researcher, and paranormal consultant". Her forte is books about ghosts. http://fionabroome.com/about-the-author (accessed July 13, 2017).

61 Trump's figure has been – equally predictably – also inserted into memes inspired by *The Beauty and the Beast*.

62 A circular (and curved?) mirror display that recalls Hugo Gernsback's TV fantasies of the 1920s was introduced by Samsung in 2015 at the Internationale Funkausstellung Berlin (IFA 2015) as part of its Smart Signage Portfolio presentation. It used the company's OLED transparent display technology, unveiled for the first time. The round mirror display was a concept model for the store of the future. The design also resonated with the round displays of smartwatches, including Samsung's own Gear S2, which was released in 2015.

63 Rina Raphael, Interactive 'Magic Mirrors' Are Changing How We See Ourselves – And Shop, in: *Fast Company*, April 6, 2017, article 3066781, http://fastcompany.com (accessed July 13, 2017). The writer evokes Disney's Evil Queen as the point of comparison. As an indication of the current attention deficit disorders, a note states the article is a *9 minute read*.

found from online resources like the Magicmirrorcentral.com, which defines itself as a "website dedicated to DIY magic mirror making (some people refer to them as Smart Mirrors)".[64] The most common technical solution to power the magic mirror is the inexpensive Raspberry Pi mini computer with custom programming in Linux. Personal enthusiasm notwithstanding, there is something puerile in the concept. In earlier manifestations of the magic mirror topos one's own reflection was normally effaced and replaced by the *data* emanating from a distance. In an era of uncertainty, superficiality and rampant narcissism, it now seems necessary to constantly monitor one's external appearance (and body functions) and to be reminded that one still exists. The obsessive selfie snapping with any place one visits as background and anyone one meets and greets as *pals* serves the same goal. The magic mirror tradition further looms behind augmented reality applications. Whatever *everyday life* used to mean is no longer enough; it has to be enhanced and boosted, modified and converted.

64 On the *home* page, http://Magicmirrorcentral.com (accessed July 13, 2017).

Tristan Thielmann

Early Digital Images

A Praxeology of the Display

> *The digital image is non-existent. If anything is responsible for missing the point here, then it is inappropriate essentialism. What there are, are innumerable analog images that illustrate the data that are present in a digital form: on monitors, televisions or paper, on movie screens, displays and so on.*[1]

Shedding Light

Media studies have postulated consistently that digital images do not exist.[2] This has not prevented the visual studies and arts from continuing with the attempt at proclaiming a phenomenology of images that refuses to negate their digital transformation or even origin.[3] The fact that we perceive images on digital displays has led to an analytical imprecision in the development of theory in the visual studies, in that the technical conditions underlying displays are referred to the materiality of images. This contribution attempts to re-introduce the separation between image and image carrier, in order to make a statement on the specific properties of digitally generated pictorial worlds. By shedding light on the historical and, simultaneously, the practice-theoretical contribution of the display in the discourse on digital imagery, we can show that the scientific reflection on images in general has been unfoundedly loaded with meaning in the truest sense of the word.

The "material turn"[4] that is currently being diagnosed in the analysis of digital media practices as well as the discourse on "soft images"[5] reveals that we are increasingly dealing with dynamic, transparent and malleable displays adapting themselves to the individual user, context and situation. This raises the question as to what contribution a practice theory can make toward conclusive media esthetics, media history and a media theory of the display.

Such a practice theory must be assessed on the basis of "the practical procedures being given precedence over all other explanatory parameters".[6] Under reference to Harold

1 Claus Pias, Das digitale Bild gibt es nicht. Über das (Nicht-)Wissen der Bilder und die informatische Illusion, in: *Zeitenblicke* 2.1 (2003), http://zeitenblicke.de/2003/01/pias/ (accessed November 1, 2017).
2 See ibid.; Wolfgang Hagen, Es gibt kein digitales Bild. Eine medienepistemologische Anmerkung, in: Lorenz Engell, Bernhard Siegert, Joseph Vogl (eds.), *Archiv für Mediengeschichte. Licht und Leitung*, Weimar: Bauhaus-Universität Weimar, 2002, pp. 103–110.
3 See Gundolf S. Freyermuth, Lisa Gotto (eds.), *Bildwerte. Visualität in der digitalen Medienkultur*, Bielefeld: transcript Verlag, 2013.
4 Bill Brown, Materiality, in: W. J. T. Mitchell, Mark B. N. Hansen (eds.), *Critical Terms for Media Studies*, Chicago: The University of Chicago Press, 2010, pp. 49–63; Sarah Pink et al. (eds.), *Digital Materialities. Design and Anthropology*, London/New York: Bloomsbury, 2016.
5 Ingrid Hoelzel, Rémi Marie, *Softimage. Towards a New Theory of the Digital Image*, Bristol/Chicago: The University of Chicago Press, 2015.
6 Erhard Schüttpelz, Skill, Deixis, Medien, in: Christiane Voss, Lorenz Engell (eds.), *Mediale Anthropologie*, Paderborn: Wilhelm Fink, 2015, pp. 153–182.

Garfinkel, we could also formulate this as follows: "Praxeology seeks to formulate statements of method, and to extend their generality, seeking as wide a domain of applicability as possible."[7]

The aim of a media practice theory of the display must be to unveil the methods of the medium.[8] Given the diversity and multiplicity of displays, this essay therefore pursues the question of what sociotechnical properties are exhibited phenomenologically by digital displays. What constitutes their specific media characteristics that distinguish them from all other forms of electronic monitors and screens? We initially need to take a step back to help us to better estimate the scope of the current development. How long have we actually been in a position of referring to the *display* as an independent medium?

The Top View of the Display

In 2003 for the first time, more LC displays were sold in Germany than conventional monitors with cathode ray tubes (CRT).[9] Since then, a fundamental change in screens can be diagnosed: from the stable, fixed CRT monitor to the flexible, mobile LC display;[10] from the heavy, furniture-like, three-dimensional object that reveals an image on its *open* side to the thin and fluid two-dimensional digital surface that appears to be one with what it depicts.

While the term monitor (Latin: an overseer, instructor or guide) still expresses specifications of an observing subject, at first sight, the display shifts what is being presented and exhibited into the center. The term "screen", which primarily emphasizes protection from electronic radiation (electronic images), already points to this fundamental difference between it and the display.[11] The type of visibility appears to be the key to understanding the display culture.[12] The materiality of the display will therefore be at the focus of a historical genealogy in the following, which reveals a series of media practice-theoretical determinants.

When contemplated from a technical and historical perspective, the term display does not originate from the medium of the television or computer but rather from the military medium of the radar: the radar display refers to the radar screen. The first field experiment using radar was conducted in Great Britain as early as February 26, 1935, during which a test airplane produced an additional illuminated dot on the screen of a cathode ray oscilloscope through the radio waves emitted by the BBC transmitter in Daventry being reflected off the body of the plane.[13]

7 Harold Garfinkel, Some Sociological Concepts and Methods for Psychiatrists, in: *Psychiatric Research Reports* 6 (1956), pp. 181–198, p. 191.

8 For visual media methodologies see: Gillian Rose, *Visual Methodologies*, London: Sage Publications, 2001.

9 See Andreas Wilkens, Erstmals mehr LC-Displays verkauft als Röhrengeräte, in: *heise.de*, 16.03.2004, http://heise.de/newsticker/meldung/45600 (accessed November 1, 2017).

10 Liquid crystal display (LCD) is an umbrella term for liquid crystal screens. The thin film transistor (TFT) refers to specific LCD technology that is used to created large-scale electronic circuits. TFT technology is currently the dominant flat screen technology, which is why LCD and TFT are used almost synonymously. More modern LCDs are also called LEDs as they use light-emitting diodes for background lighting. This allows a more compact construction and thus thinner displays. LED technology is currently becoming more and more prevalent, also in the form of its organic variant (OLED) that possesses a lower luminance density and therefore no longer relies on the use of monocrystalline materials.

11 The French écran also originally means visual protection. The term screen that is generally used in English refers to the projection surface in movies and to television, video and computers. Gunther Kress, "Screen": Metaphors of Display, Partition, Concealment and Defence, in: *Visual Communication* 5.2 (2006), pp. 199–204.

12 See Sean Cubitt, *The Practice of Light. A Genealogy of Visual Technologies from Prints to Pixels*, Cambridge, MA, 2014.

13 An oscilloscope detects changes in voltage in an electric circuit using a light trace. This becomes visible on an analog computer based on the same

During this experiment, the radar display was already calibrated so that distances were on a linear scale and it was thus easy to take a reading.[14] A ruler was depicted on the oscilloscope, from which the distance of a flying object could be read off, based on where the pulse of the echo signal amplitude was produced. However, in spite of their scaling, the distances that were measured were not yet georeferenced. Cartographic projections were not yet possible with the oscilloscope. The first radar displays simulated the practice of reading a ruler (fig. 1).[15]

This changed with the plan position indicator (PPI), proposed in 1935 and used for the first time in 1940, that allowed a top-down view of events.[16] This is the classical form of the radarscope with a panoramic display (fig. 2).[17]

"Such a system with a rotating, or sweeping, line is what most people continue to associate with a radar display."[18] The PPI display indicates the distance and direction for all altitudes through a *sweep* (a scan line that corresponds to the position of the radar antenna in a given moment) that rotates around the center of the cathode ray tube and depicts the echo blips as bright dots: "With this form of display, the airplanes were represented as dots in a two-dimensional representation of the actual airspace. Here, for the first time perhaps, a bridge is constructed between the representation of technoscientific data, such as offered by the oscilloscope, and the mimetic representation made possible by the television."[19]

In Germany, the viewing tube of the all-round sensing system was called a "*Sternschreiber*".[20] The first decades of displays are not characterized by a (4:3 or 16:9) rectangular section of *reality*, but by circular screens. Using radar systems ("*Rundsuchanlagen*", lit. "circular search apparatus") required panoramic viewing instruments.

All-round sensing with a panoramic display only became available in Germany in 1944. A distance indicator, EAG 62 "Emil", was used to locate enemy bombers and to guide the German fighter jets.[21] However, joint depiction of the fighter jets and enemy bombers was required to guide the fighter jets, leading to the development of the "*Himmelbett*" method in 1941, led by Josef Kammhuber, to project the measurements for distance, azimuth and flight altitude as spots of light onto the glass disc of the "Seeburg plotting table".[22]

In this case, the display is a window onto the turntables behind it that combine the different representations produced by the pixel projector ("*Bildpunktwerfer*") into an indexed picture (fig. 3). The foundations for the display as an independent playback medium were thus laid, and the term display was transformed from the *indicated* to the *indicator*,

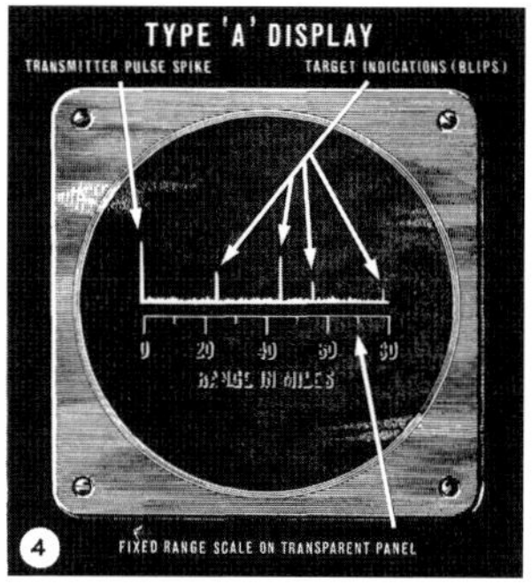

1 Radar oscilloscope before 1940.

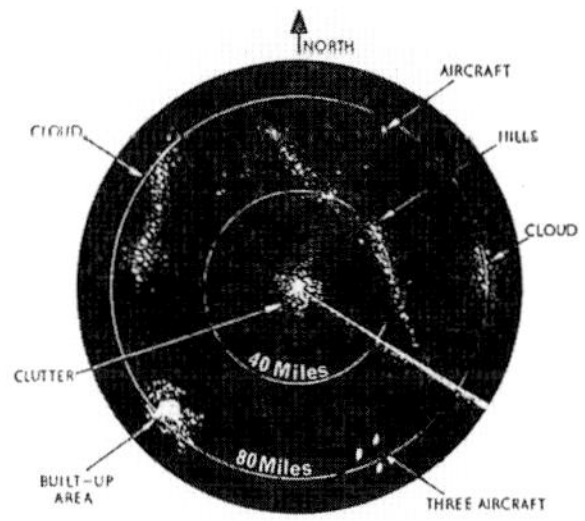

2 Schematic illustration of a classical displays on a plan position indicator.

principle as the Braun tube, as is also still the case for the image signal in current televisions. Robert Watson-Watt was already using an oscilloscope as a display for locating storms in 1923.

14 See Robert Watson-Watt, *Three Steps of Victory*, London, 1957, p. 471.

15 The History of Flight Radar, http://planefinder.net/about/the-history-of-flight-radar/ (accessed November 1, 2017).

16 Robert Watson-Watt, The Evolution of Radiolocation, in: *Journal of the Institution of Electrical Engineers* 93 (1946), pp. 374–382, p. 379.

17 Manfred von Ardenne had already developed a precursor to the panoramic display tube in the mid-1930s. See Manfred von Ardenne, Ein neuer Polarkoordinaten-Elektronenstrahl-Oszillograph mit linearem Zeitmaßstab, in: *Zeitschrift für technische Physik* 17 (1936), pp. 660–666. However, Göring rejected further development after looking at the drafts and photos of "figures on a fluorescent screen". Manfred von Ardenne, *Ein glückliches Leben für Technik und Forschung*, Munich, 1976, p. 131.

18 Wikipedia, Radar in World War II, http://en.wikipedia.org/wiki/Radar_in_World_War_II (accessed November 1, 2017).

19 Charlie Gere, Genealogy of the Computer Screen, in: *Visual Communication* 5.2 (2006), pp. 141–152, p. 146.

20 Fritz Trenkle, *Die deutschen Funkführungsverfahren bis 1945*, Heidelberg: Hüthig, 1987, p. 200.

21 David Pritchard, *Durch Raum und Zeit. Radarentwicklung und -einsatz 1904–1945*, Stuttgart, 1992, p. 63.

22 Trenkle 1987 (as fn. 20), pp. 193–194.

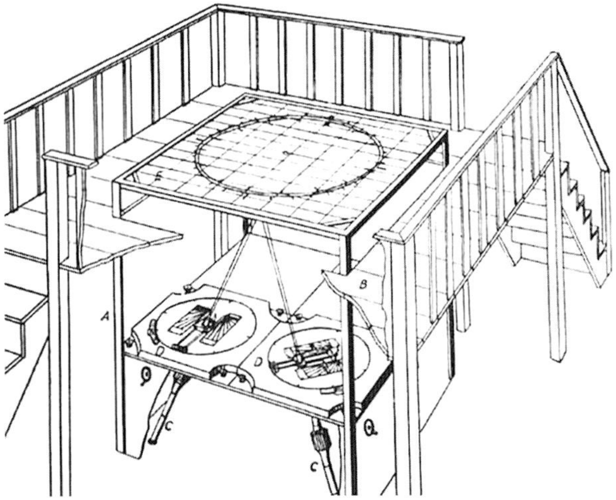

3 Sketch of the *Seeburg plotting table*.

from the *(re)presented* to the *(re)presenter*, from the image to the medium.

Instead of the now common sight of vertical screens, the dispositif of one of the first displays was determined by people moving around a table in the *Himmelbett* method. The compact all-round viewing instrument ("*Rundsichtgerät*") in the "*Jagdschloß*" system also required observation from above.[23] Even the frontally attached panoramic display in the 1944 mobile "*Rundsuchanlage*", "*Jagdwagen*" or "*Panotwiel*", still makes reference to this: it was called "*Drauf*" (from above).[24] It is this special dispositive structure, in particular, that characterizes the *point of departure* for the display and its history (fig. 3).

There were also enormous map tables in the English command centers of the pre-display era, on which, however, bits of cardboard cut into the shape of airplanes were still being used to indicate their current position (fig. 4). While the male officers studied the map, female soldiers continually changed the position of the miniature airplanes based on the incoming radar information, by moving them with long sticks.[25] The human actors were also moved about in the same way as the model airplanes. The display of the *Seeburg plotting table* has this in common with map tables and paper maps (fig. 4).

The methods that are used to process and depict the information collected from the radar have not changed the *puppet strings* guiding the supposed beneficiaries. The movement of distant objects determines the motion of the person in front of the display. The display renders the location of a distant object into externally guided movement and location of the person, as the position of the person in front of the display is determined by the position of the objects detected by radar. The display creates a uniform interaction space in which distant objects that are out of view are aligned with a subject making the observation. Coordination thus occurs without (visual) contact.

Radar technology creates semipermeable spaces in which object coordinates from an external space penetrate into an internal space, and resultant actions in the internal space have effects on the external space. A sociospatial duplicity of vehicular coordinates is produced by the simultaneity of movement in geographical space and in the map room.

23 Ibid., pp. 106–107.
24 Ibid., p. 112.
25 Lev Manovich, *The Language of New Media*, Cambridge, MA/London: MIT Press, 2001, p. 100.

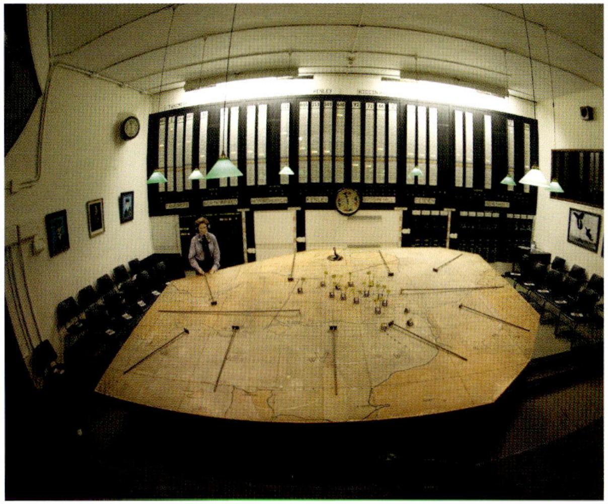

4 Plotting Room, Uxbridge, GB (1939).

The *Himmelbett* method renders the display dispositif evident. It opens up three spaces to us that can be subjected to a comparative investigation within the meaning of media praxeology. (1.) Firstly, we are dealing with objects that are out of sight, that are also moving rapidly and over great distances, and thus are scattered over a correspondingly large external space. (2.) The internal space of the display is also not visible to the observer. In this respect, it represents a miniaturized model of the external space, in that different actors (fighter jets and bombers) are being represented 1:1 here by their own medium (a ray of light). (3.) Finally, there is also the space in front of the display. This is where the actors behind and outside the display are transformed into objects that are represented together on one area. In this interaction space, alliances are formed between different observable objects and the different observers.

What is depicted, how it is depicted and the dispositive position of the subject are all still contingent here. The view from above results in a unification that documents both the dispositive structure of human and display, the relationship between significats in the external space and signifiers in the internal space, as well as all intermediaries,[26] bringing about the translation from things to signs. The *how* of the unification remains hidden, even though it takes place neither electronically, nor algorithmically, but optically and electromechanically.

If we understand technology to be a distributed action,[27] then this opens up a space that separates the space in front and below the display from each other; but the causal relationship between significat and signifier still remains preserved. The external and internal worlds of the display behave in a homomorphic fashion in relation to each other.

In the case of the Seeburg plotting table, we are still primarily dealing with interface agents who are standing around a display. In contrast, in the case of the first digital displays, we start to talk about the particular importance of coordination agents, the active, no longer passive, movement in space that was central to the functionality of the first computer and its display.[28]

26 The actor-network theory defines an intermediate link or binding agent in a sociotechnical translation chain as an intermediary, which simply transports meaning, while a mediator, or go-between, simultaneously also transforms a meaning that is to be transported. See Bruno Latour, *Reassembling the Social. An Introduction to Actor-Network-Theory*, Oxford: Oxford University Press, 2005; Ann-Sophie Lehmann, Das Medium als Mediator. Eine Materialtheorie für (Öl-)Bilder, in: *Zeitschrift für Ästhetik und Allgemeine Kunstwissenschaft* 57.1 (2012), pp. 69–88. A media practice theory that places digital displays in its center no longer needs to differentiate between *intermediaries* and *mediators*.

27 See Werner Rammert, *Technik – Handeln – Wissen. Zu einer pragmatistischen Technik- und Sozialtheorie*, Wiesbaden: Springer, 2007, pp. 79–81.

28 See Ibid., p. 83.

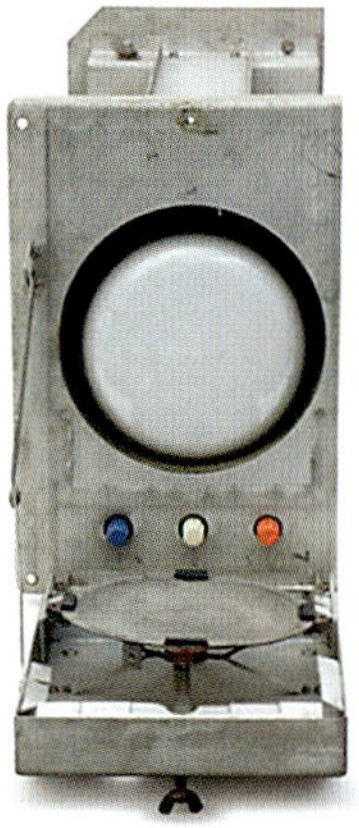

5 The Manchester Mark 1 Williams tube with the pickup plate.

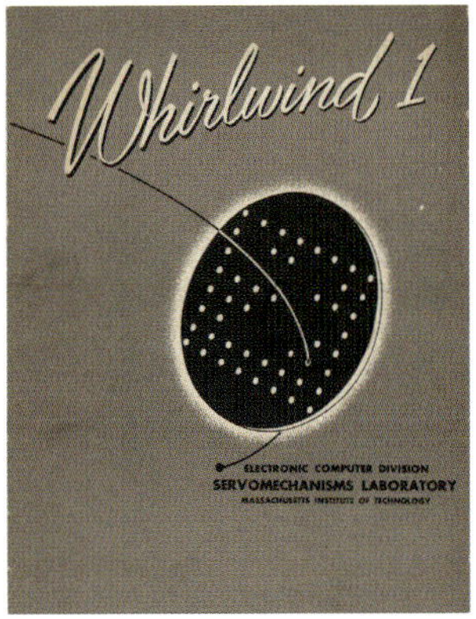

6 Cover of the Whirlwind I brochure with a schematic illustration of a Williams tube.

The Discrete Nature of the Display

For the first time in the history of computing and optical media, the radar screen allows for directly addressing an individual pixel on an illuminated area without having to pass through "precursors" and "successors".[29] This is of relevance to the optical media, as follows: In contrast to the dispositif of the television, it is not only the rows but also the columns of an image that are decomposed into single elements. In terms of the computing media, this means: a number is broken down into its decimals and is no longer depicted in its semantic unity, but instead distributed across the area of a decimal digit(al) display.

This discrete nature of the geometric location and chromatic value distinguishes the radar display from the movie/television picture and, long before the computer was to conquer media, already pointed toward our current media era in which displays have become the signature of (full) digitality. While calculating with decimal digits was still associated with the analog world, this now changes for the computer-illiterate observer, through the use of binary digits.

The 1948 Manchester Mark 1 is regarded as the first binary digital computer. Computer CRTs in this computer also acted as an accumulator, as well as control and arithmetic registers. The computer pioneers Tom Kilburn and Frederick Williams used modified cathode ray tubes, called "Williams tubes", as random access memory (RAM) for the 1948 Manchester Mark 1 (fig. 5).[30] Such vacuum tubes were widely used as RAM up until the 1950s, for example, in the mainframe computers ERA 1103 (with a Williams tube memory of 1024 words) and IBM 701 and 702[31] (fig. 5–6).

What is decisive in relation to the Williams tube is that, with it, a new form of visibility or image processing appears in the true sense of the word. However, the data that are to be processed are not simply intended for the visualization of the invisible or the absent during completion of their processing, but the dot images themselves *are* the data in the RAM. They do not represent, but instead index. They are images that are not intended for the eyes of the users, but are observed by the computer itself.[32]

The Whirlwind computer was one computer that worked with Williams tubes, the first computer that processed data in real time and was simultaneously capable of depicting anumeric data on a display.

An early computer advertisement for the Whirlwind computer shows how a cathode beam hits the phosphor layer in a cathode ray tube and produces an illuminated dot (fig. 6). An illuminated point (digit) represents a 1 and a non-illuminated point a 0 in this rectangular grid in the Williams tube. In practice, however, as a pickup plate was fixed opposite the data screen to protect the data storage tubes from electromagnetic radiation, the data storage contents could not be read off directly during normal operation. People in media studies therefore reached the conclusion that the digital *per se* is withdrawn from perception and that the hidden nature forms the condition of possibility for

29 Friedrich Kittler, Computergrafik. Eine halbtechnische Einführung, in: Herta Wolf (ed.), *Paradigma Fotografie. Fotokritik am Ende des fotografischen Zeitalters*, Vol. 1, Frankfurt/M.: Suhrkamp Verlag, 2002, pp. 178–194, p. 179.

30 Simon H. Lavington, Computer Development at Manchester University, in: Nicholas Metropolis et al. (eds.), *A History of Computing in the Twentieth Century*, New York/London: Acdemic Press, 1980, pp. 433–443, p. 433.

31 The Airspace Company Convair later developed a 7 inch tube monitor for the ERA 1103 which could display 6 × 6 characters. See Paul E. Ceruzzi, *Eine kleine Geschichte der EDV*, Bonn: mitp-Verlag, 2003, pp. 63–66.

32 Claus Pias, *Computer-Spiel-Welten*, Munich: Sequenzia Verlag, 2002, p. 75.

the function of digital data storage.[33] Full separation of data and display, which can be regarded as the central adage of digitality, was only achieved with Whirlwind I.

In 1945, project leader Jay Forrester started with the development of the Whirlwind computer at the Massachusetts Institute of Technology (MIT) with the words: "We are no longer building an analog computer; we are building a digital computer."[34] "One of the things that I think we did first was to connect a visual display to a computer",[35] reports Robert R. Everett, the engineer on the Whirlwind project at that time.

In this case, displays also initially served the purpose of checking for errors and carrying out tests, not for complex data output or input.[36] For example, in 1949, only one *ball* (dot) jumps across the 5" Tektronix oscilloscope of the Whirlwind computers to demonstrate the speed and graphics capabilities of the computer (fig. 5–6). When the Bouncing Ball Program is referred to as the first demo program in the history of software,[37] "the first display program ever written"[38] and "the first significant use of the computer display screen",[39] this is due to a perspective that attributes far greater importance to the media specifics of moving characters than the (admittedly abstract) illustration of a trajectory. Otherwise, the ENIAC demo program would have to have been ranked first in a media history of the display – especially if we consider the fact that the mediality of the ENIAC targets the contingent visualization and calculation of trajectories, just like the Whirlwind. These displays reveal an immediate continuity in computing practices – at least when Whirlwind I is considered in the context of the development of the Semi-Automatic Ground Environment (SAGE).

> *The obvious updating of SAGE related to [...] the status of the picture. The radar systems that were connected supplied the position of an object with the aid of angular coordinates that were converted into Cartesian coordinates based on the location of the radar and were indicated on the screen. The separation of data and display creates an arbitrariness of the depiction, such that it is no longer the screens doing the work (as is the case for the* Williams tube*) but users working on them.*[40]

During the Bedford Tests in 1950, the Whirlwind computer was used as the central control station for the Cape Cod Experimental Air Defense System, a prototype for the aerial defense and early warning system, SAGE. An additional "computer-generated visual display"[41] not only depicted

33 See Wolfgang Ernst, Den A/D-Umbruch aktiv denken – medienarchäologisch, kulturtechnisch, in: Jens Schröter, Alexander Böhnke (eds.), *Analog/Digital – Opposition oder Kontinuum? Zur Theorie und Geschichte einer Unterscheidung*, Bielefeld: transcript-Verlag, 2004, pp. 49–65; Pias 2002 (as fn. 32), p. 75.

34 Robert R. Everett, Whirlwind, in: N. Metropolis et al. (eds.), *A History of Computing in the Twentieth Century*, New York/London: Acdemic Press, 1980, pp. 365–384, p. 365.

35 Ibid., p. 375.

36 "All we used the displays for was testing the various parts of the system so displays were ancillary completely to the main event." Norman Taylor, as cited in Jan Hurst et al., Retrospectives I. The Early Years in Computer Graphics at MIT, Lincoln Lab and Harvard (Panel Proceedings of SIGGRAPH '89), in: *Computer Graphics* 23.5 (1989), pp. 19–38, p. 22.

37 Claus Pias, Die Pflichten des Spielers. Der User als Gestalt der Anschlüsse, in: Martin Warnke et al. (eds.), *Hyperkult II*, Bielefeld, 2004, pp. 313–341, p. 321.

38 Jan Hurst et al., Retrospectives I. The Early Years in Computer Graphics at MIT, Lincoln Lab and Harvard(Panel Proceedings of SIGGRAPH '89), in: Computer Graphics 23.5 (1989), pp. 19–38, p. 21.

39 Jan Hurst et al., Retrospectives II. The Early Years in Computer Graphics at MIT, Lincoln Lab and Harvard (Panel Proceedings of SIGGRAPH '89), in: *Computer Graphics* 23.5 (1989), pp. 39–74, p. 40.

40 Pias 2002 (as fn. 32), p. 77.

41 IBM Corporation – Military Products Division, On Guard! The Story of SAGE, https://archive.org/details/OnGuard1956 (accessed November 1, 2017).

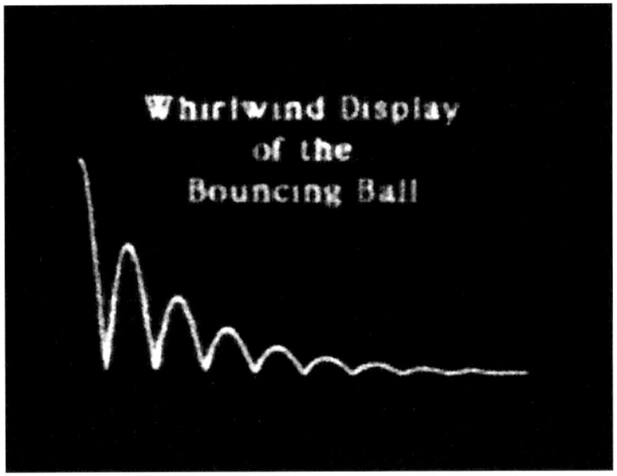

7 Trajectory of a bouncing ball and a rocket on the Whirlwind I 5" Tektronix oscilloscope.

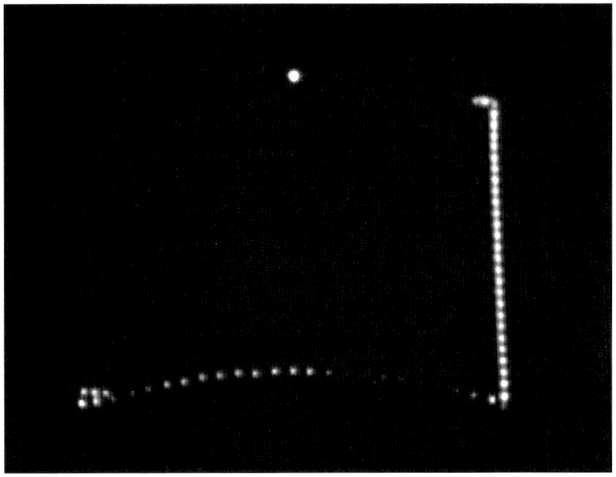

8 Trajectory of a rocket (without afterglow). The x-axis of the coordinate system is slightly curved on the right display. The dot of light (the rocket) is located at the highest point, the apex of the trajectory. The right bar indicates the velocity of the rocket.

dots but also depicted and georeferenced lines and text as Cartesian target coordinates in the visualization of airplane movements (fig. 7–8):

> *Such screens, referred to as 'vector screens' […] had no filled areas, but potentially only lines, such as those indicating borders and travel routes. The alphanumeric text of coordinates also had the status of a map. Letters and numbers were composed of dots in the blackness of the monitor that were connected by lines like an astronomical constellation.*[42]

Even so, practices relating to the display still had a long way to go before it could be used like a map. The initial difficulties in depicting flying objects on electronic maps included the fact that the radar system at Bedford airport that was used to carry out the first Cape Cod tests allowed two different modi of data representation.[43] The radar antennae rotated four times a minute without suppression of interference, while the radar only achieved two rotations when suppression of *clutter* was switched on – the deletion of undesirable data was required for improved legibility of the display (fig. 7–8).

42 Claus Pias, Punkt und Linie zum Raster, in: Markus Brüderlin (eds.), *Ornament und Abstraktion*, Cologne: Dumont, 2001, pp. 64–69, p. 66.

43 Kent Redmond, Thomas M. Smith, *From Whirlwind to MITRE. The R&D Story of the SAGE Air Defense Computer*, Cambridge, MA: MIT Press, 2000, pp. 81–82.

In practice, the airplanes thus easily disappeared from the perceptual horizon as their blips only lit up twice a minute. Continuous *tracking* was therefore almost impossible.[44]

The determination of the location of mobile airborne objects and the depiction of their location on the display were strictly coupled to each other. Even though the digital PPI screen therefore primarily visualizes the function of the rotating radar antenna itself and thus renders apparent the media methodology of the display, all practical steps that are taken are tied to it as if the basis of the display image was formed by a continuous "track-while-scan ability"[45].

On March 13, 1953, Robert Wieser explained in a lecture to visitors to the Cape Cod Experimental Air Defense System what the different data processing steps were in order for the Whirlwind computer processes to make a *special display* available that monitors and controls the paths, path guidance and path guidance aids:

> *The radar data is fed into the Whirlwind I computer at the Barta Building in Cambridge, which processes the data to provide 1) vectoring instructions for mid-course guidance of manned interceptors and 2) special displays for people who monitor and direct the operation of the system.*
>
> *In processing data, the computer automatically performs the track-while-scan function, which consists of l) taking in radar data in polar coordinates, 2) converting it to rectangular coordinates referred to a common origin, 3) correlating or associating each piece of data with existing tracks to find out which pieces of data belong to which aircraft, and 4) using the data to bring each track up-to-date with a new smoothed velocity and position, and 5) predicting track positions in the future for the next correlation or for dead reckoning if data is missed. Once smoothed tracks have been calculated, the computer then solves the equations of collision-course interception and generates and displays the proper vectoring instructions to guide an interceptor to a target.*[46]

Even though the entire process was predetermined, such that the operators are denied independent coordinating practices on the displays in the literature, the interpretation that is required in relation to the displays must not be forgotten in a sociotechnical consideration of the issue.[47]

The radar screen shows radial snapshots in time, at regular intervals, of the continuous movement of an object in real space. The conditions underlying the visualization are clear and apparent to the participating actors; they require no justification. Nobody would assume that the airplane that is being tracked in this way was moving discontinuously, as depicted on the display. A significant praxeology of the display has developed based on the presentation of incomplete path information and its completion by the actors, as practices relating to interpretation and actions have arisen due to this medium that require no further justification, given the strict adherence to instructions and incorporation into the rules of the system.

44 See Servomechanism Laboratory, Air Traffic Control Summary Report 7, July 25, 1950 – October 25, 1950, Massachusetts Institute of Technology, Cambridge, MA, 1950.

45 Redmond, Smith 2000 (as fn. 43), p. 77.

46 C. Robert Wieser, Cape Cod System and Demonstration. Memorandum VI – L-86, MIT Lincoln Laboratory, Division 6, Cambridge, MA, March 13, 1953, p. 2, http://dome.mit.edu/bitstream/handle/1721.3/41510/MC665_r28_L-86.pdf (accessed November 1, 2017).

47 See Kjeld Schmidt, *Cooperative Work and Coordinative Practices. Contributions to the Conceptual Foundations of Computer-Supported Cooperative Work (CSCW)*, London et al.: Springer, 2011, p. 318.

Precisely because the computer users sitting in front of the displays were not in a position to change the situation or the program,[48] practices relating to displays require a not entirely insignificant ability to normalize, as is also demonstrated in studies on current GPS navigation practices.[49]

Lev Manovich's reference to "visual nominalism", the automated "use of vision to capture the identity of individual objects and spaces by recording distances and shapes",[50] must therefore be contrasted with an equivalent *visual norminalism*. Manovich reduces the radar to "seeing without eyes",[51] whereby the praxeological changes that arise from the radar display are overlooked: a modified *seeing with eyes* under the conditions of *remote sensing*.

In addition, the central change that is associated with the digital radar display occurs at the interactional level: in mid-1950, a joystick was initially used to pursue a target object, with which a mobile dot of light was moved until it covered the radar blip of the target object. The target thus selected could then be pursued automatically.[52]

In addition to this manual targeting, simultaneous work was being carried out on a) how targeting can be carried out automatically within a selected corridor, b) how target objects can be separated from each other, even if they are close together, and c) how data can be smoothed – something that was required due to the inaccuracy of the positional data obtained by radar.[53] The unified planimetric depiction on the display required "rationalization of sight"[54] – a modification to adapt the technically perceived reality to the conditions of the display.

Robert Everett, who had already made a significant contribution with Jay Forrester toward driving forward the construction of the Whirlwind computer, developed a light gun as a computer interface – "a photoelectric device which is placed over the desired spot on a display scope" – as the joystick had proven too slow and cumbersome to operate.[55] The joystick was then discarded at the end of 1950. The medium of the light gun corresponded to the planned media practice of the Air Defense computer "to zero in on a selected target",[56] not only on a screen but also in the air with the help of fighter jet interceptors.

The programing foundations for the interception tests were laid in the second half of 1950. The criteria for a successful computer test were a) the simultaneous tracking of the target of the attack and the interceptor on the display, b) the calculation of the target guidance data, and c) the transmission of the control command to the fighter jet interceptor or rocket interceptor.[57] However, one problem in relation to tracking flying objects was that stationary targets were not yet suppressed on radar displays until mid-1950.[58] This only changed in October 1950, with the patenting of the "Moving

48 Judy E. O'Neill, *The Evolution of Interactive Computing Through Time-sharing and Networking*, University of Minnesota, 1992, p. 21.

49 See Barry Brown, Eric Laurier, The Normal Natural Troubles of Driving with GPS, in: *Proceedings of the ACM CHI 2012. Conference on Human Factors in Computing Systems*, Austin (2012), pp. 1621–1630.

50 Lev Manovich, The Mapping of Space: Perspective, Radar, and 3-D Computer Graphics, in: Thomas Linehan (ed.) *Computer Graphics Visual Proceedings, Annual Conference Series, ACM SIGGRAPH '93*, New York, 1993, pp. 143–147, http://manovich.net/content/04-projects/003-article-1993/01-article-1993.pdf (accessed November 1, 2017), p. 2.

51 Ibid., p. 4.

52 See Servomechanism Laboratory, Air Traffic Control Summary Report 6, April 25, 1950 – July 25, 1950, Massachusetts Institute of Technology, Cambridge, MA, 1950, p. 4, p. 8.

53 See Redmond, Smith 2000 (as fn. 43), p. 80.

54 William M. Ivins, *On the Rationalization of Sight. With an Examination of Three Renaissance Texts on Perspective*, New York, 1973 [Original published as *Metropolitan Museum of Art Papers* 8 (1938)].

55 Servomechanism Laboratory 1950 (as fn. 52), p. 1.

56 Redmond, Smith 2000 (as fn. 43), p. 81.

57 See Ibid., pp. 83–84.

58 See Ibid., p. 79.

Target Indicator", a module for the suppression of stationary targets.[59]

Before radar targeting of mobile objects turned into the suppression of stationary targets, two 5" oscilloscopes were used to separate mobile from stationary objects. A 12" PPI screen only became available to the MIT Digital Computer Laboratory toward the end of 1950.[60] One display was used to select objects with a light gun and a second display for the presentation of the marked objects. Input and output screens were still separated from each other.

> *Tracking would be initiated manually by applying the light gun to a selected target on the main scope. So far, use of the test patterns indicated that the light-gun technique should work. Holding the light gun on the location of the target long enough to detect the computer's next scanned display would transfer the target's display to the second scope. Doing the same to the interceptor's spot on the first scope would select it too for display on the second scope. From that time on, the two selected blips would be tracked in isolation on the second scope, without further need for the light gun. Their courses would be predicted on the basis of the history of preceding sightings. A collision course would be computed, proper heading instructions for the interceptor would be displayed, and the scope operator would pass on the information by voice to the pilot in the interceptor.*[61]

59 Charles T. Baker Jr., Moving Target Indicator Radar, Patent-No. US 2811715 A, October 2, 1950.

60 See Redmond, Smith 2000 (as fn. 43), p. 83.

61 Ibid., p. 84.

The scope operator's task was thus reduced to the selection and passing on of information. The operator fulfilled a relay function[62] that also could have been automated.

In this sense, selection was part of a "distributed cognition" process,[63] because the power behind the action of selecting the target with the *light gun* is limited to the separation of a mobile object from the stationary objects, to extracting it and transferring it to a second display using copy/paste. This second display is based on a different construction of reality: on a world that only knows vehicles.

The production of coordinates that had been conducted previously outside the display was shifted to translation steps between two displays. With reference to the narrative and visual complexity of the overall context that is being depicted, this therefore threw the development of displays back to before the Seeburg plotting table, which was oriented to an even greater extent based on mimetic procedures. The *Himmelbett* method required a consensual interpretation by the actors standing around the display. From an action-theoretical perspective, the development of the radar display that occurred in 1950 simultaneously constituted a step in sociotechnical innovation: Through the separation into a data input and a data output display, "cooperation

62 See Antoine Hennion, Cécile Méadel, In the Laboratories of Desire. Advertising as an Intermediary between Products and Consumer, in: *Reseaux. The French Journal of Communication* 1.2 (1993), pp. 169–192. Within the scope of the actor-network theory, a relay can also be understood as a team that takes over and triggers a relay race of further actions. See Tristan Thielmann, Digitale Rechenschaft. Die Netzwerkbedingungen der Akteur-Medien-Theorie seit Amtieren des Computers, in: Tristan Thielmann, Erhard Schüttpelz (eds.), *Akteur-Medien-Theorie*, Bielefeld: transcript-Verlag, 2013, pp. 377–424, p. 382.

63 Edwin Hutchins, *Cognition in the Wild*, Cambridge, MA: MIT Press, 1995.

without consensus"[64] became possible for the first time in graphic human-computer interaction, in that targeting (enemy flying objects) and target guidance (of the fighter jet interceptors) were separated media-praxeologically and became an action distributed across displays.

At this point, it is interesting to note the use of the term *scope*, used colloquially to refer to a) an oscilloscope and b) a viewfinder or a telescopic sight in military terms, and c) the word more generally refers to a frame for movement and latitude.

These three different levels of meaning also actually emerge in the SAGE Air Defense Computer and its prototypes. On the one hand, this display is a converted measuring instrument – an oscillograph that reduces waveform graphs to the depiction of dots. The *scope* is one such measuring instrument and indicates that it once simply served to control the computer, as was the case for the ENIAC. On the other hand, it also served as an instrument to search for enemy objects. At the same time, the only demand that could be made was the displaying of a section of reality: a reality that is solely determined by the movement of the objects that have been reduced to a dot.

Equally, the *scope* no longer serves the sole purpose of monitoring the internal and external world of the computer. With the advent of digitality, the internal world increasingly closes up; additional translation steps are incorporated by new interfaces; the external world is simultaneously represented in a more media-differentiated way – limited to a circular excerpt, to mobile objects and discrete characters.

The display reveals its media methodology in the first computer applications, as demonstrated by these elaborations. This shows that a technical component (an oscilloscope) is used in a different way from what was originally intended – as an optical snapshot in time of the location of dots, instead of one or more courses of waveforms. Put in graphical terms, a loophole is left that must be closed by the user through the determination of a mobile final destination. Both at the technical and at the practical levels, this display is all about producing an endpoint coordinate. Furthermore, a *scope* is not only etymologically linked to the media practice of searching, from a genealogical perspective, a *scope* is also focused on applications that try to fix movement.

The different Whirlwind computer displays thus have a media-technical, practical, normalizing and nominalist dimension, all of which are also of importance for our current understanding of displays. This section has demonstrated that the separability and addressability of the individual pixel is a variable that takes precedence over the question of whether data are used for display or storage. The procedure of discretization must therefore be considered as an essential component of a praxeology of the display.[65] The reason this is so important is that the discretization comprises both the fabrication of image and location dots. This step in the development in display media praxeology is therefore characterized by a sociotechnical duplicity of endpoint coordinates.

64 Susan Leigh Star, Cooperation Without Consensus in Scientific Problem Solving. Dynamics of Closure in Open Systems, in: Steve M. Easterbrook (ed.), *CSCW. Cooperation or Conflict?*, London: Springer, 1993, pp. 93–106.

65 On understanding media praxeology, see special issue of *Digital Culture & Society* 2 (2017) on *Mobile Digital Practices*; Erhard Schüttpelz, Epilogue. Media Theory Before and after the Practice Turn, in: Ulrike Bergermann et al. (eds.), *Connect and Divide: The Practice Turn in Media Studies. The 3rd DFG conference of Media Studies*, Zürich/Berlin: diaphanes (forthcoming).

Flood Lights

The current analysis demonstrates that representations and their arbitrary dimension can only form part of a media-esthetic investigation. The decisive media-methodological occurrence does not take place on the display, but behind it from a technical perspective and in front of the display from a practical perspective. A combined technical and practice theory is therefore required to understand the specific mediality of the display.

This kind of approach shows how the practice of coordination and the materiality of coordinates each determine the media-specific nature of the display through their duplicity. This socio-technology of the duplicity of image-space coordinates has the capacity to more closely define the dispositive structure of digital displays and to present them in their different variants. In this process, the innovative steps that emerge in the media history of the display are characterized by the elimination and inclusion of mediators, through which the perceptibility of co-coordination is newly materialized in each case.

Up to now, the duplication of spatial and image-related coordinates was simply declared to be a cultural-geographic characteristic of *augmented reality* applications.[66] This essay shows that this scope of analysis that is driven by intervention is not far-reaching enough. Spatial co-coding not only characterizes the layering technologies through which the location-relevant internet information is merged with the live camera image on mobile consumer devices. Co-coding of online and offline spaces already occurred with radars. In this respect, a far more comprehensive heuristic continuum can be described based on the "duplicity of code",[67] through which image and spatial production are constituted interactively and are reified in displays.

In addition, the praxeology of the display shows how historical media methods reach into the present and are still having a formative effect on current manifestations of displays. The display not only harbors within it the intrinsic persistence of guidance and bearings, it cannot deny its genealogy, which stems from the medium of the radar. The media methodology of the display aims at a mediated seeing over distance and the depiction of discrete and addressable pictorial symbols in the form of co-existing light and location dots. Even when considered in light, the dispositif of the display remains seeing in the dark, inherent in surroundings and in proximity. Displays show the immediate mediate. That is its media practice-theoretical dimension.

Displays can therefore by all means be called visible objects as per the meaning of Thomas Elsaesser, giving light a spatial form and materiality that goes beyond the artist's flat and framed canvas.[68] They represent an a-modern return in the cloak of the digital modern, "that returns the fixed spectator facing the fixed rectangular screen to being a historically contingent actor"[69] and thereby also allocates spatio-temporally limited valence to the dispositif of the movie.[70]

66 See Mark Graham et al., Augmented Reality in Urban Places. Contested Content and the Duplicity of Code, in: *Transactions of the Institute of British Geographers* 38.3 (2013), pp. 464–479.

67 Ibid.

68 Thomas Elsaesser, The 'Return' of 3-D. On Some of the Logics and Genealogies of the Image in the Twenty-First Century, in: *Critical Inquiry* 39.2 (2013), pp. 217–246.

69 Ibid., p. 244.

70 See Hermann Kappelhoff, Der Bildraum des Kinos. Modulationen einer ästhetischen Erfahrungsform, in: Gertrud Koch (ed.), *Umwidmungen. Architektonische und kinematographische Räume*, Berlin Vorwerk 8, 2005, pp. 138–149.

The rectangular viewing window has been a "medium of visibility"[71] since Roman antiquity, which opens up elementary practical functions, such as illumination and views. The display is the first medium that opposes this image-generating dispositif in its mediality, materiality and media practice. Based on the variability and adaptability of its shape and its co-coordinating function, it represents a disillusioning feast for the eyes. Unlike the rectangular window or screen that is not subdivided, it no longer provides a view in or out – is thus no longer constrained by architecture – but represents a socially canonized practice[72] of telemetry and remote sensing.

When we take a look at the media history of the display, it becomes clear that the discourse in the pictorial sciences on the discrete nature, operationality and spatial control of images precedes the methodology of displays outlined here. Stiegler has recognized this:

> *[I]t will not be much longer before we can view images analytically: screens* [l'écran] *and what is written* [l'écrit] *are not simply opposed to each other.*[73]

71 Gerd Blum, Epikureische Aufmerksamkeit und euklidische Abstraktion. Alberti, Lukrez und das Fenster als Bild gebendes Dispositiv, in: Horst Bredekamp, Christiane Kruse, Pablo Schneider (eds.), *Imagination und Repräsentation. Zwei Bildsphären der frühen Neuzeit*, Paderborn: Wilhelm Fink Verlag, 2010, pp. 79–118, p. 80.

72 See Hans-Jürgen Horn, Fenster (kulturgeschichtlich), in: *Reallexikon für Antike und Christentum*, Vol. 7. Stuttgart: Anton Hiersemann, 1970, pp. 732–737; Achatz von Müller, Der Politiker am Fenster. Zur historischen Ikonographie eines „lebenden Bildes", in: Gottfried Boehm (ed.), *Homo Pictor*, München/Leipzig: De Gruyter, 2001, pp. 323–338.

73 Bernard Stiegler, Das diskrete Bild, in: Jacques Derrida, Bernard Stiegler (eds.), *Echographien. Fernsehgespräche*, Vienna: Passagen, 2006, pp. 162–188, p. 180.

Both are increasingly becoming one from a phenomenological perspective. This increases the lack of conceptual clarity in relation to what we understand an image to be and simultaneously moves into the foreground implicit knowledge and media methods that are solely based on displays.

In point of fact, it will indeed not be much longer, as it has always taken place and is always occurring anew: Displays allow us to view images analytically.

Figures

1–2 Royal Air Force, Standard Technical Training Notes, Part 3: Elementary Radar, Great Britain, Air Ministry, London, 1965, p. 37, p. 87.

3 Gyges Publishing Company, http://gyges.dk/Seeburg%20(Small).jpg (accessed November 1, 2017).

4 Danielstirland, https://en.wikipedia.org/wiki/Battle_of_Britain_Bunker (accessed November 1, 2017).

5 http://virtualtravelog.net/images/2003-08-Williams-Kilburn-tube.jpg (accessed November 1, 2017).

6 Curtesy MIT Museum, image curtesy Computer History Museum.

7 Claus Pias, Die Pflichten des Spielers. Der User als Gestalt der Anschlüsse, in: Martin Warnke et al. (eds.), Hyperkult II, Bielefeld: transcript-Verlag, 2004, pp. 313–341, p. 321.

8 Diego Hernan Sanhueza Martínez, See It Now: Jay W. Forrester and the WHIRLWIND Computer - (1951), https://youtube.com/watch?v=5ZQP4G3Qwb4 (accessed November 1, 2017).

Jan Distelmeyer

Carrying Computerization

Interfaces, Operations, Depresentations

Indispensable and Invisible[1]

Interaction is dismissed. In the end of 2016 the cover of the *Interactions* magazine, published by the Association for Computing Machinery since 1994, crossed out the last word in human computer interaction and replaced it with *integration*. The "era of human-computer interaction", the cover story stated, "is giving way to the era of human-computer integration – integration in the broad sense of a partnership or symbiotic relationship in which humans and software act with autonomy, giving rise to patterns of behavior that must be considered holistically" (fig. 1).[2] After a summary how "the nature of our interaction has continuously evolved" from "switches, cards, and tape to typing, mice, and styluses, adding speech and gesture"[3] and a forecast on "brainwave interaction"[4] (recalling Vannevar Bush's thoughts on "a couple of electrodes on the skull"[5]), the most interesting question was raised:

> *We can see these changes, but the most dramatic change affecting human-computer interaction was invisible:* what the computer does when we are not interacting with it.[6]

This correlation between integration, (autonomous) activity and hiddenness was accompanied by another marketing-related connection between computers and invisibility, published at the same time. The international video campaign "Feel connected all over Europe" of the German telecommunications company Deutsche Telekom presented the singer Andrea Bocelli praising what is called laconically the network: "It gives me freedom. Reliably wherever I am. It transcends boundaries. It's indispensable and invisible."[7] Bocelli's statements are decorated and elaborated by images of him walking, riding, and boating in several iconic places in Europe (fig. 2). The fact that Andrea Bocelli is blind and that this promotional video about network technology and digitization shows no form of computer technology or infrastructure at all, is important for the message with which

1 Some parts of this paper have been published in the journal *Cinéma & Cie* (Vol. XVII). I would like to thank the participants and organizers of the workshop "Screen Operations. Conditions of Screen-based Interaction" at Humboldt University Berlin, 2016, for discussions and comments.
2 Umer Farooq, Jonathan Grudin, Human-Computer Integration, in: *Interactions* 23.6 (2016), pp. 27–32, p. 27.
3 Ibid., p. 28.
4 Ibid.
5 Vannevar Bush, As We May Think, in: *The Atlantic Monthly* 176 (1945, https://theatlantic.com/magazine/archive/1945/07/as-we-may-think/303881 (accessed September 20, 2017).
6 Farooq, Grudin 2016 (as fn. 2), p. 28.
7 Serviceplan Group, Telekom Connecting Europe – TV commercial 2016, https://youtube.com/watch?v=6No-bDXIdEE (accessed September 20, 2017).

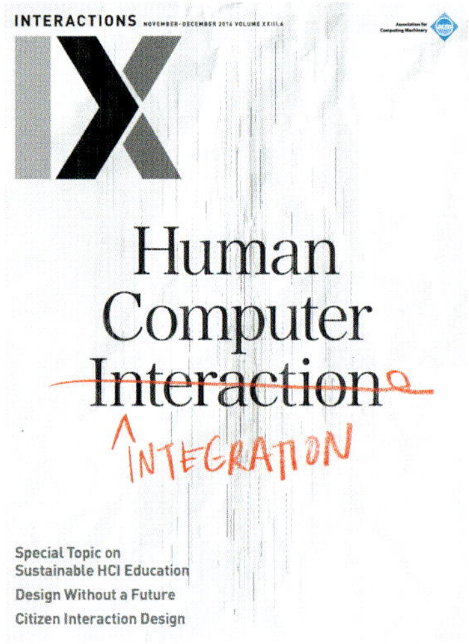

1 Cover of Interactions, 23/6 (2016). Association for Computing Machinery.

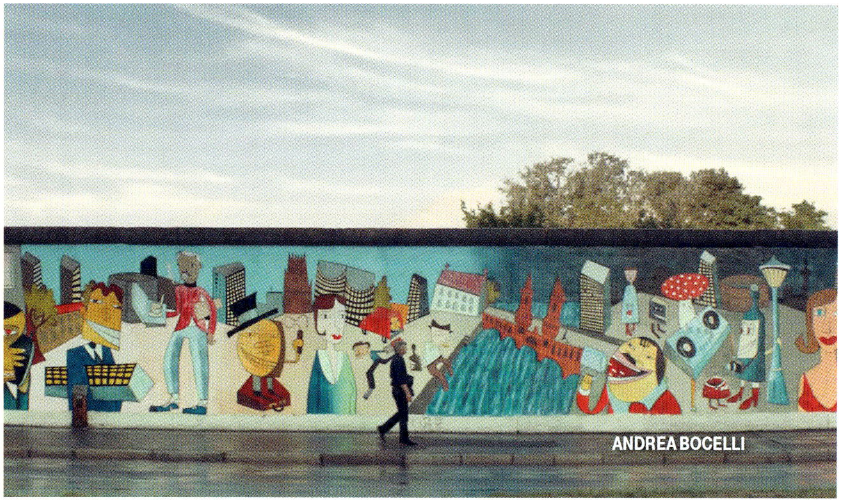

2 Still from Telekom commercial *Feel connected all over Europe*

the video concludes: "The network – it's the present and the future. You can't see it. But you can feel it."[8]

Both contributions to the (inexisting) presence of computers are expressions of an important consilience: On the one hand, the current discussions and developments of concepts like *Ambient Intelligence*, *Internet of Things*, and *Smart Environments* promote the powerful and consequential omnipresence of computers, which is, on the other hand, understood and promoted as indiscernible, unobservable, embedded, and (nonetheless or therefore) effective.[9] A magical match – (omni-)present and hidden at the same time. From early plans of *Ubiquitous Computing* up to current concepts of "calm" and imbedded technologies the linkage between ubiquity, efficiency, and invisibility is important, especially in concepts and presentations of developers.[10]

One of the most famous examples of the living contradiction of invisibility and readiness-to-hand was given by Steve Jobs' farewell performance as CEO of Apple in San

8 Ibid.

9 "It is still a matter of some debate whether and how interaction designs for or ubicomp culture are to tend towards the transparent and calm invisibility of an infrastructure for interactivity or towards a more personalized, attention-getting, even exciting unfolding, and mediatory laying bare of potential paths for technocultural interrelations and interactivities." Ulrik Ekman, Individuations, in: Ulrik Ekman, Jay David Bolter, Lily Diaz, Maria Engberg, Morten Søndergaard (eds.), *Ubiquitous Computing, Complexity, and Culture*, New York: Routledge, 2016, pp. 77–90, p. 83.

10 Natascha Adamowsky, Vom Internet zum Internet der Dinge. Die neue Episteme und wir, in: Florian Sprenger, Christoph Engemann (eds.), *Internet der Dinge. Über smarte Objekte, intelligente Umgebungen und die technische Durchdringung der Welt*, Bielefeld: transcript, 2015, pp. 231–265, p. 245.

Francisco on 6 June 2011. Introducing the service iCloud, Jobs illustrated his idea of an autonomous service, for which "I don't even have to take the devices out of my pocket" or "to be near my Mac or PC".[11] Since

> *all these new devices have communications built into them; they can all talk to the cloud whenever they want. [...] And now everything's in sync with me not even having to think about it. [...] [S]o everything happens automatically and there's nothing new to learn. It just all works. It just works.*[12]

Obviously today's interface culture is shaped very much by various forms of interfacing with computers that cannot be reduced to user interfaces, we address by touching or clicking, gestures or voices. N. Katherine Hayles remark, "[m]obile phones, GPS technology, and RFID (radio frequency identification) tags, along with embedded sensors and actuators, have created environments in which physical and virtual realms merge in fluid and seamless ways",[13] sums up some forms of interfaces building and organizing *seamless* processes of connectivity. But this development, mirrored by the term "Post-Interface"[14] and Mark B. N. Hansen's perspective on "twenty-first-century media" ("no longer a delimited temporal object that we engage with focally through an interface such as a screen, media become an environment that we experience simply by being and acting in space and time"),[15] should not be misunderstood as a disappearance of human computer interfaces or even of interfaces at all.

Interfaces Carry

Why do interfaces matter? Firstly, because the term refers to different modes and processes of connectivity that ensure the functionality of (all forms of) digital computers in the first place. It is important to remember that the term *interface*, introduced by the physicists James and William Thomson in late 19th century, was originally used to describe the transmission of energy.[16] Their usage of the term "would define and separate areas of unequal energy distribution within a fluid in motion, whether this difference is given in terms of velocity, viscosity, directionality of flow, kinetic form, pressure, density, temperature, or any combination of these".[17] With this in mind, the question of the pursued ubiquity and networked embeddedness of computing, relying basically on transportation of signals and the carrying of electricity, is inevitably a question of interfaces. The term interface helps to describe the "interior telegraphy" of the computer as well as all forms of its networks, its relations to us and its incorporations.[18] Hence, the ongoing development

11 EverySteveJobsVideo, Steve Jobs introduces iCloud & iOS 5 – WWDC (2011), https://youtube.com/watch?v=gfj7UgCMsqs (accessed September 20, 2017).
12 Ibid.
13 N. Katherine Hayles, Cybernetics, in: W. J. T. Mitchell, Mark B. N. Hansen (eds.), *Critical Terms for Media Studies*, Chicago: University of Chicago Press, 2010, pp. 145–156, p. 148.
14 Michael Andreas, Dawid Kasprowicz, Stefan Rieger, Technik | Intimität. Einleitung in den Schwerpunkt, in: *Zeitschrift für Medienwissenschaft* 15 (2016), pp. 10–17, p. 12.
15 Mark B. N. Hansen, Ubiquitous Sensation. Towards an Atmospheric, Impersonal and Mircotemporal Media, in: Ulrik Ekman (ed.), *Throughout. Art and Culture Emerging With Ubiquitous Computing*, Cambridge: MIT Press, 2013, pp. 63–88, p. 73.
16 See Peter Schaefer, Interface. History of a Concept, 1868–1888, in: David W. Park, Nicholas W. Jankowski, Steve Jones (eds.), *The Long History of New Media. Technology, Historiography, and Contextualizing Newness*, New York: Peter Lang, 2011, pp. 163–175.
17 Branden Hookway, *Interfaces*, Cambridge: MIT Press, 2014, p. 59.
18 Hartmut Winkler, *Prozessieren. Die dritte, vernachlässigte Medienfunktion*, Munich: Fink, 2015, p. 294.

of an increasingly hidden dissemination, interconnection, and implementation of computers cannot be understood without asking about interface processes.[19]

Graphical user interfaces are but one of the multilayered aspects characterizing interfaces in terms of digital computing. These "symbolic handles", as Florian Cramer and Matthew Fuller have called them, "which [...] make software accessible to users" depend on and are connected to other interface aspects and processes, such as hardware connecting humans/bodies to hardware, hardware connecting hardware to hardware, software connecting software to hardware, and software providing software to software connections.[20]

Secondly, our encounter with computers in all its forms by use of programmed and designed user interfaces is not superseded but accompanied by "pervasive" and "ubiquitous" computing that Ulrik Ekman has described as "a sociocultural *and* technical thrust to integrate and/or embed computing pervasively, to have information processing thoroughly integrated with or embedded into everyday objects and activities, including those pertaining to human bodies and their parts".[21] Mark B. N. Hansen's description of the "experiential shift" by "twenty-first-century media" depicts the diversity of interconnected interface politics:

> *Thus, well before we even begin to use our smart phones in active and passive ways, the physical devices we carry with us interface in complex ways with cell towers and satellite networks; and preparatory to our using our digital devices or our laptops to communicate or to acquire information, the latter engage in complex connections with wireless routers and network hosts.*[22]

While these devices are constantly and so-called calmly interfacing with networks and servers, we also do use *our smart phones in active ways*, which is why we pay for and update them. Even today, graphical user interfaces are so obviously omnipresent that this manifestation of software is still "often mistaken in media studies for 'interface' as a whole".[23] Screen operations belong to the chief activities in large parts the world; work and leisure activities are increasingly involving screen activities, just as three most popular websites worldwide – Google, YouTube and Facebook – bank on our interactions with their offerings on different kind of screens.[24] Despite this, media studies analyses of common user interfaces are still not common.[25] This must change if we are to better understand of our relationship with (previous, current, and upcoming) forms of computing.

19 "But alongside and interwoven with computational and networked digital media, more than one 'environmental' system of calculation, slipping in and out of direct perception, and the multiple interfaces between them are to be reckoned with." Matthew Fuller, Foreword, in: Ekman 2013 (as fn. 15), pp. xi–xxvi, p. xx.

20 Florian Cramer, Matthew Fuller, Interface, in: Matthew Fuller (ed.), *Software Studies. A Lexicon*, Cambridge: MIT Press, 2008, pp. 149–152, p. 149.

21 Ulrik Ekman, Introduction, in: Ekman 2013 (as fn. 15), pp. 1–59, p. 22.

22 Mark B. N. Hansen, *Feed Forward. On the Future of Twenty-First-Century-Media*, Chicago: University of Chicago Press, 2015, p. 62.

23 Cramer, Fuller 2008 (as fn. 20), p. 149.

24 Alexa web analytics, http://alexa.com/topsites (accessed September 20, 2017).

25 For exceptions, see Matthew Fuller, It looks like you're writing a letter. Microsoft Word, in: Matthew Fuller (ed.), *Behind the Blip. Essays on the Culture of Software*, New York: Autonomedia, 2003, pp. 11–37; Christian Ulrik Andersen, Søren Pold (eds.), *Interface Criticism. Aesthetics Beyond Buttons*, Aarhus: Aarhus University Press, 2011; Margarete Pratschke, Interacting with Images. Toward a History of the Digital Image: The Case of Graphical User Interfaces, in: Horst Bredekamp, Vera Dünkel, Birgit Schneider (eds.), *The Technical Image. A History of Styles in Scientific Imagery*, Chicago: University of Chicago Press, 2015, pp. 48–57; Teresa Martínez Figuerola, Jorge Luis Marzo (eds.), *Interface Politics*, Barcelona: BAU, 2016; Florian Hadler, Joachim Haupt (eds.), *Interface Critique*, Berlin: Kadmos, 2016; Jan Distelmeyer, *Machtzeichen. Anordnungen des Computers*, Berlin: Bertz + Fischer, 2017.

These analyses are necessary because interfaces define today's reality in manifold ways. Understood as the complex of various processes of connectivity and conduction, interfaces do carry – on all levels of its acceptation – the worldwide computerization, in which graphical user interfaces still build the real but underestimated blockbusters of today's visual politics.

Graphical user interfaces inform us (to some extent) of the real and the imaginary, the well-prepared and consequential relations between humans and computers as applied in computers. They mediate interrelations between humans and computers. Studying its interface politics allows for the computer to be realized as a particular "power machine",[26] which enables us to examine a key component of computers and computerized media/things/beings: programmability.

The fact that graphical user interfaces work so differently from, for instance, cinematic or televisual appearances and do inevitably rely on other interface processes between all sorts of hard- and software makes the task of interface analysis and critique so urgent. The example I would like to comment on here is the YouTube interface – those immensely popular conditions with which we upload, search, identify, organize, tag, encounter, and negotiate the occurrence of video material on this second most popular website worldwide. To turn towards these special screen operations, it is important to consider their operative images as depresentations.

26 Distelmeyer 2017 (as fn. 25), pp. 82–92.

Operative Images and Depresentation

The interdependence of aesthetics and dispositifs demands attention be paid to the special status of these images and signs that – to quote a Windows 10 commercial from 2015 – "help you do your thing".[27] Of course, these so-called *computer icons* could likewise be symbolic, depending on the specific interface design. Regardless of the potentially iconic or symbolic character of these images and signs, all clickable or touchable appearances correspond to Peirce's idea of indices.[28] These images and signs must have a physical relation to the somehow presented processes of computing, to the interior telegraphy of the computer. They "show something about things, on account of their being physically connected with them"[29]; otherwise they simply would not work.[30]

Hence, images and videos on YouTube could combine different indexical qualities. Dealing with the YouTube interface by clicking/touching on a video thumbnail offered on the screen or by clicking/touching on a running video to stop or to enlarge it, is an expression of a certain semiotic shift. Because indexicality of these images and videos is no longer only generated by a potential trace to a pre-filmic reality but also by the trace to the interior telegraphy of the networked computer that makes its existence and usage possible. Since we are invited and enabled to interact with these

27 Windows, Windows 10 Highlights Reel, https://youtube.com/watch?v=j-3ZLphVaxkg (accessed September 20, 2017).

28 See Marianne van den Boomen, *Transcoding the Digital. How Metaphors Matter in New Media*, Amsterdam: Institute of Network Cultures, 2014, pp. 37–41.

29 Charles S. Peirce, What is a sign, in: The Peirce Edition Project (ed.), *The Essential Peirce. Selected Philosophical Writings* Volume 2 (1893–1913), Bloomington: Indiana University Press, 1998, pp. 4–10, p. 5.

30 To specify this indexicality, it is helpful to remember the difference between what Peirce called a genuine index and a degenerated index, because graphical user interfaces combine both forms of Peirce's indexicality.

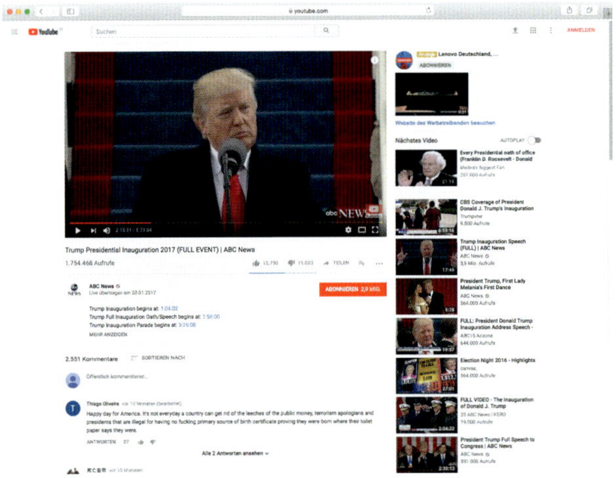

3 Screenshot of the ABC video Trump *Presidential Inauguration* on YouTube.

images, they refer to another realm – to the processuality of my networked computer (fig. 3).

The video *Trump Presidential Inauguration 2017* and its thumbnail image is photographically indexical because of its "physical relation between the object photographed and the image finally created".[31] The additional indexical qualities of the video – which we have learned to stop and continue by clicking/touching on it, as well as of the thumbnail, which we click/touch to select and start the video in the first place – are relying on the programmability of computers. This programmatic indexicality is reliable, because these images and signs "materially refer to an act of executing machine code"[32]: "They refer to existential, physical chains of causation, to machine processes to be executed in order to yield a specific result."[33]

Graphical user interfaces visualize what the computer offers to do in a particular way without, showing what is actually happening *inside* the machines. "Software, or perhaps more precisely OS," as Wendy Chun has stated, "offer us an imaginary relationship to our hardware: they do not represent the motherboard or other electronic devices but rather desktops, files, and recycling bins."[34] Unsurprisingly the YouTube interface does not represent any of the hardware of the servers or network processes necessary to realize my search request.

Nevertheless at the same time this hardware based relationship offered by software – depresented by symbolic or iconic signs – offers more than just an *imaginary relationship* to the working hardware of the computer, for example, in the form of the motherboard. These clickable or touchable signs are simultaneously linked electronically to the inner processes of the machine, to its interior telegraphy, whose flow of electronic signals connects, among others, the motherboard to the indexical signs of the graphical user interface. In terms of YouTube the click/touch on the offered thumbnail of the video *FULL VIDEO – The Inauguration of Donald J. Trump* instructs the interior telegraphy of my computer to use the exterior and protocol-driven telegraphy of the World Wide Web to stream the requested data. An imaginary and at the same time real and physical relationship enables us to click/touch these images, to start the prom-

31 Tom Gunning, What's the Point of an Index? Or, Faking Photographs, in: *Nordicom Review* 25.1–2 (2004), pp. 39–49, p. 40.

32 van den Boomen 2014 (as fn. 28), p. 40.

33 Ibid.

34 Wendy Hui Kyong Chun, *Control and Freedom. Power and Paranoia in the Age of Fiber Optics*, Cambridge: MIT Press, 2006, p. 20.

ised and hidden algorithmic processes, which is why Frieder Nake calls them "algorithmic images".[35]

The contradictory character of these images and signs has led Marianne van den Boomen to introduce the very fruitful term of depresentation. They show what we can do without showing the "procedural complexity" and the multitude of requirements and consequences attached: "[T]he icons on our desktops do their work by representing an ontologized entity, while depresenting the processual and material complexity involved. This is the way icons manage computer complexity, this is the task we as users (in tacit conjunction with designers) have delegated to them."[36]

To address the special quality of these "symbolic handles",[37] I have discussed them as "operative images", adopting a concept coined by Harun Farocki to describe the production of images by machines for machines.[38] This term – translating "*operative Bilder*" Farocki has called them "operative images" as well as "operational pictures" and "operational images" – is driven by the interest in processes not represented by these operative images but rather of which operative images are part of themselves.[39] "These are images", Farocki explained, "that do not represent an object, but rather are part of an operation."[40]

Volker Pantenburg has emphasized that operative images "aren't intended to be released separately, and strictly speaking don't need to appear as images at all but emerge as the intermediate product of a wider technical process".[41] Farocki described them as differentiated by purpose: "In my first work on this subject, Eye/Machine (2001), I called such pictures, made neither to entertain nor to inform, 'operative images'."[42] This last point is crucial and marks a productive difference between Farocki's concept and my application of it.[43] Whereas the operative images of a graphical computer interface may not be made for edification, information or instruction in the classical sense ("*Erbauung oder Belehrung*"[44]), they do (and must) instruct users on what can be done. What they instruct and are part of through depresentation is a kind of knowledge about computers, about their usage, and about us – it forms an "implicit memory".[45]

The interdependency with technical execution ("*technischen Vollzug*"[46]) differentiates this form of operativity from others, as for instance the operative imagery, operative writing, and diagrammatic operations of Sybille Krämer's approach to diagrammatology.[47] Operative images as depre-

35 Frieder Nake, The Semiotics Engine. Notes on the History of Algorithmic Images in Europe, in: *Art Journal* 68.1 (2009), pp. 76–89.
36 van den Boomen 2014 (as fn. 28), p. 36.
37 Cramer, Fuller 2008 (as fn. 20), p. 149.
38 See Distelmeyer 2017 (as fn. 25), pp. 92–98.
39 Harun Farocki, Quereinfluss/Weiche Montage, in: Christine Rüffert et. al. (eds.), *Zeitsprünge. Wie Filme Geschichte(n) erzählen*, Berlin: Bertz + Fischer, 2004a, pp. 57–61, p. 61; Harun Farocki, Phantom Images, in: *Public. Art, Culture, Ideas* 29 (2004b), pp. 12–22; http://harunfarocki.de/installations/2000s/2003/eye-machine-iii.html (accessed September 20, 2017).
40 Farocki 2004b (as fn. 39), p. 17.
41 Volker Pantenburg, *Farocki/Godard. Film as Theory*, Amsterdam: Amsterdam University Press, 2015, p. 210.
42 Farocki 2004b (as fn. 39), p. 17.
43 For other approaches to the term, see Werner Kogge, Lev Manovich. Society of the Screen, in: Alice Lagaay, David Lauer (eds.), *Medientheorien. Eine philosophische Einführung*, Frankfurt/M.: Campus, 2004, pp. 297–315; Ingrid Hoelzl, *The Operative Image. An Approximation*, http://mediacommons.futureofthebook.org/tne/pieces/operative-image-approximation (accessed September 20, 2017).
44 Farocki 2004a (as fn. 39), p. 61.
45 Jan Distelmeyer, An/Leiten. Implikationen und Zwecke der Computerisierung, in: *Navigationen. Zeitschrift für Medien und Kulturwissenschaften* 17.2 (2017), pp. 37–53.
46 Farocki 2004a (as fn. 39), p. 61.
47 See Sybille Krämer, Operative Bildlichkeit. Von der Grammatologie zu einer 'Diagrammatologie'? Reflexionen über erkennendes Sehen, in: Martina Heßler, Dieter Mersch (eds.), *Logik des Bildlichen. Zur Kritik der ikonischen*

sentations of computer performance are parts and thresholds of (at least) four types of mutually connected operations – that is, interface operations within the meaning of the multilayered interface facets:

1. Operations as the various interrelations between hardware and software ensuring that these general-purpose machines and universal symbolic machines fulfill their tasks.
2. Operations as the interrelations of several computers, leading to further co-action of hardware and software by protocol-driven networks.
3. Operations as the connections and communications between computers and non-computer forms of interconnected materiality – such as human bodies or technical artifacts – that lead to the issues of surveillance and cybernetization of beings and (an internet of) things under programmed control.
4. Operations as *us* dealing with *them* – operations as the handling of and dealing with computers, hence: operations understood as technical, physical, and cognitive processes, including questions of the relationship between software and ideology raised by Wendy Chun,[48] Alexander Galloway,[49] and Cynthia and Richard Selfe.[50]

I would like to highlight just two aspects of the last type: The first aspect is related to the special indexicality of these operative images, which leads back to the question of how analyzing graphical user interfaces could help address the dicey character of computerization. Addressing this indexicality inevitably confronts us with consequences of programmability, which I understand as perhaps the most thought-provoking characteristic of computers and computerized media, things and beings. Graphical user interfaces always propose ideas and depresentations of more than just the computer; instead, "[i]nterfaces and operating systems produce 'users' – one and all."[51] And since all of our computer use has to be envisaged and enabled by programming, computer interfaces always empower users to regulate while at the same time forcing them to be regulated.[52] Hence, the interface mise-en-scène – the available structure of operative images and depresentations – shapes the aesthetic appearance of the computer as an *aesthetics of regulation* (Ästhetik *der Verfügung*).[53]

This aesthetics is marked by a particular power structure – a logic of regulation: Actively regulating users are being regulated in a system, in which they have to play under the default rules with the provided tools and prerequisites.

Vernunft, Bielefeld: transcript, 2009, pp. 94–123; Sybille Krämer, Christina Ljungberg (eds.), *Thinking with Diagrams. The Semiotic Basis of Human Cognition*, Boston/Berlin: Mouton de Gruyter, 2016.

48 Wendy Hui Kyong Chun, *Programmed Visions. Software and Memory*, Cambridge: MIT Press, 2013.

49 Alexander Galloway, *The Interface Effect*, Cambridge: MIT Press, 2012.

50 Cynthia L. Selfe, Richard J. Selfe, The Politics of the Interface. Power and Its Exercise in Electronic Contact Zones, in: *National Council of Teachers of English* 45.4 (1994), pp. 480–504.

51 Chun 2013 (as fn. 48), pp. 67–68.

52 I would like to stress the point that the common distinction between *users* and *programmers* is highly problematic – especially when it comes to interfaces. As Wendy Chun has pointed out, "programmers are users" since "they create programs using editors, which are themselves software programs": "The distinction between programmers and users is gradually eroding, not only because users are becoming programmers (in a real sense programmers no longer program a computer; they code), but also because, with high-level languages, programmers are becoming more like simple users. The difference between users and programmers is an effect of software." Wendy Hui Kyong Chun, On Software. Or the Persistence of Visual Knowledge, in: *Grey Room* 18 (2004), pp. 26–51, p. 38.

53 See Distelmeyer 2017 (as fn. 25), pp. 65–126.

But this is not a one-way street: Precisely because every computer operation relies on programs, all programmed functions, regulations, barriers, and presets are principally alterable and expandable by users or hackers. This processuality identifies dealing with computers as a power struggle with which its political issues may begin. It confronts us with controllability resulting from programmability.

The second aspect of operations in terms of human handling of computers is related to knowledge, informing our actions. Criticized by various media scholars, the mythical term *digital* has been an extremely powerful buzzword and sales argument at least since the early 1990s.[54] To mark *the digital* as a myth and to keep in mind the problems of coping with mythical terms as shown by Roland Barthes, I arranged myself some years ago with another not yet mythical term: the neologism *digitalicity.*[55]

In Western-European and US-American discourse since the early 1990s digitalicity is shaped to a special degree by promises (and fears) of interactivity, flexibility, control, freedom, and empowerment. Celebrated as a victory of digital media's acclaimed elasticity as opposed to rigid, inflexible, passive, and hierarchy-based predecessors, the same programmatic linkage between flexibility and control is now – at the latest since the Snowden disclosures and the debates about dominating corporations and algorithmic regulation – also an object of criticism.[56] As just one example I would like to quote maybe the most influential protagonist of digitalicity from the 1990s, Nicolas Negroponte:

> *[M]ore than anything, my optimism comes from the empowering nature of being digital. The access, the mobility, and the ability to effect change are what will make the future so different from the present.*[57]

Understanding digitalicity as one important discursive aspect of computerization and – not least – the hopes of and investments in the "fourth industrial revolution" as for instance shown in the European Commission's "path to digitise European industry"[58] – the question arises, how a given interface mise-en-scène corresponds to the promises and fears that have shaped digitalicity. With this question I would like to turn to YouTube as an example.

YouTube Operations

If you enter the URL www.youtube.com or follow a corresponding link, bookmark, or presetting, the front page of YouTube presents a deployment of selectable operative images, depresenting potentially upcoming video events (fig. 4).[59] Even if you have no personal account to log in, the personalizing *you* of YouTube is taken seriously right from the start: Thanks to recorded, evaluated, and conjugated former visits and dealings with YouTube, every front page should be a customized performance. This personalization is the outcome or yield of *my* work within the YouTube

54 See Hartmut Winkler, *Docuverse. Zur Medientheorie der Computer*, Munich: Fink, 1997; Lev Manovich, *The Language of New Media*, Cambridge: MIT Press, 2001; Chun 2006 (as fn. 34).

55 See Jan Distelmeyer, *Das flexible Kino. Ästhetik und Dispositiv der DVD & Blu-ray*, Berlin: Bertz + Fischer, 2012.

56 See Distelmeyer 2017 (as fn. 25), pp. 98–126.

57 Nicolas Negroponte, *Being Digital*, New York: Alfred A. Knopf, 1995, p. 230.

58 European Commission Directorate General for Communications Networks, Content & Technology, *The Fourth Industrial Revolution*, https://ec.europa.eu/digital-single-market/en/fourth-industrial-revolution (accessed September 20, 2017).

59 I am describing the YouTube interface performed by a browser; the interface designed for the YouTube app is a different formation.

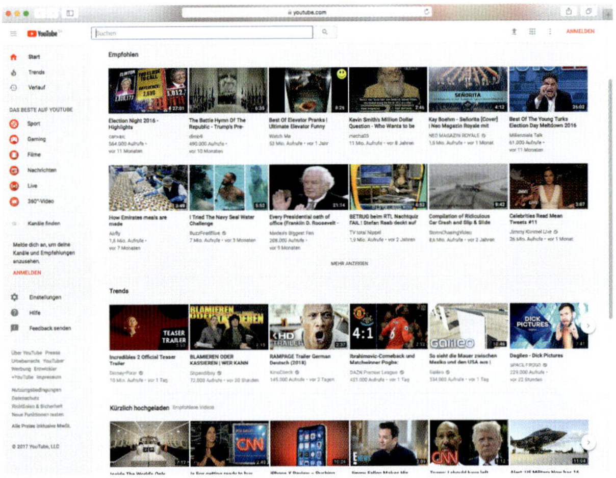

4 Lists on the front page of YouTube.

interface that Till A. Heilmann has described as data labor in current capture-capitalism.[60]

The moment I make my selection, the former depresented video starts in a frame, in which the video is a working as an operative (moving) image in its own right. If I click on the running video, it pauses until another click on the now freezed operative image starts the movement and sound again. A double-click leads to the full screen mode, another double-click brings back the YouTube website interface. Here the expandable video frame is escorted by another arrangement of selectable operative images to the right of the frame. This arrangement of thumbnails could be described as a remaining gesture of wealth and richness – a power of control related to a variety of depresented audiovisual material classified by taglines, genres, categories, and other visualized metadata. It keeps up the empowerment gesture and the *ability to effect change*: Even though I have already chosen a video, this choice is accompanied by a selection of another to-be-selected material.

This choice-empowerment relies heavily on a mode of presentation that dominated and still is dominating more than a few interface formations. This tradition presents the aesthetics of regulation as an "order of selectivity"[61] – offering options and reassuring usability as a freedom of choice in the form of menus, buttons, lists, and the like. This "freedom as control" is a question of strictly defined and prepared choices.[62]

We encounter this traditional and surprisingly long-lasting WIMP (abbreviation for windows, icons, menus, and pointer that denotes an interface design paradigm in human-computer interaction) cosmos by, for instance, using popular online shops like iTunes or Amazon, the grid-apposition of apps on multi-touch devices like Google Nexus and Samsung Galaxy, iPhone and iPad, the "active apps" and "ideal apps" arrangement on the Fairphone 2, the Launchpad of MAC OS Yosemite, Windows 10, and the Linux-Interface GNOME 3 with its "activities overview" described by the GNOME Project as "an easy way to access all your basic tasks. A press of a button is all it takes to view your open windows, launch applications or check if you have new messages."[63]

60 Till A. Heilmann, Datenarbeit im 'Capture'-Kapitalismus. Zur Ausweitung der Verwertungszone im Zeitalter informatischer Überwachung, in: *Zeitschrift für Medienwissenschaft* 13 (2015), pp. 35–47.

61 Jan Distelmeyer, Freiheit als Auswahl. Zur Dialektik der Verfügung computerbasierter Medien, in: Jan-Henrik Möller, Jörg Sternagel, Leonore Hipper (eds.), *Zur Paradoxalität des Medialen*, Munich: Fink, 2013, pp. 69–90.

62 Chun 2006 (as fn. 34).

63 GNOME Project, https://gnome.org/gnome-3/ (accessed September 20, 2017).

Considering the familiarity with this widespread freedom as prepared choice-control, other widespread aesthetics of regulation could easily be overlooked. Especially computer games challenge and play with this dominant overview order. Examples can be found in different sorts of games, most obvious maybe and long-lasting in first-person shooters like the popular Tom Clancy's Rainbow Six Siege (Ubisoft, 2015), in which the crucial point is not to know but to explore what actually is offered and waiting around the corner. Nevertheless this exploring mode of aesthetics is quite often supplemented by another order of selectivity, showing available weapons, equipment, maps, and the like.

Hence, an order of selectivity, invoking our wealth of choice by menus and similar arrangements, is not in the least determined by technology. Instead this order of selectivity is a cultural construction and just one, yet dominant mode of aesthetics of regulation. It presents the computer as an empowering decision devise and shapes YouTube to a special degree (fig. 5).

The aforementioned flexibility of the video appearance in the YouTube frame is increased by the offer to transform the running video appearance in terms of language, subtitles and resolution, which can be adjusted using the operative image of a gearwheel on the bottom right of the video frame. Furthermore since 2012 each YouTube video is presented in a paradigmatic way: When the cursor moves the progress bar, the video blurs and a collection of somehow representative single frames pop up as a preview, offering a navigation aid through the whole video by means of this frame collection.

Hereby the video is not playing but displayed as an area, as a visible set of not yet operative images. This YouTube approach to the order of selectivity touches upon fundamental questions of moving images elucidated by an even more

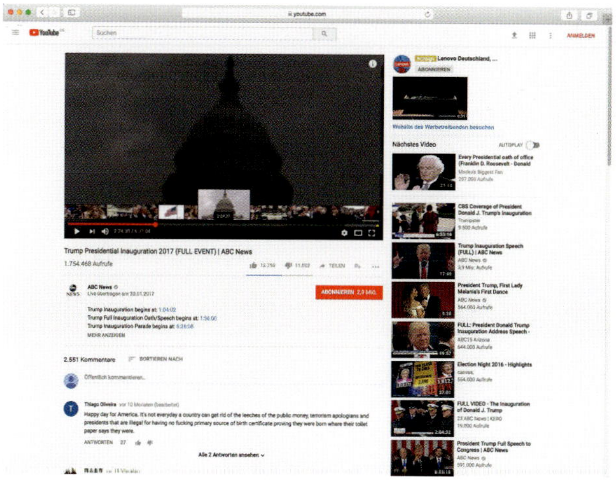

5 YouTube, navigation with progress bar.

obvious and radical programming that changed the look of YouTube already shortly after it has been sold to Google at the end of 2006. In the early days of YouTube, right after a video had been played, it still filled the whole video frame with one somehow representative image, ready to start anew. Since 2007 a finished video is replaced by a thumbnail collection of selectable videos: a new grid order of choice in exactly the frame supposedly reserved for moving images (fig. 6). This programmatic displacement becomes peculiar picturesque when the video is watched in full screen mode. Regarding this familiar mise-en-scéne – this grid of selectivity – Geert Lovink's résumé about YouTube from 2008 could be loaded with a new intention: "We no longer watch films or TV; we watch databases."[64]

64 Geert Lovink, The Art of Watching Databases. Introduction to the Video Vortex Reader, in: Geert Lovink, Sabine Niederer (eds.), *Video Vortex Reader.*

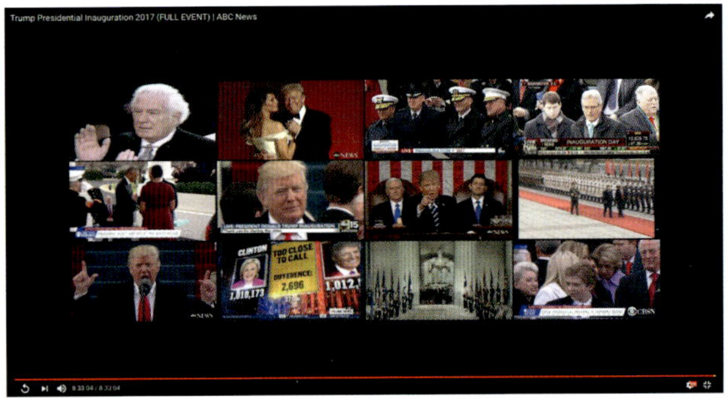

6 YouTube, grid collection of recommended videos.

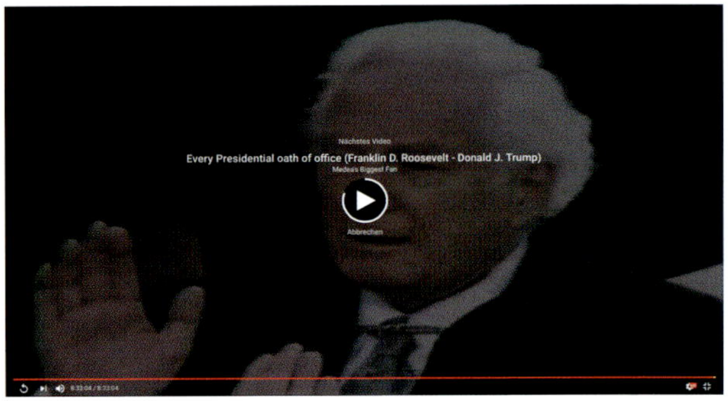

7 YouTube, autoplay feature.

Instead of the video appearance – that is, the chosen succession and process of moving images and sounds as a syntagmatic gesture – now the exact opposite is taking over: the invitation to select among replaceable images as a paradigmatic gesture, which consists of operative images. Thus YouTube's magic – the additional transformation of moving images into operative images – is demonstrated once more insistently. The programmatic indexicality of these images moves to the front.

Bearing in mind the second type of interface operations, this programmatic indexicality of the collected videos is based not only on the fact that they "all refer causally and physically to a set of software instructions to be executed" but also because of the operative trace to the processing of recorded and algorithmically evaluated data labor, with which these appearances are causally and physically linked.[65]The grid collection of recommended videos – that is, the idea and promise of this reference – is referring to the recorded viewing and search history. Because these operative images are therefore both depresenting and acting, i. e. part of an agency and an agenda, these aesthetical questions are also and unavoidably political.

With this in mind, a displacement and respectively a diversification of film/video aesthetics by aesthetics of regulation could be witnessed here. The logic of the filmic syntagm gets involved in the paradigmatic logic of digitalicity and its performed freedom as choice-control. To this, I would like to add, another potential relationship: the connection of this exhibited flexibility, a crucial promise of digitalicity, with the sociocultural ideal and pressure of flexibility in today's formations of flexible and communicative capitalism. Jodi Dean and Franco Berardi describe "a key contradiction of communicative capitalism" – if you "want to survive you have to be competitive and if you want to

Responses to YouTube, Amsterdam: Institute of Network Cultures, 2008, pp. 9–13, p. 9.

65 Marianne van den Boomen, Interfacing by Material Metaphors. How Your Mailbox May Fool You, in: Marianne van den Boomen et al. (eds.), *Digital Material. Tracing New Media in Everyday Life and Technology*, Amsterdam: Amsterdam University Press, 2009, pp. 253–264, p. 257.

be competitive you must be connected, receive and process continuously an immense and growing mass of data".[66]

The preliminarity and replaceability of the selected video can be interpreted as visualization and maybe familiarization of what Dean calls "the competitive intensity of neoliberal capitalism".[67] This aesthetic fate of chosen videos may be understood as a reminder of the competitive pressure, analyzed by Boltanksi and Chiapello,[68] and as an echo of Gilles Deleuze's *Societies of Control*[69]: Even these or them, which may have been chosen once among the many, always have to face a new competition right after the very selection. Ongoing flexibility and changeability is to learn and to rely on (fig. 7).

I would like to conclude with the observation that even this well-established paradigmatic logic of YouTube is subjected to changes. The installation of the "autoplay" mode, switched on by default since 2015, forms a counterpart to the order of selectivity: "The Autoplay feature on YouTube makes it easier to decide what to watch next. After you watch a YouTube video, we'll automatically play another related video based on your viewing history."[70] Thereby YouTube creates a new emphasis of flow that can be discussed from various perspectives: for instance, both in terms of YouTube's acclaimed reputation as *the new television* and in terms of the "data stream", estimated by Lev Manovich as a cultural form of presenting data in web-based social network services, heightening "the experience of the 'data present'".[71] Another form of flexibility is performed here – an ongoing flow of change that seems to be no longer under our (prepared and advised) control but that is controlled by information processing like a showcase for "algorithmic governmentality".[72]

Monitoring

This deserves a closer look and further steps. My remarks here are intended as starting points for an analysis that takes account of the complex procedures enabling and pursuing the options of uploading, searching, watching/hearing, classifying, valuing, and exposing data in the form of videos, requests, comments, clicks, and all sorts of metadata. In the end all the options depend on processes that ask for a new attention for intertwined interface operations.

How I operate with the YouTube interface is wedded to other interface operations by which a request for a video finds its way from, for instance, my processing smartphone to the responding server located in one of the European Google server farms in Dublin, Eemshaven, Hamina, or St Ghislain, relying on what Florian Sprenger has discussed as "politics of micro-decisions"[73] and producing traffic that is recorded both to customize my next visit on YouTube (and other informed websites) as well as to profit from my

66 Jodi Dean, The Limits of Communication, in: *Guernica*, http://guernicamag.com/features/the-limits-of-communication/ (accessed September 20, 2017).

67 Ibid.

68 See Luc Boltanski, Ève Chiapello, The New Spirit of Capitalism, New York: Verso, 2007.

69 See Gilles Deleuze, Postscript on the Societies of Control, in: *October* 59 (1992), pp. 3–7.

70 YouTube, Autoplay videos, https://support.google.com/youtube/answer/6327615?co=GENIE.Platform%3DAndroid&hl=en (accessed September 20, 2017).

71 Lev Manovich, Data Stream, Database, Timeline, in: *Software Studies Initiative*, http://lab.softwarestudies.com/2012/10/data-stream-database-timeline-new.html (accessed September 20, 2017).

72 Antoinette Rouvroy, Bernard Stiegler, The Digital Regime of Truth. From the Algorithmic Governmentality to a New Rule of Law, in: *La Deleuziana. Online Journal of Philosophy* 3 (2016), pp. 6–27.

73 Florian Sprenger, *Politik der Mikroentscheidungen. Edward Snowden, Netzneutralität und die Architekturen des Internets*, Lüneburg, 2015.

ascending browsing record by customized advertising.[74] These constructed, programmed, instructed, and performed interface operations built the processual character of YouTube videos that Yuk Hui has described as "digital objects", focusing "on data and metadata, which embody the objects with which we are interacting, and with which machines are simultaneously operating".[75] Interface operations characterize them as "new", that is "dynamic and energetic", forms of "industrial objects".[76]

Precisely because the question raised by *Interactions* – what computers are doing when we are not consciously interacting with them – is of prime importance, the complex of human and automatized interface operations needs to be explored. Interface operations include humans in front of monitors as well as, for instance, sensor-based *Ambient Intelligence* monitoring human or any other activities. Interface analyses should therefore be interested in aesthetics and a specific interface mise-en-scène as well as in the heterogeneous mesh of conditions enabling and determining it together with the practices of use and understanding. Their relevance is increasing the more interfaces and operations between beings, things, and computers are built and relied on – and the more these operations are planned and mediated as *indispensable and invisible*, heading towards new forms of depresentation.

74 In the case of YouTube interface analyses overlap with "platform studies" insofar as "platform" is understood as "a broad enough category to capture a number of distinct phenomena, such as social networking, the shift from desktop to tablet computing, smart phone and 'app'-based interfaces as well as the increasing dominance of centralised cloud-based computing". Joss Hands, Introduction. Politics, Power and 'Platformativity', in: *Culture Machine* 14 (2013), pp. 1–9, p. 1.

75 Yuk Hui, *On the Existence of Digital Objects*, Minneapolis, 2016, p. 48.

76 Ibid., pp. 49–57.

Figures

1 Cover of *Interactions* 23/6 (2016).

2 Deutsche Telekom, Telekom Europe TV commercial with Andrea Bocelli, https://youtube.com/watch?v=eZu6oWfrieo (accessed November 20, 2017).

3 ABC News, Trump Presidential Inauguration 2017 (FULL EVENT) | ABC News, https://youtu.be/Nieiu8tmLIM?t=2h13m31s (accessed November 20, 2017).

4 YouTube, https://youtube.com (accessed November 20, 2017).

5 ABC News, Trump Presidential Inauguration 2017 (FULL EVENT) | ABC News, https://youtu.be/Nieiu8tmLIM?t=2h24m30s (accessed November 20, 2017).

6–7 ABC News, Trump Presidential Inauguration 2017 (FULL EVENT) | ABC News, https://youtu.be/Nieiu8tmLIM?t=8h33m4s (accessed November 20, 2017).

Luisa Feiersinger

Spatial Narration

Film Scenography Using Stereoscopic Technology

Situating the Screen in Stereoscopic Practice

If the screen is a plane on which images are displayed in order to view them, then what precisely happens to the screen in those films that are presented under the catchphrase *3-D*?[1] The visual impression created by such *stereoscopic* films is based on a technical imitation of binocular vision, which allows for the perception of three-dimensionality. Through an intricate arrangement of projectors, filter foils, specially coated screens, and distinctive eyeglasses, two film tracks – each taken from a slightly different perspective – are simultaneously delivered to the eyes of the spectator. The viewer is able to perceive a visual space that is not merely restricted to the plane of the screen, but expands in front and behind it. This *negation of the screen*, in which the abstraction of two-dimensionality appears to dissolve, has always been been understood within film theory as a means of achieving greater immediacy.[2] Miriam Ross, in her recent discussion on stereoscopic visuality, actually begins her analysis by asking whether the screen is even still present.[3] She reaches the conclusion that the screen dissolves into a "field screen," thereby facilitating a "fundamentally different viewing experience."[4] That a change in the filmophanic space[5] occurs is indisputable, but if one wishes to situate the screen in visual practice, as this volume suggests, then the bigger picture – so to speak – must be considered. As such, the stereoscopic image space, having dispensed with planar limitations, can only be generated through particular filming techniques and mise-en-scène strategies. Fully in the tradition of narrative cinema,

1 This essay was first published in 2015, and has been translated and revised for this volume. I thank Deborah J. Curtis and Julia Sittmann. For the original text see Luisa Feiersinger, Räumliches Erzählen. Filmszenographie in stereoskopischer Technik, in: Annette Dorgerloh, Marcus Becker (eds.), *Alles nur Kulisse?! Filmräume aus der Traumfabrik Babelsberg*, Weimar: VDG, 2015, pp. 140–145.

2 A comprehensive introduction to the historical background is beyond the scope of this essay and, therefore, reference is only made here *pars pro toto* to Arnheim, *Film als Kunst*, Frankfurt/M.: Suhrkamp, [1932] 2002, and to André Bazin, Der Mythos vom totalen Film, in: Robert Fischer (ed.), André Bazin, *Was ist Film?*, Berlin: Alexander Verlag [1946] 2009. Although their basic approaches are almost diametrically opposed, both film theoreticians ascribe an "immediacy of appearance" and thus a direct influence on the viewer to the three-dimensional film. Arnheim 2002, p. 266; Bazin 2009, p. 47.

3 Miriam Ross, Stereoscopic Visuality. Where is the Screen, where is the Film?, in: *Convergence. The International Journal of Research into New Media Technologies*, 19.4 (2013), pp. 406–414, p. 406.

4 Ibid., p. 413. It must be mentioned, that she also discusses the aesthetic changes in stereoscopic film based on its technical conditions.

5 This text draws on the *Vocabulary of Filmology* used by Etienne Souriau to make a distinction between the reality that is independent of the film ("afilmic reality"), the reality that pertains to the film ("profilmic reality"), and the narrated world ("diegesis"), as well as to separate the processes and characteristics of film projection ("filmophanic reality") from those of the film material ("filmographic reality"). Etienne Souriau, Die Struktur des filmischen Universums und das Vokabular der Filmologie, in: *montage/av*, 6.2 (1997), pp. 140–157.

the mediation itself (the complex technological and cinematographic arrangement that produces the image) must be rendered invisible to the viewer.[6]

The ostensibly *negated screen* is made expressly manifest in the practices and techniques involved in producing the immediacy of this perception, and stereoscopic films – just as any other screen-based images – are impossible to conceive of outside their production and reception possibilities. An investigation of the film-image in films produced since 2009 must be based on its interconnection with its production techniques – not because the visual space created through stereoscopic technology is new,[7] but because the combination of stereoscopic alignments with digital recording, processing and playback techniques is. Although the shift from analog to digital techniques may not have been apparent to the untrained eye, this transition was essential to the development of the aesthetic qualities inherent to contemporary stereoscopic visual imagery.[8] In order to investigate how narration can take place through and within stereoscopic film spaces, three vital questions must be addressed: firstly, how does this diegetic space interact with the boundaries of its images, both on the plane of the screen, and within the projection that extends beyond it? Secondly, how is the profilmic space prepared and translated for the shot, using cinematographic techniques? Finally, how has the understanding of these cinematographic practices and the existing technical requirements favoured the formulation of specific narrative structures and motifs? In the following, Alfred Hitchcock's film, *Dial M for Murder* (1954),[9] produced using analog technology, and *The Three Musketeers*, an exemplar of digital stereoscopic films, directed by Paul W. S. Anderson (2011),[10] will be compared in order to illuminate these issues.

Translating Diegesis into Stereoscopic Film Space

In *The Three Musketeers*, Alexandre Dumas' well-known story is re-packaged as a action movie spectacle, meant to satisfy modern sensibilities: With the help of inordinate amounts of weaponry, the three musketeers and the young D'Artagnan foil a conspiracy by Cardinal Richelieu to rob the inexperienced King Louis XIII of the throne. Central to the plot is a necklace belonging to the Queen, which – in the wrong hands – could trigger war between England and France. This diegetic 17th Century France is located in the profilmic spaces of Bavarian castles and Babelsberg green-screens, where a new world was created, which fulfilled the visual requirements of stereoscopic films, while

6 This structure of technical images is known as the "principle of disjunction" in the discipline of *Bildgeschichte*, Horst Bredekamp, Angela Fischel, Birgit Schneider, Gabriele Werner, Bildwelten des Wissens, in: *Bildwelten des Wissens. Kunsthistorisches Jahrbuch für Bildkritik*, 1.1: Bilder in Prozessen (2003), pp. 9–20. Not by name but in principle, it was established for the two-dimensional film by Bordwell, Thompson and Staiger in their seminal study on the classical Hollywood cinema. David Bordwell, Janet Staiger, Kristin Thompson, *The Classical Hollywood Cinema. Film Style and Mode of Production to 1960*, London: Routledge, 1994.

7 For in-depth information on the history of stereoscopic films and their occurrence in waves, see Ray Zone, *Stereoscopic Cinema & the Origins of 3-D Film, 1838 –1952*, Lexington: University Press of Kentucky, 2007; Ray Zone, *3-D Revolution. the History of Modern Stereoscopic Cinema*, Lexington: University Press of Kentucky, 2012.

8 Thomas Elsaesser highlighted this link in an essay on the re-establishment of stereoscopic films since 2009, emphasising the benefits of the visibly *different* stereoscopic film image for the dissemination of digital technology. Thomas Elsaesser, The 'Return' of 3-D. On Some of the Logics and Genealogies of the Image in the Twenty-First Century, in: *Critical Inquiry* 39 (Winter 2013), pp. 217–246, pp. 221–225.

9 Alfred Hitchcock, *Dial M for Murder*, USA 1954, in: 3-D Blu-ray, Warner Bros. Entertainment Inc. 2012, 105 Min.

10 Paul W. S. Anderson, *The three Musketeers*, Germany/France/UK/USA 2011, in: 3-D Blu-ray, Constantin Film 2011, 111 Min.

simultaneously serving as a visual subtext for the plot and its characters. Richelieu's room and its furnishings, for example, convey his tactical cleverness: Our first encounter with the Cardinal is cinematically staged over a chessboard, as the camera pans upwards from a close-up of the chess pieces towards a medium close-up shot of his face.[11] In this sequence, the game of chess and the Cardinal's face are arranged not only on a vertical axis, but also positioned separately in the depth of space produced by the stereoscopic film. The elements of the image are distributed visually in this stereoscopically-created space (depending on where the optical axes of the two image tracks intersect) either in front or behind the plane of the screen, the so-called *zero parallax*.[12] While the chess pieces are in front of this plane – referred to as *negative parallax* – the Cardinal's face is in *positive parallax*, namely behind it. The game of chess is thus spatially accentuated, through its position directly in front of the eyes of the audience. That the game serves as an allegory for the Cardinal's political manoeuvres – which he plans like chess moves – becomes abundantly clear as the scene continues, and Richelieu reveals to his interlocutrice, *Mylady*, that he only ever plays against himself – no other suitable challenger exists.[13] Standing behind the table with the chessboard, the protagonists are shown in a two shot – wherein the frame encompasses a view of two people (fig. 1). Once again positioned in slight negative parallax, the chess game continues to occupy the front of the image space, framed symmetrically between two candlesticks and two small ornate cases. The two individuals dominate the shot, while the room spreads out in positive parallax in the background around them.

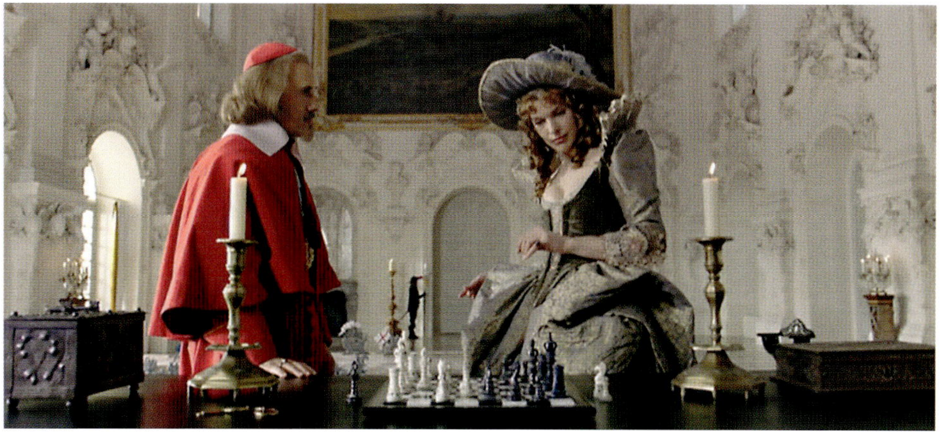

1 Semantics of space in *The Three Musketeers* (2011), screenshot, TC: 00.22.21.

For attentive film audiences, this specific mise-en-scène of objects in the foreground, actors in the middle ground, and a room in the background will already be familiar from shots in numerous stereoscopic films, including Hitchcock's *Dial M for Murder*.[14] In this 1954 film, a husband attempts to have his unfaithful wife murdered in his absence. Even though – or perhaps precisely because – the murder attempt fails, the husband is found out by dint of a key, crucial to

11 Ibid., TC: 00.21.52–00.22.02.

12 For an in-depth description of the stereoscopic production of space and on the associated terminology, see the contribution by Shannon Benna and its glossary in this volume, pp. 133-145.

13 Anderson 2011 (as fn. 10), TC: 00.22.02–00.23.04.

14 The film was produced using stereoscopic technology, but has generally been listed as a 2-D film due to the rapid decline of the 3-D boom in the 1950's. For a history on the screening of the film, see R.M. Hayes, *3-D Movies. A History and Filmography of Stereoscopic Cinema*, Jefferson/London: McFarland & Co 1989, pp. 171–173 and Zone 2012 (as fn. 7), pp. 35, p. 42. David Bordwell discusses this particular mise-en-scène in the entry *Dial M for Murder: Hitchcock frets not as his narrow room* on his blog *David Bordwell's website on cinama*, David Bordwell, Kristen Thompson, Observations on film art, http://davidbordwell.net/blog/2012/09/07/dial-m-for-murder-hitchcock-frets-not-at-his-narrow-room/ (accessed May 12, 2015).

his ingenious plan, that ultimately betrays him. Almost all of the scenes in *Dial M for Murder* are filmed using the image composition mentioned above, with only a few crucial elements jutting out into the movie theatre in strong negative parallax: first the wife's hand, which she desperately stretches towards the viewers at the moment of her attempted murder, and then the key, which the police inspector displays in an equally dramatic fashion.[15] These narrative moments are foregrounded – quite literally – as exceptions to Hitchcock's stereoscopic formula. Similarly, in *The Three Musketeers*, the Queen's necklace, in addition to the chessboard, also often appears in the visual foreground, thereby marking its narrative importance in the film.

However, spatiality is deployed at other levels as well. While the depth of space is relatively flat in dialogue scenes (such as the one previously mentioned between Richelieu and Mylady), it is extended in more dramatic moments, as the so-called *depth budget* is enlarged. As the *inter-axial* distance (the space between the cameras recording the images) is increased, the physical expansion of the image elements is heightened. The stereoscopically produced space is not dependent on the expansion of the actual space being recorded by the cameras, but on specific cinematographic strategies and conditions. This fact applies equally to analog and digital cinematography, although greater control can be exerted over digital shots, since they can both be assessed on the spot during filming, and corrected later in the production process. In addition, the necessary manipulations – equally possible in analog films – appear easier to achieve and can be implemented more rapidly by digital means. As such, continuous minimal adjustments become feasible, permitting – for instance – for the space to be flattened to spare the eyes of the viewer in a scene with rapid cutting. In *The Three Musketeers*, these adjustments, the exaggeration and the flattening of the visual space, can be observed in the sequences where the three musketeers encounter Rochefort, chief of the Cardinal's guardsmen, on airships.[16]

Elements in the film that move towards the audience must be handled with the same care, as they entail an intrinsic contradiction: They are both expected to appear in a stereoscopic film, but when they do, are often condemned as both cheap gimmickry and hard on the eyes.[17] In addition, they harbour the danger of destroying the illusion of physicality produced in stereoscopic films, and thus laying bare the technical sleight of hand that brought them into existence. The visual elements in negative parallax practically force themselves onto the viewer. But were they to follow their natural impulse to test the image's physicality, the viewer would reach into nothingness, reinforcing the

15 Hitchcock 1954 (as fn. 9), TC: 00.44.04 and TC: 01.39.29. The film director confirms the positioning of these image elements in his interview with François Truffaut, although he has little praise for his only 3D project. François Truffaut, *Mr. Hitchcock, wie haben sie das gemacht?*, München: Heyne, 2003, pp. 207–210, p. 208.

16 Anderson 2011 (as fn. 10), TC: 01.25.41–01.29.19. While the space is flattened in the battle sequences it is exaggerated in the sequences opening up the view into the landscape. Glen MacPherson, who worked on this film as a camera man, as well as on numerous other projects by Anderson, confirms these techniques for another joint 3-D project in the interview with R. Emmet Sweeney. R. Emmet Sweeney, Interview: Glen MacPherson, 3D DP, http://filmcomment.com/entry/interview-glenn-macpherson-3d-dp-resident-evil (accessed January 23, 2015).

17 The critics' response to the film was mixed, mostly highlighting the excessive use of visual effects in a flat literary adaptation. For one example, see Mark Feeny, The Three Musketeers Movie Review, in: *The Boston Globe* October 22, 2011, http://archive.boston.com/ae/movies/articles/2011/10/22/three_musketeers_when_swords_meet_cgi/ (accessed March 23, 2018). Elsaesser highlighted this type of criticism as a general trend in the discussion of 3-D in his essay on the genealogy of stereoscopic films and pointed out the contradictory demands placed on them. Elsaesser 2013 (as fn. 8), p. 237.

images' lack of corporality.[18] Furthermore, even if the audience accepts the optical illusion as is, it is precisely these forward-moving elements that can produce perceptual conflicts, through their positioning in the visual space relative to the screen's boundaries. If, due to negative parallax, an object appears to be placed in front of the screen, but is simultaneously intersected by the framing of the film, then this results in competing and contrasting depth references, since such an overlap normally indicates, by convention, that the object is positioned in the background.[19] The visual space at these points does not extend forwards or backwards, as is characteristic for stereoscopic technology, but moves to and over the sides. While the expansion of the diegetic space over the side boundaries of the visual space is unproblematic in two-dimensional films,[20] the frame appears more fundamentally to be recognized as a border in stereoscopic films: In *Dial M for Murder*, table lamps, which are placed at the front of the image space often produce such a conflict. Specifically in the longer takes, the viewer perceives the intersection between objects in negative parallax and the frame as breaking the illusion of corporality that stereoscopic films try to convey.[21] Even if these lamps are only slightly in front of zero parallax, they are visually irritating, since they exceed the full height of the image. Even when the objects do not produce any perceptual conflicts, their positioning in the foreground often distracts from the main action, which is in part covered up by them.[22] Image composition and framing must therefore be re-conceived and re-learned for stereoscopic filming. The placement of these

18 In the essay on stereoscopic visuality by Miriam Ross, already mentioned above, the author focuses, in particular, on image elements presented in negative parallax. The potential of the stereoscopic film to dissolve its illusion of reality would be *concentrated* in these elements. She therefore refers to these elements as destabilising the screen and its illusion, Ross 2013 (as fn. 3), p. 409. They simultaneously expand the sensory potential of the stereoscopic film in its own fashion, as she demonstrates in reference to the discourse on the haptic film. Jennifer Barker, *The Tactile Eye. Touch and the Cinematic Experience*, Berkeley, CA: University of California Press, 2009; Guiliana Bruno, *Atlas of Emotion. Journeys in Art, Architecture and Film*, New York: Vers, 2002; Laura Marks, *The Skin of the Film*, Durham: Duke University Press, 2000; Anne Rutherford, Cinema and Embodied Affect, in: *Senses of Cinema* 25 (March 2003), http://sensesofcinema.com/2003/feature-articles/embodied_affect/ (accessed January 23, 2018); Steven Shaviro, *The Cinematic Body*, Minneapolis: University of Minnesota Press, 1993; Vivian Sobchack, *The Address of the Eye. A Phenomenology of Film Experience*, Princeton: Princeton University Press, 1992; Vivian Sobchack, *Carnal Thoughts. Embodiment and Moving Image Culture*, Berkeley: University of California Press, 2004; Christiane Voss, Film experience and the formation of illusion. The spectator as 'surrogate body' for the cinema, in: *Cinema Journal* 50.4 (2011), pp. 136–150.; Ross 2013 (as fn. 3), p. 412. In parallel with this argument on the disruptive potential of image elements in negative parallax, I have interpreted these elsewhere as revenants of philosophical toys. Luisa Feiersinger, Berührung im stereoskopischen Film. Über das Ergreifen und Ergriffenwerden von optischen Illusionen, in: Steffen Haug, Thomas Helbig, Tina Zürn (eds.), *„Don't touch! Touch screen!" Das Bild, der Blick und allerhand Formen taktiler Wahrnehmung und Erkenntnis. Eine Tagung für Michael Diers*, Munich: Fink, in preparation.

19 Raymond and Nigel Spottiswoode were already working on this problem in the 1950's. They therefore propose a *stereo window* that, printed around the film image in the form of a black frame, also floats in space visually as an image element and thus eliminates the irritations produced by the overlap, Zone 2012 (as fn. 7), pp. 268–269. This frame, which is incorporated, but not perceived as such, just as is demanded by the tradition of narrative cinema, is used much more frequently in digital cinema, predominantly in individual shots, mainly thanks to the simplicity of the production of these stereo windows with digital techniques. Once again, see Benna 2018 (as fn. 12), pp. 135-136.

20 Instead of referring to the numerous publications that discuss the onscreen-offscreen relationship from specific points of view, it should be emphasised here that David Bordwell and Kristin Thompson name framing, i. e. the relationship between what is depicted and its frame, as a central category for analysis in their seminal book on the analysis of films. David Bordwell, Kirstin Thomson, Narrative as a Formal System, in: David Bordwell, Kirstin Thomson, *Film-Art. An Introduction*, New York: MacGraw-Hill, 2010, pp. 186–212.

21 Barbara Flückiger provides a clear discussion on this dissolution of the physicality of objects when they are intersected by the margin. Barbara Flückiger, Aesthetics of Stereoscopic Cinema, in: *Projections* 6.1 (2012), pp. 101–122, pp. 116–117.

22 This happens very often troughout the entire film, for one examplary instance, see Hitchcock 1954 (as fn. 9), TC: 00.15.03.

visual elements in *Dial M for Murder* raises the question as to their function, whereby the likelihood is high that their purpose was simply to stagger the depth of space, but that the chosen lamps were simply too large for the task.[23]

In the later film, *The Three Musketeers*, greater attention was paid to the relationship between the larger objects shown in negative parallax and the frame. They are never truncated by more than one visual edge, and especially not by the upper one, and then, only briefly. In addition, the viewer can observe a greater focus on the main action within the general composition of the scene. There is also a clear attempt to better integrate elements already in strong negative parallax into the image as a whole, while simultaneously maintaining the invisibility of the techniques used in the medial transmission. In concrete terms, this means that an effort was made to ensure that elements entering into the viewer's space respected the frame of the screen. In one sequence, which follows the flight of a cannonball, the projectile is staged in a complex manner within the depth of space:[24] the warhead initially hurtles straight towards the viewer, crossing through the entire stereoscopically created space, from positive into negative parallax. But, before the cannonball reaches the viewer, the camera rotates around it, subsequently following it in slow motion from the side, as it now, all of a sudden, floats in negative parallax in front of the audience, tantalizingly within reach. Safely out of range of any overlaps, its physicality appears beyond question. The camera then pans around again, this time behind the cannon ball, and follows its flight, back at normal speed, until impact.

The specific constraints that shots with effects in negative parallax must adhere to, in order to avoid irritating the viewer, are also liable to influence the narratives of stereoscopic films. The historically inconsistent re-imagination of Dumas' *The Three Musketeers* to include airships is likely a consequence of those conventions, insofar as flying elements are particularly well suited to the medium. Setting the action at height, with the protagonists hovering in the air, facilitates not only the emergence of image elements in negative parallax, without the danger of encroaching on the image frame, but also the subliminal introduction of the motif of falling. A common theme in stereoscopic films, falling, with its ability to depict spectacular views into the depths below, produces a potent vacuum-effect which pulls at the viewer, and is, as such, a favoured cause of death in *The Three Musketeers*, despite the plethora of actual weapons available. The final battle between the adversaries Rochefort and D'Artagnan takes advantage of precisely this danger, impressively displayed through stereoscopic techniques.[25] The duel on the gables of Notre Dame Cathedral in Paris, with its steeply pitched roof, opens up numerous opportunities for shots from above, looking down into the depths below. Rochefort ultimately falls into the abyss – effectively staged in positive parallax, emphasizing the dramatic nature of the location and his death.

Means of Constructing Stereoscopic Space

It goes without saying that the risk to the actors in this scene was minimal, since the gables were located no more than half a metre above the ground on soft mats in a film-studio in Babelsberg (fig. 2a,b). The musketeers' airship also flew in

23 On the production of space in *Dial M for Murder*, see also Jesco Jockenhövel, *Der digitale 3D-Film. Narration, Stereoskopie, Filmstil*, Wiesbaden: Springer, 2014, pp. 60–64.

24 Anderson 2011 (as fn. 10), TC: 01.26.44–01.26.48.

25 Anderson 2011 (as fn. 10), TC: 01.32.31–01.36.00.

front of the green screen there, negating the need to reconstruct Notre Dame in Babelsberg, which thus only existed in virtual space (fig. 3a,b),[26] similarly to all elements lending structure to the aerial space. What Hitchcock attempted with the help of table lamps, can now be accomplished in digital films with much smaller elements. Water vapour and clouds, for instance, demarcate the spatial expansion in numerous scenes in *The Three Musketeers*. This "stereoscopic debris" is so easy to produce with CGI that it has become seemingly ubiquitous in recent films.[27] Independent of the fact that digital techniques have resulted in a simplification, and thus in an increase, of these types of cinematographic manipulations,[28] film space has nonetheless always been a synthetic space. Right from the beginning, film space was untethered from physical reality, as shots of small-scale models (standing in for larger cityscapes) or even black and white, and deep focus shots, manipulated our visual perception.[29]

Within stereoscopic film techniques, the crucial difference between analog and digital manipulations remains the ability to control the outcome. With analog techniques, the success or failure of the artificial creation of space can only be assessed once the celluloid has been developed. Digitally manipulated space can be checked on the control screen during production and sometimes even instantaneously on-set. If the visual spaces are entirely digitally generated,[30] control over the image is extended even further: Every aspect of the various components of the simulated image can be controlled and arranged.[31] In a stereoscopic set-up, the 'cameras' are essentially viewpoints onto intricately calculated generated worlds: their alignment, as well as their various stereo-parameters, can be perfectly synchronized. Light reflections, for example, that present themselves differently to analog cameras taking the shot from different positions, can result in contradictory images, which dissolve the spatial effect.[32] In CGI, they are introduced individually, and as such become easily manageable. This element of control in digital film space facilitates its use in both two-dimensional and stereoscopic films. Whereas the creation of space was possible with analog techniques, implementing it with two image tracks was far more difficult. The construction of artificial spaces was more noticeable in shots taken with two instead of one camera. Due to their planar nature *matte paintings*, used to introduce foreign environments into the backdrop, just like *rear projections*, provided the cameras filming them from different perspectives with no

26 With reference to work on the virtual spaces, see the interview with Eric Robinson, the head of the VFX team in *The Three Musketeers*. Vincent Frei, The three Musketeers: Eric Robinson – Digital Effects Supervisor – Mr. X., http://artofvfx.com/?p=1713 (accessed May 12, 2015).

27 Ross calls these elements "stereoscopic debris". It is precisely this debris that is capable of producing the "thick, tactile field screen" that is typical for stereoscopic films in the 21st Century and she attributes a prominent role to it in the construction of a "field screen". Nonetheless she notes, that these elements are not limited to the current productions techniques, but the simplicity with which they can be controlled, made it easier to integrate them. Ross 2013 (as in fn. 3), pp. 409–410.

28 On the construction of these worlds and on their persuasive powers through the simulation of photographic appearance, see Stephen Prince, True Lies. Perceptual Realism, Digital Images and Film Theory, in: *Film Quarterly* 49.3 (1996), pp. 27–37.

29 An overview on the advanced production of artificial worlds using analogue techniques is provided by Thomas G. Smith, *Industrial Light & Magic. The Art of Special Effects*, New York: Ballantine Books, 1986.

30 Nowadays, the default construction of digital worlds is that of 3D animations. These are characterized by their volumetric figures in spatial settings. They do not, however have an intrinsic connection to 3-D projection.

31 Prince 1996 (as fn. 28), as well as the discussions of the possibilities of the camera in digital film. Jessica Aldred analyzed these considering their effects on the viewer and their immersion into the film. Jessica Aldred, All Aboard *The Polar Express*. A 'Playful' Change of Address in the Computer-Generated Blockbuster, in: *animation: an interdisciplinary journal* 1.2 (2006), pp. 153–172.

32 Flückiger 2012 (as fn. 21), pp. 106–107.

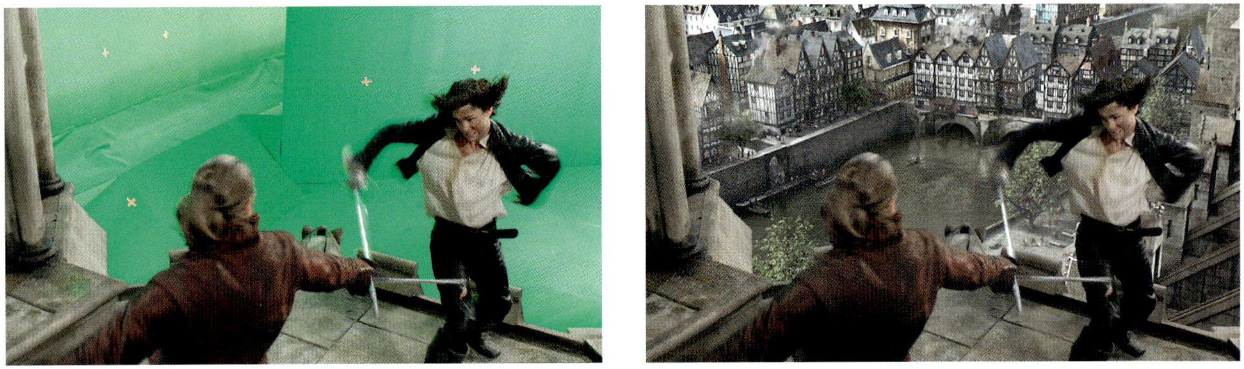

2a,b Duel on the roofs of Notre Dame in *The Three Musketeers* (2011), working photography before and after insertion of the digital background.

differentiating information. As can be observed, for example, in *Creature from the Black Lagoon* (1954), they thus appear strangely flat in stereoscopic set-ups.[33] It is possible that for this very reason, Hitchcock decided to stick to a chamber play in his stereoscopic film, thereby avoiding the techniques he otherwise favoured for the incorporation of any external environment into his films.[34] The production possibilities thus effect the options available for the setting. In addition to the construction of virtual worlds and the possibilities inherent to post-production, digital filming devices produce a liberty within their scenographic circumstances that stereoscopic films did not previously have. While Hitchcock still had to build a gigantic model of a telephone to film a close-up,[35] today, digital camera can film such a scene normally, as cameras have shrunk, permitting a shorter inter-axial distance.

Opening the Window

Overall, digital stereoscopic film is subject to fewer technical limitations than its analog predecessor, permitting a greater measure of control over profilmic objects and cinematographic staging. The wide scope for manipulation, inherent to worlds that are stereoscopically reproduced, allows spatial constructions to communicate the narrative. Despite this ostensible freedom for film producers, certain conventions have established themselves, such as reducing the effects of negative parallax and the flattening of space during rapid cut-sequences, for instance. Some of these stan-

33 Jack Arnold, *Creature from the Black Lagoon*, USA 1954.

34 The entire film, apart from two short scenes outside, takes place in two rooms in the couple's apartment. Hitchcock states that the play the film was based on was pivotal to this decision, see Truffaut 2003 (as fn. 15), pp. 208–209. Jockenhövel also highlights that it is precisely the selection of a piece that requires no outside space, which Hitchcock preferred to incorporate through matte paintings and rear projections, can be viewed as associated with the stereoscopic techniques. Jockenhövel 2014 (as fn. 23), p. 64.

35 David Bordwell shows this in his already mentioned in-depth analysis of the stereoscopic variant of the film that he published on his blog Bordwell, Thompson 2012 (as fn. 14).

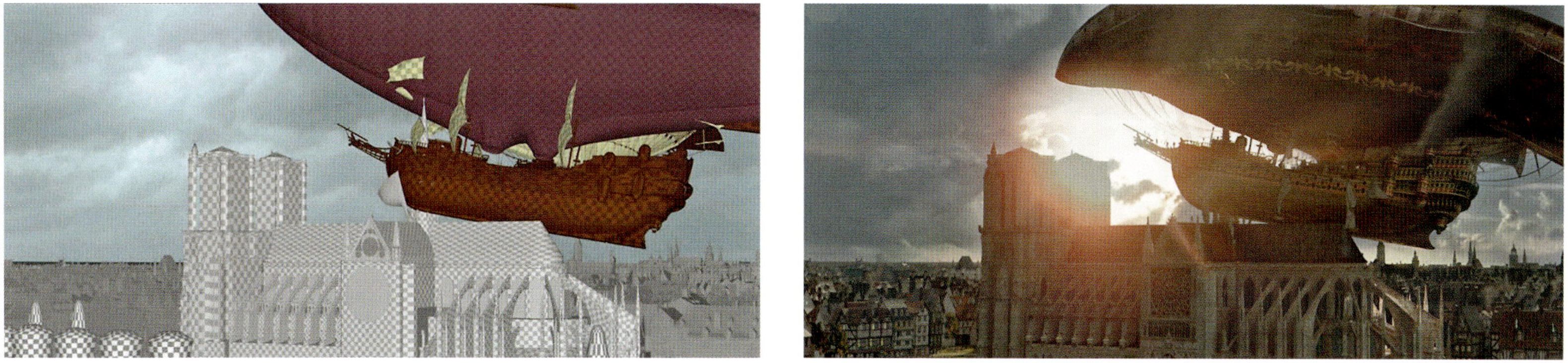

3a,b Airship above Notre Dame in *The Three Musketeers* (2011). CAD working photography in the raw and fully rendered version.

dards are also directly incorporated into film plots, which are adjusted to create stereoscopically suitable scenes and motifs. Above all, the mise-en-scène conventions outlined here serve the purpose of imitating natural perception: The technical mediation (between cinematographic manipulations and the viewer) remains as invisible as possible[36] – a principle in line with a tradition that reaches as far back as Renaissance painting, with its emphasis on a central pictorial perspective.[37] This classic art historical concept of the picture as an *open window* binds the viewer and the image to a single point, in order to convey an illusionistic space. In its two-dimensional alignment, film – even though it sets its images in motion – continues in this tradition, to a certain degree.

The title sequence of *The Three Musketeers*, invokes this concept only to leave it behind.[38] The opening credits start with a view of a map that initially appears flat, positioned slightly behind the plane of the screen (fig. 4). The map is framed by a dagger, a revolver and some coins, all elements reminiscent of *trompe l'œil* paintings. Such paintings, usually depicting flatware, present themselves as illusionistic expansions of the viewer's space. Precisely this space is then burst open when the camera moves first towards and then into the map. Its flatness dissolves into different planes within the stereoscopic depth of space: the grid that delineates the map is revealed, floating in front of the map, not dissimilar to the gridlines used in image composition. Moving through the transparent grid, and thus, quite lit-

36 Benna points out the natural depth method, developed by stereoscopic filmmakers Alan & Josephine Derobe, that mimics human binocular vision. Benna 2018 (as fn. 12), p. 142.

37 In the 15th Century, the image in central perspective was described as an open window by Leon Battista Alberti in his treatise on the art of painting. Leon Battista Alberti, *On Painting, De Pictura*, New Haven: Yale University Press, 1966. This idea of immediacy has been discussed so comprehensively as a metaphor in the discourses on the history of art and imagery, at least since Panofsky's *Perspective as Symbolic Form* (1927), that even an illustration of the central positions alone cannot be given here. In its place, reference is therefore made to the discussion of these metaphors specifically in relation to the film. Sobchack 1992 (as. fn 18), pp. 14–25.

38 Anderson 2011 (as fn. 10), TC: 00.00.35–00.01.37.

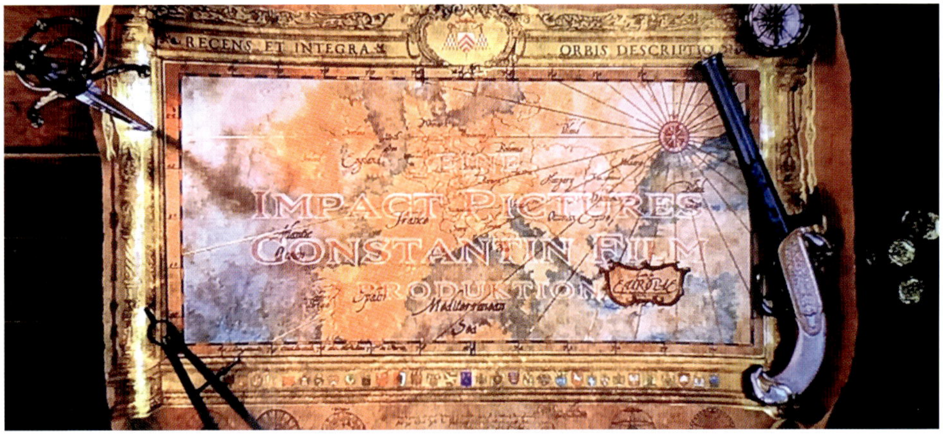

4 *Trompe-l'œil* in the opening credits for *The Three Musketeers* (2011), screenshot, TC: 00.00.37.

erally, leaving it behind, the camera then opens up a visual space that is only possible in digital stereoscopic films. Flying through an artificial space that is populated by figurines (familiar from re-enactments of historical battles), the camera's movement is reminiscent of a physical camera, moving freely within the space. But the zoom through the sky and the clouds was created digitally. By making use of stereoscopy, the film aims to produce sensation rather than realism. The figurines and their stereoscopic viewpoints are rendered in such a way that their three-dimensionality equates to human size, with the camera's flight up and through the space causing a kind of a roller coaster sensation for the viewer.

The *being there* in an artificial world, which succumbs to the screen as mediator, is nonetheless a worthy successor in this longer tradition, if one considers both the invisible screen-plane, and the technical and practical set-up that produces the screen-based image. The various viewpoints as well as the artificial world with its population of figurines have to be rendered on numerous screens. The impression of human-sized three-dimensionality is then possible through a careful arrangement of those viewpoints in *hypostereo*. Furthermore, elements – such as the clouds – are chosen for their ability to be displayed stereoscopically in arresting ways. The dissolution of the screen and the abstraction of two-dimensionality can thus only be left behind through intricate technical and narrative alignments. The screen-based image remains fixed in the mechanics and techniques of its production and reception, even when the screen is negated in a narrative sense.

Figures

1, 4 Paul W. S. Anderson, *The Three Musketeers*, Germany/France/UK/USA 2011, in: 3-D Blu-ray, Constantin Film Verleih GmbH 2011, 111 Min., TC: 00.22.21, 00.00.37.

2–3 Digital Effects Supervisor, http://artofvfx.com/?p=1713 (accessed January 23, 2015).

Lasse Scherffig

From Action Capture to Interaction Gestalt

How can moving a small physical object on a table, observing *apparent motion* on a computer screen and pressing a finger onto one part of the object be experienced as one integrated action: clicking on something? This text proposes a novel answer to this question that goes back to the beginning of interactive computing and before, re-activating ideas from cybernetics to arrive at a new understanding of screen-based interaction.[1]

In this paper, I will briefly introduce cybernetics and its role in interactive computing and show how at the transition from analogue to digital computing screen-based interaction was introduced. I will then explain how screen-based interaction is subject to questions regarding the perception of motion that were first raised by gestalt psychology, explain how these questions relate to the idea of direct manipulation and how we might have to rethink the gestalt of an interface as an effect of interaction.

Cybernetics and Interaction

While cybernetics played an important role in the formation of early computer science, in recent times it has mainly been discussed in the humanities.[2] Here, the focus often is on the epistemological implications of this discipline, which, from its very beginning, proclaimed it would erase the boundaries between animal and machine, living and non-living systems.[3] One of its core tenets is the application of negative feedback to the description and control of any process that can be described as goal-directed behaviour.[4]

Negative feedback implies that the output of a system is fed back to its input as a negative quantity, resulting in a system that operates on the difference between its output and a desired goal. Systems using negative feedback hence use their own deviation from a given goal as a means of correcting this error.

Cybernetics to some extend can be understood as a science undertaking a "totalization" of feedback control.[5] Its importance for answering the question about *clicking*

1 For a much broader and more detailed development of this argument see Lasse Scherffig, *Feedbackmaschinen. Kybernetik und Interaktion*, Köln: Kunsthochschule für Medien Köln, 2017.

2 Kathryn Hayles, *How we Became Post-human. Virtual Bodies in Cybernetics, Literature, and Informatics*, Chicago, IL: Chicago University Press, 1999; Claus Pias, Zeit der Kybernetik. Eine Einstimmung, in: Claus Pias (ed.), *Cybernetics/Kybernetik – The Macy Conferences. Volume II*, Zürich/Berlin: Diaphanes, 2004, pp. 9–41.

3 This is already apparent in the title of Norbert Wiener's *Cybernetics. Or Control and Communication in the Animal and the Machine*, Cambridge, MA: MIT Press, 1948.

4 Arturo Rosenblueth, Norbert Wiener, Julian Bigelow, Behavior, Purpose and Teleology, in: *Philosophy of Science* 10 (1943), pp. 18–24.

5 Peter Galison, The Ontology of the Enemy. Norbert Wiener and the Cybernetic Vision, in: *Critical Inquiry* 21.1 (1994), pp. 228–266, p. 233.

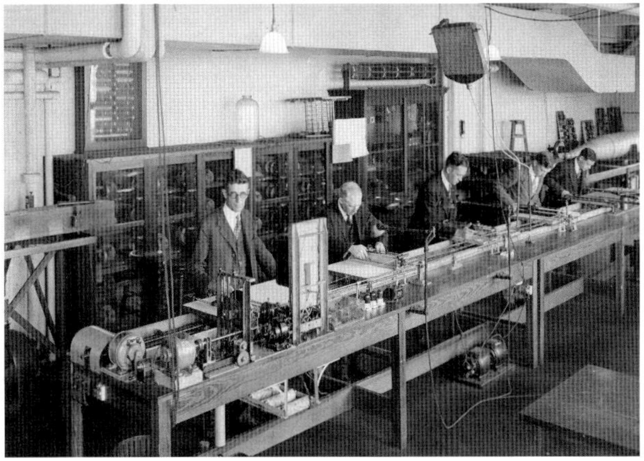

1 The Product Integraph, an electro-mechanical analogue computer built at the Servomechanisms Laboratory.

on something is two-fold: on one hand, the idea of negative feedback can help us to understand how in clicking on something, hands on the table and motion on the screen come together in one integrated sensorimotor act that, although involving a variety of distinct processes and locations, is perceived as one. On the other, historically, the MIT Servomechanisms Laboratory was behind much of the rise of negative feedback and thus the emergence of cybernetics as a field, as well as the construction of the first digital computer that was interactive in today's sense.

This computer started as an analogue computer for flight simulation, the Aircraft Stability and Control Analyzer (ASCA). The machine originally was planned as a continuation of the laboratory's successful work in analogue computing. Especially Vannevar Bush's famous Differential Analyzer had made clear that analogue computing can be applied to a variety of problems,[6] paving the way for the idea of building a flight simulator for arbitrary (existing and future) airplanes as "a cockpit or control cabin connected, somehow, to an analog computer"[7] (fig. 1).

Analogue computing, in this context, denotes a form of computation where a physical system is built in analogy to a phenomenon under study.[8] At the Servomechanisms Laboratory, during the early twentieth century, this practice led to the construction of a series of feedback-based electro-mechanical devices to study the dynamics of the electrical power grid and other complex systems (fig. 1). Likewise, the ASCA was conceived as an electro-mechanical system whose kinetic and electrical dynamics would resemble the dynamics of flying. Crucially, this meant that the cockpit would be an integral part of the computer – as the motion of its instruments and controls would be inseparable from the motion of computation.

However, halfway through its construction, the computing part of the machine was turned into a digital computer, because the project leads had realized the future potential of this emerging technology.[9] This change meant that a digital computer was to take over the role of an electro-mechanical device intrinsically connected to an environment (a cockpit, in this case). It hence had to be a special kind of digital computer: a computer that operates in real-time and allows for the exchange of data with its environment while operating.

6 David A. Mindell, *Between Human and Machine. Feedback, Control, and Computing before Cybernetics*, Baltimore: Johns Hopkins University Press, 2004, pp. 157–158.
7 Kent C. Redmond, Thomas M. Smith, *Project Whirlwind. The History of a Pioneer Computer*, Bedford, MA: Digital Press, 1980, p. 32.
8 Charles West Churchman, *Operations research. Eine Einführung in die Unternehmensforschung*, München: Oldenbourg Verlag, 1971, pp. 151–152.
9 Redmond 1980 (as fn. 7), pp. 27–44.

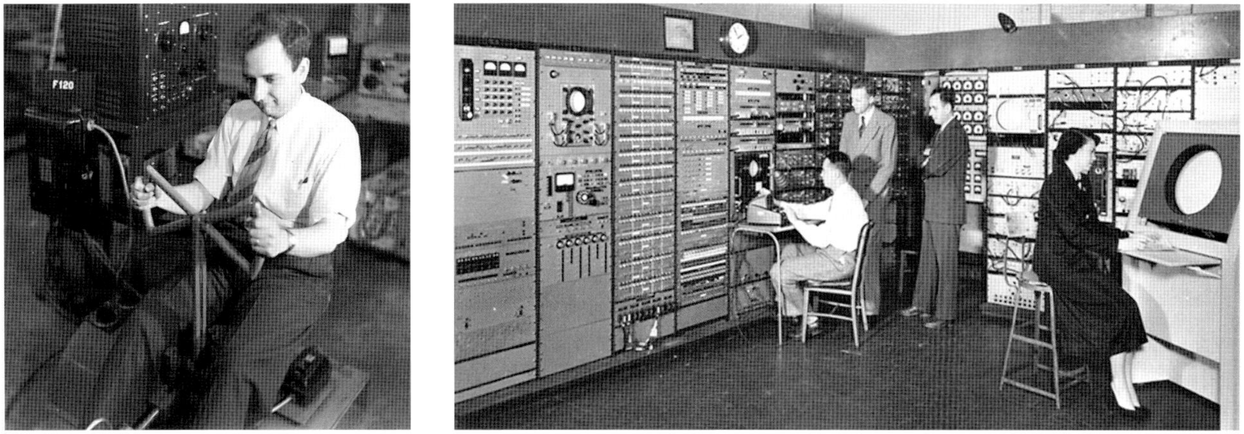

2 ASCA 1947 (left) and Whirlwind 1950 (right), a cockpit whose moving parts are part of a computational process versus the shape of computation to come.

During the construction of this novel machine, however, the task of building a versatile and fast digital computer became so dominant, that the engineers involved in the project increasingly neglected the cockpit portion of the system. As this cockpit still constituted a system of analogue instruments and moving parts, it later became clear that connecting these instruments to a digital computer posed fundamental problems that had never been dealt with before: "These problems were not impossible, but neither did established solutions exist. The digital computer was too new."[10]

In consequence, the simulator's cockpit was scrapped in late 1948 and the result was named Whirlwind the first interactive computer ever built and no longer a flight simulator (fig. 2).[11]

Reciprocal Visibility

What was *too new* to make a digital ASCA possible comes down to two questions: how to make digital data and processes visible to human viewers, and how to make the viewer's actions and reactions, in turn, *visible* to the computer?

Both problems are rooted in the nature of digital computation: the visibility of a representation in analogue computing is determined by the relationship between a physical system and the system it was made to model. Bush's Differential Analyzer, like the other analogue computers of the Servomechanisms Laboratory, was not so much a computer that solved differential equations as it was "an elegant, dynamical, mechanical model of the differential equation" that did "kinetically act out the mathematical equation".[12]

10 Ibid., p. 49.
11 Ibid., p. 60, pp. 43–44.

12 Larry Owens, Vannevar Bush and the Differential Analyzer. The Text and Context of an Early Computer, in: *Technology and Culture. The International*

Likewise, the ASCA would have been an electro-mechanical model of the aerodynamics of flight. This model would seamlessly integrate the instruments and controls of the cockpit, as well as any human action bearing on them, simply because the motion of instruments and controls would be part of *acting out* the computation. In contrast, digital computation has no *a priori* relationship to the systems it models.[13] It is marked by discrete states, represented by the symbols of a formalism, and "carefully chosen rules that dictate how one symbol succeeds another".[14] In order to be visible, digital computation must be translated into representations that "stand in an arbitrary relationship to the objects they represent."[15]

In addition, digital computation from the very beginning was conceptualized as a time and context free process. The idea of the Turing Machine (and equivalent definitions of computation) assumes that computation starts with a fixed input, operates on this input according to a fixed set of rules and terminates after a finite number of steps (or goes into an infinite loop of repetition).[16] Hence "[t]uring machines cannot handle the passage of time".[17]

Originally conceptualized as a machine in constant dialogue with a crew of flight operators in training, Whirlwind had to deviate from this assumption. The fact that indeed almost every computer we use today does so – by constantly waiting for new input from its environment while producing output that may affect future inputs – has only relatively recently been acknowledged by theoretical computer science.[18]

During the transition from an analogue ASCA to a digital Whirlwind, both problems were addressed pragmatically. The problem of the visibility of digital data was approached by establishing the mode of representation that is still dominant today: the computer drew arbitrary symbolic representations on the screen. To that end, Whirlwind's data registers were linked to the x/y-position of the electrode beam of a cathode-ray tube (CRT, fig. 3).[19] By so doing, the discrete states of machine computation were translated into representations that are readable by human observers, and the computer screen was introduced.

Within the project, the establishment of this new form of connecting people and computation was not seen as a great leap. Robert Everett, one of Whirlwind's engineers, simply noted later: "One of the things that I think we did first was to connect a visual display to a computer."[20] It was understood as something *I think we did first* because the engineering practice of the Second World War had already established the possibility of thinking (and building) this connection. With the Williams Tube a combination of digital computation and CRT was already in use. As the Williams Tube was a form of digital memory that *drew* zeros and ones onto a CRT screen in order to store them for a few milliseconds, it was not intended to be looked at by a human observer.[21] But in analogue radar technology, CRTs

Quarterly of the Society for the History of Technology (1986), pp. 63–95, p. 75.

13 Gerard O'Brien, Jon Opie, The Role of Representation in Computation, in: *Cognitive Processing* 10.1 (2008), pp. 53–62.

14 Ibid., p. 56.

15 Ibid., p. 58.

16 Georg Trogemann, Jochen Viehoff, *Code@Art. Eine elementare Einführung in die Programmierung als künstlerische Praktik*, Wien/New York: Springer, 2005, p. 85.

17 Peter Wegner, Why Interaction is More Powerful Than Algorithms, in: *Communications of the ACM* 40.5 (1997), pp. 80–91, p. 83.

18 Ibid.

19 Robert Everett, Whirlwind, in: J. Howlett, Gian Carlo Rota, Nicholas Metropolis (eds.), *A History of Computing in the Twentieth Century*, Orlando: Academic Press, 1980, pp. 365–384, p. 365.

20 Ibid, p. 375.

21 Claus Pias, *Computer Spiel Welten*, Dissertation, Weimar: Bauhaus-Universität, 2000, pp. 55–56.

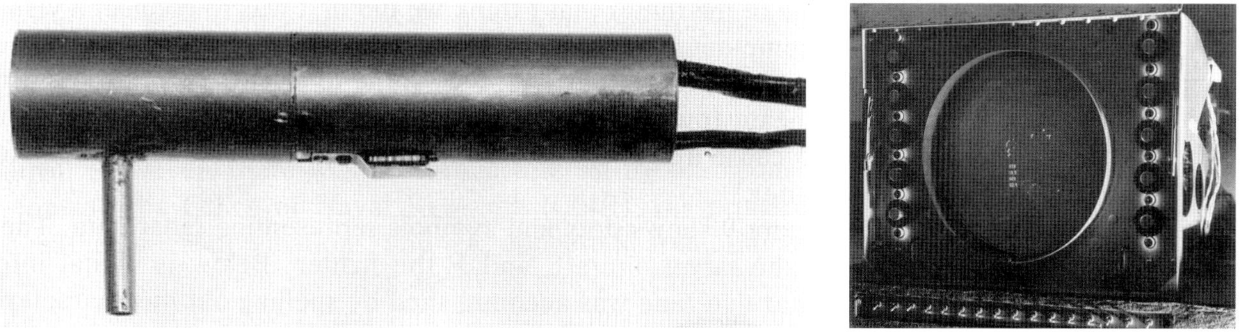

3 Light-gun and symbolic representations on an early screen of Whirlwind.

had already been employed as visual displays.[22] Finished after the war, even radar CRTs leftover from the war could be used in the construction of Whirlwind.[23] The project thus simply had to connect both pre-existing practices (CRT-based digital memory and analogue radar displays) to create the arrangement of computation and representation we now refer to as the computer screen.

The problem of the *visibility* of human action to the process of computation was addressed by interrupting this process. A light-gun allowed for a literal *handling* of computation, as it made it possible to touch symbolic representations by pointing at them (fig. 3). This was achieved by placing a light sensor at the tip of the gun that would interrupt the computer's drawing process. As Whirlwind did not draw a rasterized image (organized in rows and columns of pixels), but drew one representation after the other, interrupting this process entailed that the light picked up at the moment of interruption would be emitted from the very object the gun was pointed at. It could thus be interpreted as a selection to be taken into account for further computation.[24]

With this setup, Whirlwind was ready to become the origin of SAGE, the Semi-Automatic Ground Environment air defence system – the largest computer built to date, managing American air defence until 1983.[25] More importantly, however, it established a feedback loop between screen-based representation and action. In consequence, the visibility of what was represented on screen became subject to interactions between motor activity and visual perception.

Direct Manipulation

The closing of this loop, in which action would be taken on a screen-based representation that in turn would react to that action, preconfigured how we interact with computers

22 Axel Roch, Die Maus. Von der elektrischen zur taktischen Feuerleitung, in: *Lab. Jahrbuch 1995/96 für Künste und Apparate*, Köln: Verlag der Buchhandlung Walther König, 1996, pp. 166–173, p. 170.

23 Everett 1980 (as fn. 19), p. 379.

24 C.R. Wieser, *Cape Cod System and Demonstration*, Cambridge, MA: MIT Lincoln Laboratory, 1953, p. 2.

25 Redmond 1980 (as fn. 7), p. 206.

until today. It established a remarkably stable dispositive of interaction, sustainably structuring large parts of the field of human-computer interaction (HCI), which would later refer to the combination of (mostly screen-based) representation with the capability to act on these representations as an *interface*. Nevertheless, it took the field until the 1980s to conceptualize the closed loop between representation and action as *direct manipulation*.

This discussion initially was framed by cognitive science and computational theories of the mind that treat interaction as a process of rule-based problem solving. For Ben Shneiderman, who introduced the term "direct manipulation", the phenomenon can accordingly be explained by assuming a difference between non-physical "semantics" of human problem solving and the physical "syntax" of representation and action at an interface.[26] While, according to this view, any form of HCI has to mediate between these two domains, direct manipulation reduces the difference between them by having users act in the world of semantics as opposed to syntax: direct manipulation, the argument goes, allows a writer to, for instance, directly interact with a paragraph of text (by marking it with the mouse) as opposed to decomposing high-level semantic intentions into low-level abstract commands whose syntax is largely unrelated to the paragraph itself and the act of manipulating it.[27]

Later, Edwin Hutchins, James Holland and Donald Norman expanded on Shneiderman's work, providing a seminal discussion of direct manipulation from a cognitive science perspective.[28] Starting from the assertion that "[w]e see promise in the notion of direct manipulation, but as of yet we see no explanation of it",[29] they develop an explanation that follows Shneiderman's path by distinguishing between the physical reality of an interface and the non-physical "model-world" of what it represents.[30] Direct manipulation, in this view, implies acting with the metaphors of that model-world, while well-chosen metaphors align this model with a user's problems. It is thus a function of the cognitive or information processing "distance" between the model-world and intention.[31]

Surprisingly, however, this does not seem to account for the whole phenomenon. Direct manipulation for the authors seems to possess a qualitative or experiential component that is hard to grasp in the terms of cognition and problem solving. In addition to cognitive *distance*, direct manipulation relies on emotional *engagement*, resulting from the feeling of being causally effective in that world – a phenomenon that cannot be understood in terms of goal-directed problem solving. The authors thus admit that direct manipulation seems like an "atavistic [...] return to concrete thinking".[32] It may, however, be precisely the messy concrete thinking of our hands engaged in syntactic activities (or sensorimotor loops) that can help us to understand direct manipulation, as will become apparent later in this paper.

26 Ben Shneiderman, Direct Manipulation. A Step Beyond Programming Languages, in: Noah Wardrip-Fruin, Nick Montfort (eds.), *The New Media Reader*, New York, NY/London: W.W. Norton & Company, 2001, pp. 486–498.

27 Ibid. This argument alone is enough to cast doubt on the supposed directness of direct manipulation, as manipulating a paragraph with the mouse still presupposes a decomposition into low-level hand movements and button presses – only that this low-level syntax is *different* from, say, a command line interface.

28 Edwin L. Hutchins, James D. Hollan, Donald A. Norman, Direct Manipulation Interfaces, in: *Human-Computer Interaction* 1 (1985), pp. 311–338.

29 Ibid., p. 316.

30 Ibid., p. 317.

31 Ibid., p. 311.

32 Ibid., p. 337.

Gestalt and Apparent Motion

The motion we see on computer screens is what the psychology of perception calls *stroboscopic* or *apparent* motion, an illusionary impression of motion created by the succession of static frames. Historically, the systematic investigation of apparent motion is closely connected to gestalt psychology,[33] as one of the texts defining the field is Max Wertheimer's (still untranslated) *Experimental Studies about the Perception of Motion.*[34]

Wertheimer's seminal study tries to understand how stroboscopic stimuli that are objectively not moving create the subjective *percept* of motion. For the study, Wertheimer employs the Schumann Tachistoscope as a stroboscope (fig. 4). This device uses rotation to quickly cover and uncover stimuli. A setup using two stimuli, a and b, and a prism allows Wertheimer to use the apparatus in a way that, to a viewer, presents both stimuli in quick alternating succession.

Focusing on those cases of apparent motion that do not yield a perfect illusion of seeing moving objects but, for instance, fractured and partial motion percepts[35], Wertheimer arrives at a remarkable conclusion that ultimately reverses the relation of movement and object as it was understood by his contemporaries (fig. 5). These, he argues, assume that the perception of motion presupposes the perception of a moving object, understanding the moving object as a primary and its motion as a secondary feature ascribed by perception. Wertheimer, instead, sees motion, named "pure φ" or "pure motion", as a primary object of perception, even reconstructing the identity of moving objects as a limiting case of motion.[36] In this view, perception of motion happens directly and immediately, preceding and enabling the perception of the gestalt of an object. The latter is hence conceived as a "short-circuit" of motion perception as a "duo-in-uno" when, for instance, two lines, a and b, in

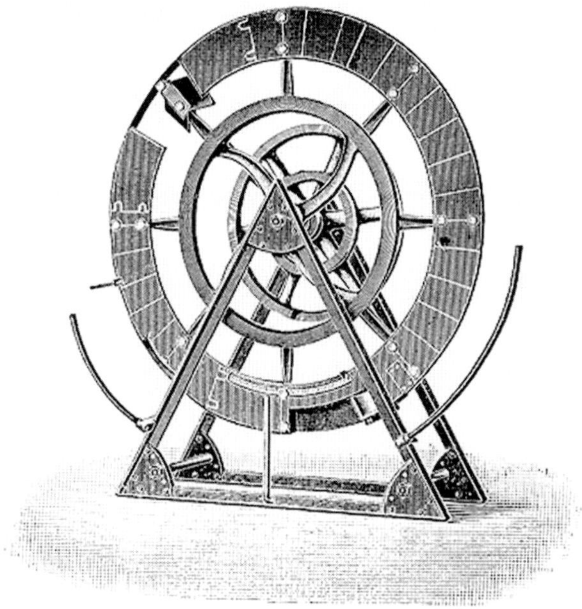

4 The Schumann Tachistoscope.

33 Robert M. Steinman, Zygmunt Pizlo, Filip J. Pizlo, Phi is not beta, and why Wertheimer's discovery launched the Gestalt revolution, in: *Vision Research* 40 (2000), pp. 2257–2264.

34 Max Wertheimer, Experimentelle Studien über das Sehen von Bewegung, in: *Zeitschrift für Psychologie und Physiologie der Sinnesorgane* 61 (1912), pp. 161–265.

35 Ibid., p. 191.

36 Ibid., p. 221.

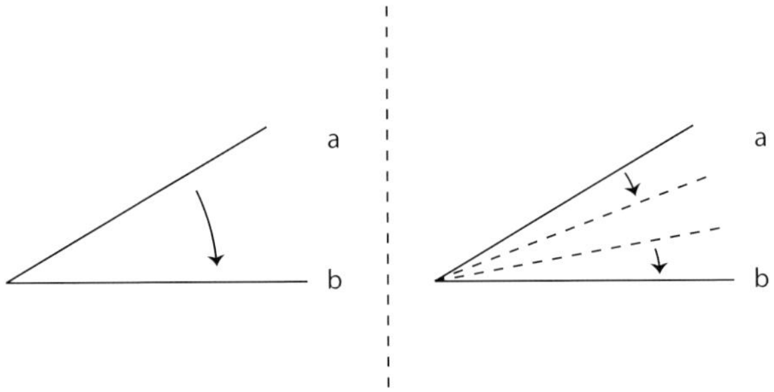

5 Apparent motion of a line from a to b (left) and partial apparent motion if the time-interval between frames a and b becomes too long (right), as described by Wertheimer.

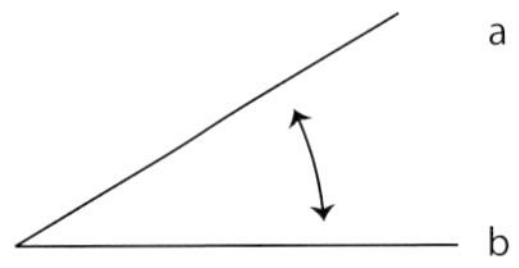

6 Two lines in rapid succession forming an angle composed of two sides, as described by Wertheimer.

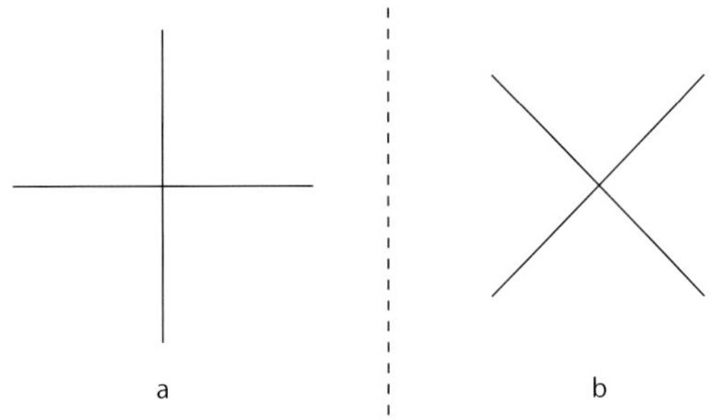

7 Ambiguous motion as described by Linke.

rapid succession lead to the perception of one object: an angle composed of two sides (fig. 6).[37]

In special cases, the stroboscopic stimuli causing the perception of apparent motion may be ambiguous. Whenever, for example, two or more concurring interpretations of one stimulus are possible, their perception becomes multi-stable: Subjects perceive one possible percept at a time, while perception alternates between possibilities over time. This was first demonstrated by Paul Linke using a cross that is rotated by 45° from stimulus to stimulus and that, as a bistable stimulus, can be perceived as clockwise or counterclockwise rotation (fig. 7).[38] Termed "ambiguous motion", this effect was later studied by Paul von Schiller, who tried to isolate the factors that determine which possible percept is perceived at a time, trying to establish the laws of how ambiguous motion is disambiguated to distinct percepts.[39] During this study, von Schiller made a remarkable observation: His subjects were able to control the perceived direction of motion most effectively by actively moving their hands and heads. This, he writes in a footnote, constitutes a case of motor activity having a gestalt influence on visual perception.[40] Because the experimental systems of experimental psychology of that time, such as the tachistoscope, only allowed for the precise control of the presentation of stimuli without connecting it to human action, this effect seemed too hard to control for him to warrant further investigation.[41]

Action Capture

During the past decades, the methods of experimental psychology have changed significantly in favour of quantitative research that relies on a universal experimental system, enabling not only the precise control of the exposure of stimuli but also the measurement of human action. This system is fundamentally structured by the interactive

37 Ibid., p. 251.
38 Paul von Schiller, Stroboskopische Alternativversuche, in: *Psychologische Forschung* 17 (1933), pp. 179–214, p. 180.
39 Ibid.
40 Ibid., p. 196.
41 Ibid., p. 195.

computer and was, for instance, pioneered in the famous experimental studies by Douglas Engelbart[42], which led to the decision to replace the light-gun for computer input with the mouse.[43] Today, "the computer [...] has taken over practically all the experimental procedures used to examine the perception of space and time".[44]

It nevertheless took until 1994 for the first publication to present quantitative evidence of the influence described by von Schiller.[45] Since then, a series of studies have shown that if ambiguous motion is coupled to physical motion, then the bodily movement *captures* its perception by influencing it in the direction of motion. This capture effect is strongest for voluntary self-motion and has therefore been named "action capture"[46] (or "priming by actions"[47]), to do justice to the fact that the influence is caused by whole actions, comprised of intentions, motor planning and execution. In most studies analysing the effect, computers are used to couple the movement of the hands with ambiguous motion stimuli presented on a screen (fig. 8). Action capture, one could hence argue, has been mostly studied as inter-action capture.

The concept of action capture not only holds for visual stimuli and the motion of our hands; it also has been shown that the perception of ambiguous auditory[48] and tactile[49] stimuli can be captured by the movement of the hands, eyes[50] and the walking body[51]. For the interface this implies that we may have to understand interfaces as perceived in action. Their visual (and even tactile and auditory) qualities are influenced by our actions, and interface design may have to take into account that it is not only the functioning of an interface that depends on its use, but also its perceptual qualities.

Research examining action capture has shown that it is facilitated by a close physical and temporal distance of the action and stimulus[52], as well as a correspondence of the axes and orientation of motion between both.[53] More importantly, the correspondence of stimulus and action that drives the effect is not an *a priori*. It is context dependent, as it can be influenced by expectations: when, for instance, a button with a right arrow is pressed, perception of apparent rotation is captured in the clockwise direction, because we have learned that pushing a round object to the right will most likely cause it to rotate in the clockwise direction.[54] The effect, in addition, can be modified and even reversed by training.[55] And finally, it is already present when actions are merely planned and not yet carried out.[56]

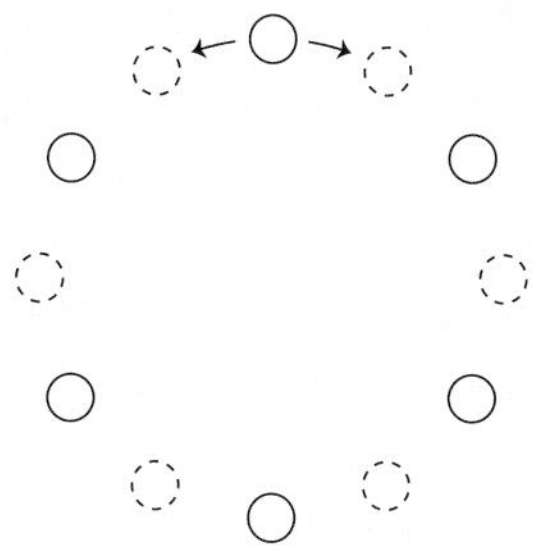

8 A typical ambiguous motion stimulus as used in experiments. Stroboscopic motion of the circles is presented on a computer screen at the same time as subjects perform physical motion on an input device, such as a keypad, knob, or mouse.

42 As a dispositive it structures the presentation of stimuli, the measurement of responses, the design and statistical analysis of experiments, and by that "the nature of the questions that can be addressed". Nicholas J. Wade, Dieter Heller, Scopes of Perception. The Experimental Manipulation of Space and Time, in: *Psychological Research* 60.4 (1997), pp. 227–237, p. 235.
43 William K. English, Douglas C. Engelbart, Melvyn L. Berman, Display-Selection Techniques for Text Manipulation, in: *IEEE Transactions on Human Factors in Electronics* 8.1 (1967), pp. 5–15.
44 Wade, Heller (as fn. 42), p. 235.
45 G. Ishimura, S. Shimojo, Voluntary Action Captures Visual Motion, in: *Investigative Ophthalmology and Visual Science (Supplement)* 35 (1994), p. 1275.
46 Ibid.
47 Andreas Wohlschläger, Visual Motion Priming by Invisible Actions, in: *Vision Research* 40 (2000), pp. 925–930, p. 929.
48 Bruno H. Repp, Günther Knoblich, Action Can Affect Auditory Perception, in: *Psychological Science* 18.1 (2007), pp. 6–7.
49 Olivia Carter, Talia Konkle, Qi Wang, Vincent Hayward, Christopher Moore, Tactile Rivalry Demonstrated with an Ambiguous Apparent-Motion Quartet, in: *Current Biology* 18 (2008), pp. 1050–1054.
50 Ibid.
51 Yoshiko Yabe, Gentaro Taga, Treadmill Locomotion Captures Visual Perception of Apparent Motion, in: *Experimental Brain Research* 191.4 (2008), pp. 487–494.
52 G. Ishimura, Visuomotor for Action Capture, in: *Investigative Ophthalmology and Visual Science (Supplement)* 36 (1995), p. 357.
53 Wohlschläger 2000 (as fn. 47), pp. 927–929.
54 Ibid., p. 928.
55 Ishimura 1995 (as fn. 52), p. 357.
56 Wohlschläger 2000 (as fn. 47), p. 929.

Accordingly, action capture seems to not only depend on how much an action corresponds with what we perceive (in terms of spatio-temporal distance and orientation), but rather seems to depend on how much a possible percept corresponds to the result we expect an action to have. If I expect (or plan) my hands to be involved in causing clockwise rotation, I am more likely to perceive an ambiguous rotation as clockwise.

This corresponds to early findings from human factors indicating that the speed and error rate of actions at an interface depend on the "compatibility" of stimulus and response.[57] This compatibility has from the beginning been understood as an acquired relationship, for which it holds that "stimulus and response sets are optimally matched when the resulting ensemble agrees closely with the basic habits or expectancies of individuals".[58]

In order to further analyse the significance of action capture for screen-based interaction, I have conducted a study linking earlier research on action capture to the compatibility of mouse action and computer response.[59] Assessing compatibility, however, is a messy task, since basic habits or expectancies do not translate well into experimental protocols. But the computational tools of cybernetics – the field whose heritage still defines the way interactivity works – at least provide ways of measuring a non-semantic similarity of stimuli and response, understood as the cross-correlation of a time series of measurements.

Coupling an ambiguous motion stimulus to mouse movements, the experiment measured how subjects moved the mouse as they were asked to perform circular motion while looking at an ambiguous rotation on screen. It thus relied on a paradigmatic case of interaction, linking the motion of the mouse with apparent motion on screen, while making their interrelation measurable by using ambiguous motion that can be captured by the body's activity. In a series of trials, mouse motion was recorded together with the perceived direction of rotation of each trial, ascertained through questions (fig. 9).

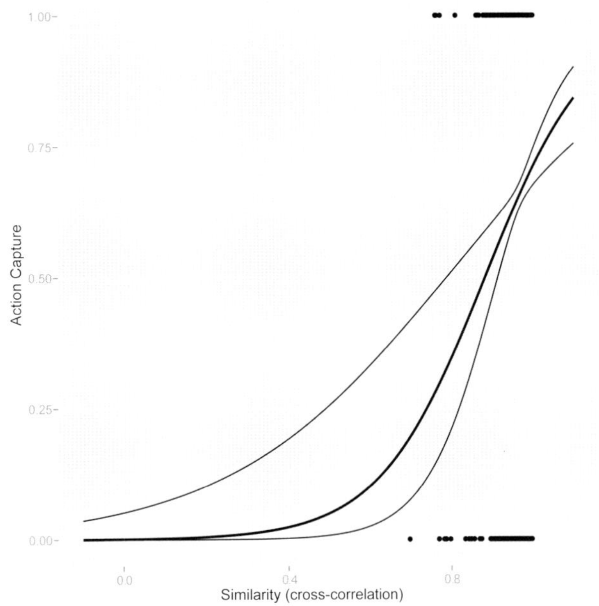

9 Logistic regression of the relationship between a) the similarity of mouse motion and a perfect rotation on screen, in the direction subjects were asked to move the mouse, and b) the coincidence of perceived direction of ambiguous motion with the direction of mouse motion (action capture). The more similar the motion of the hand is to a perfect rotation in the correct direction, the more likely is action capture. For details see Scherffig 2017 (as fn. 1), pp. 259–261.

57 Paul M. Fitts, Charles M. Seeger, S-R Compatibility. Spatial Characteristics of Stimulus and Response Codes, in: *Journal of Experimental Psychology* 46.3 (1953), pp. 199–210, p. 199.

58 Ibid., p. 208.

59 Scherffig 2017 (as fn. 1), pp. 257–262.

The result is as simple as it is statistically significant: the more similar the mouse motion is to the motion subjects were asked to perform, the higher the likelihood of action capture (the significance level of their correlation is $p<0.0001$, see also fig. 5). In other words, the more similar the action of the hand on the mouse and the reaction on the screen seem to be, the more the former captures the latter. Action capture, therefore, seems to quantitatively incorporate our actions into what we perceive. Our perception seems to *calculate* with our actions and their expected results.

Interaction Gestalt

We have seen the idea of calculating with one's actions before. It is the idea of comparing actions with changes in perception, or – in the technical terms of cybernetics – the idea of using the difference between a system's output and some goal to determine its future output.

Past and contemporary discourse in physiology, the psychology of perception and cognitive science has often identified this goal with expectations or predictions[60]: We constantly compare the change in perception induced by our activities with expected change. This is what becomes apparent (and even measurable) as action capture: our predictions regarding a physical action capture the way we perceive the results of that action.

Action capture thus suggests that at the interface, too, it is our actions that determine what we perceive. If we move a mouse and observe its cursor and the on-screen reactions to pressing a finger onto it, the motion of our hands on a physical object on a table and the perception of *apparent motion* on a computer screen is fused *internally* into a sensorimotor unity that goes beyond the mere correlation of both.

This is exactly what was observed by computer scientist Dag Svanæs.[61] Conducting experiments in which subjects interacted with abstract interactive systems consisting of black and white squares[62] he analysed their behaviour in correlation with their verbal descriptions of it, paying attention to the way the abstract black and white squares slowly became perceived as objects:

> *The objects described by the subjects in the experiments existed for them only through interaction. The objects emerged as a result of the interplay between the intentions of the users, the users' actions, and the feedback given by the system.*[63]

Observing the interaction with a simple system that would eventually be perceived as a switch, he notes:

> *When the subjects said 'It is a switch', they did not come to this conclusion from a formal analysis of the State Transition Diagram of the example. Nor did they conclude it from the visual appearance of the square, as the squares all looked the same. The switch behavior*

60 See, for instance, Karl J. Friston, Christopher Thornton, Andy Clark, Free-Energy Minimization and the Dark-Room Problem, in: *Frontiers in Psychology* 3 (2012), pp. 1–7; Jack M. Loomis, Distal Attribution and Presence, in: *Presence* 1.1 (1992), pp. 113–119; or Sarah-Jayne Blakemore, Chris D. Frith, Daniel M.Wolpert, Spatio-Temporal Prediction Modulates the Perception of Self-Produced Stimuli, in: *Journal of Cognitive Neuroscience* 11.5 (1999), pp. 551–559.

61 Dag Svanæs, *Understanding Interactivity. Steps to a Phenomenology of Human-Computer Interaction*, Dissertation, Trondheim: Norges Teknisk-Naturvitenskapelige Universitet, 2000.

62 Ibid., pp. 128–132, pp. 108–110.

63 Ibid., p. 230.

> *slowly emerged from the interaction as the square repeated its response to the subject's actions.*[64]

This implies that the hands on the mouse in dialogue with the computer's response yield the emergence of perceptual units having "gestalt properties" as their perception, once emerged, is "direct and immediate" and not a cognitive interpretation of action and perception.[65]

Svanæs therefore suggests treating the objects that compose an interface as "interaction gestalts", entities that are "similar to visual gestalts in that they are wholes, and not compositions of analytical elements".[66] The form or gestalt of an interface, understood as *interaction gestalt*, can be seen as a perceptual or experiential whole that is based on action and perception as *duo-in-uno* – as a limiting case of the loop of human action and machine reaction.

From a sensorimotor perspective, the elements that make up an interface are hence not so much the discrete entities that they have been designed and programmed to be; instead they are the results of being used. Buttons, in this sense, look like buttons because they are used as such – *and* the other way round. Their form does not imply or communicate their function. Instead, their (subjectively experienced) form and function are interdependent and are the result of their use and its context.

This suggests a *cybernetic* model of the interface and interaction, implying that what we see is enacted by how we react to it. According to this model, we can indeed understand direct manipulation in terms of *distance* end *engagement*. But distance would be reduced to simple spatio-temporal distance of stimulus and response and the perceptual similarity between both. Or more generally, it would be redefined as the distance of predicted and actual reaction, which is the negative feedback at the heart of cybernetics. Engagement, in turn, would become being engaged in sensorimotor loops that are continuously learned and exercised, forming the objects they deal with within this cyclical process. If today's touch-based interaction on mobile phone screens seems to constitute a return to Whirlwind's combination of screen-based representation and the possibility of touching it, this may be understood in light of HCI's constant effort to minimize distance and maximize engagement in these literal terms.

The simple need to establish *reciprocal visibility* between computation and human environment thus introduced a dispositive of interaction in which bodily movement at the computer screen, its predicted and its observed results together are integrated into coherent perceptions of interaction, that form the gestalt of the interface. This is the integration of hand and screen-based representation that allows us to speak of *clicking on something* while we steer a physical mouse on a table and watch apparent motion on a screen. What, according to this view, creates interfaces, is interaction.[67]

64 State Transition Diagrams are formal graphical representations of how a system of discrete states (such as combinations of black and white squares that can switch their color) can transition from one state to another. In Svanæs' experiments these diagrams describe the actual behavior of the systems used, as opposed to the perceived behavior described by his subjects. Ibid., p. 206.

65 Ibid., p. 244.

66 Ibid.

67 As in: not only enables and shapes their functioning as interface but also their appearance.

Figures

1 MIT Museum. Reprinted with permission from MIT Museum.

2 Copyright Computer History Museum. All Rights Reserved (left), The MITRE Corporation. All Rights Reserved (right). Reprinted with permission from The MITRE Corporation.

3 The MITRE Corporation. All Rights Reserved. Reprinted with permission from The MITRE Corporation.

4 Collection Jirí Hoskovec. Reprinted with permission by Simona Hoskovcová.

5–9 Author's figure.

Kathrin Friedrich

Screening Bodies

Radiological Screens and Diagnostic Operations

Screens provide the basis for actions and diagnostics in clinical radiology. From a media-historical perspective, the transformation from light boxes to today's digital screen configurations provides the opportunity to explore the epistemic and operational conditions of screens. In particular, screen architectures and screen-bound tools are central to the question of how screen operations prefigure the diagnostic screening of patients' bodies. By focusing on the exploration of the operational and epistemic relations between screens and screening, I will use the order of hanging or arranging film print-outs on light boxes as an example of a site-specific practice which was significantly altered by the introduction of digital infrastructures into clinical radiology departments. The screen-based radiological hanging protocol, which specifies the arrangement of visualizations on a light box or in a graphical user interface (GUI), reveals the extent to which media transformation and epistemic practices are mutually contingent as well as how profound and abundant the apparently flat and limited screen is.

Protocols and Practices of Light Box Hanging

Diagnostic image viewing based on analog radiograms, which are hung up on electrically illuminated light boxes, dates back to the 1910s. Later on, as an established collective and probably even instructive practice, the screen-bound dispositive of radiological diagnostics in pre-digital form (both regarding the imaging technique of radiography and the diagnostic practice at the light box) involved several main objects and actors (fig. 1). These constitute the scenery that might have been staged for the photograph to be taken. Nevertheless, it becomes obvious that the hanging of x-ray films on the wall-sized light box is neatly ordered. A series of thorax and abdomen radiograms in different perspectives of supposedly the same patient are provided. While two eagerly interested physicians, probably radiological novices, are sitting and staring at the light box in front of them, a seemingly skilled radiologist instructs their sight by using a pointing stick.

Even if this diagnostic dispositive is primarily directed towards the screen of the light box, several other media technologies and infrastructures intersect in the process of crafting a diagnosis.[1] Next to the telephone in the lower corner on the right there is also a Dictaphone to record diagnoses that are later on typewritten by busy clerks. Diagnostic viewing is presented as a collective and distributed practice rather than just an almost contemplative posture in front of

1 On the design and conception of light box dispositive in early radiology see Christian Vogel, Epistemischer Sinn und ästhetische Wirkung. Das Betrachten von Röntgenbildern im Schaukasten, 1896–1930, in: *Fotogeschichte* 138 (2015), pp. 19–28.

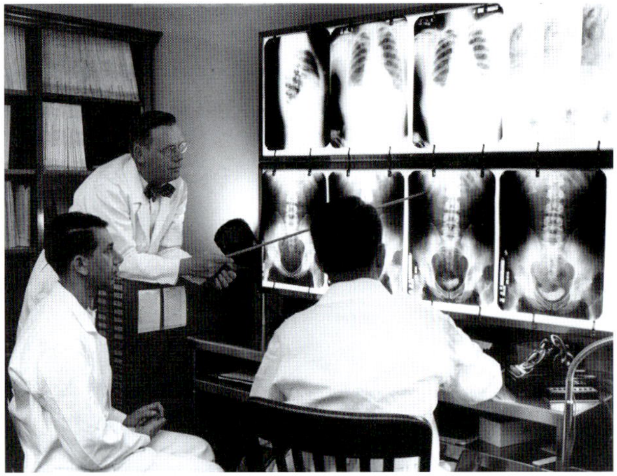

1 Photograph of x-ray diagnostics at Hermann Hospital in 1953 using a light box.

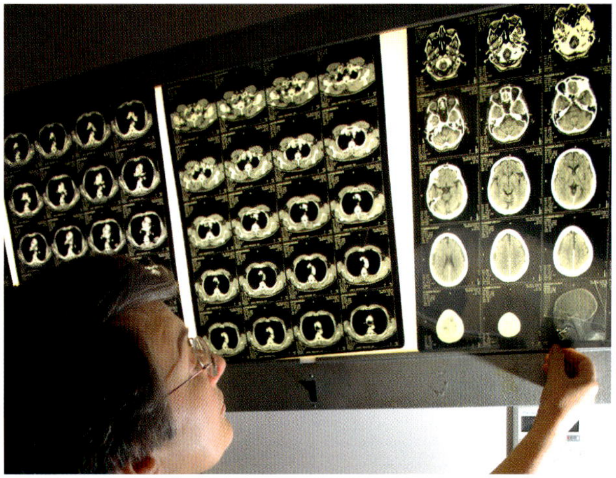

2 Example of hanging CT scans on a light box.

a bright screen. As suggested by this photograph, viewing images on the light box is associated with perceiving from a distance static images that are strictly ordered in a static frame. The x-rays could not be changed once they were printed out on film. Only their order at the light box was reconfigurable, and to a certain degree instruments such as the pointing stick or magnifying glasses could help to guide the diagnostic gaze and bridge the operational and probably epistemic gaps between distant users and static images.

From the mid-1970s, digital imaging technologies such as computer tomography (CT) and later magnetic resonance imaging (MRI) were introduced into clinical practice. These genuinely digital imaging processes complemented analog x-ray visualization, but the diagnostic viewing of images on the light box continued. Even though CT and MRI data were produced and processed digitally, their tomographic visualization for diagnostic purposes was not yet delegated to digital representation modes, such as computer screens, until the mid-1990s. One quite practical reason for this technological difference between modes of data acquisition and modalities of visualization was the need for advanced and in particular networked software and also technically advanced computer screens, such as high-resolution displays. Another, more epistemological reason was the established diagnostic routine and its persistence. The radiologists' expertise slowly adapted to the technical features and diagnostic possibilities of new digital imaging techniques, but it responded even more slowly to changing dispositives of diagnostically screening images and bodies. For another 15–20 years after digital imaging such as CT and MRI were introduced to clinical practice, the light box remained the primary place of radiological diagnostics.

The light box hanging of CT visualizations consists of individual cross-sectional images from a digital scan of a specific body region (fig. 2). These cross-sectional image series are printed on films in a matrix. Typically, these films consist of 4 × 5 matrices (4 cross-sectional images alongside each other, arranged in 5 rows). Within each film sheet, the images are then read from left to right and from top to bottom. This schema is applied to the overall arrangement of films on the light box. Anthropologist and radiologist Barry Saunders notes in his ethnographic observation of diagnostic CT reading on light boxes:

> *CT images are typically displayed in an order of magnitude or so smaller than the specimen they reference, with many images on one sheet of film. Film size is standard, but the 'matrix' of slices on each sheet [...] is variable, subject to differing conventions, even to ad hoc specification by readers.*[2]

The hanging sequence and hence the viewing sequence applied to the image rows is based on the linear writing and reading direction of Latin scripts, whereas the actual diagnostic routine, i.e. reading the images, varies among physicians depending on their operational routines, experience and the diagnostic request as well as on body region. Formally, the hanging protocol and the screen install and propose a certain order of viewing through their very own architecture. Hence, the gaps between individual cross-sectional images may be bridged in the direction of vision, but cognitively and epistemically, this bridging is performed by the radiologists themselves. Where the printouts simply leave a white space between the sequence of cross-sectional images, a cognitive bridging and spatial reformation must be performed in the viewer's mind based on radiological expertise. Finally, this imaginative reformation enables conclusions to be drawn as to the size, position and development of a problematic structure in the patient's body. The flatness of the view, the layout of the film hanging and the linear arrangement of images all play a role in the mental summing up of the slices on the light box into a body volume. At the same time, this kind of hanging films creates a form of clarity that invites an elliptical and comparative way of seeing.[3] Simply by changing position, the viewer in front of the light box can navigate between different cross-sectional planes or hang old and new images directly next to each other.

Besides the arrangement of images on films as well as their hanging on the light box, cross-sectional image viewing is structured in another media technical respect. The size of the box determines the number of films that can be examined at one time. To an extent, the available area limits the number of cross-sectional images per film sheet, as the slices as such would otherwise become too small to be examined in a detailed and diagnostically significant way. Moreover, the viewer's capacity would be challenged by an increased number of films and cross-sectional images. Fading out or integrating the gaps between individual tomographical slices may be merely a question of focus for experienced radiologists. But for the less experienced radiologists, it may cause a loss of orientation in the body volume.

To a degree, the exterior form of the box specifies an epistemic and aesthetic framework in which diagnostic

2 Barry F. Saunders, *CT Suite. The Work of Diagnosis in the Age of Noninvasive Cutting*, Durham/London: Duke University Press, 2008, p. 18.

3 Eva Cancik-Kirschbaum, Bernd Mahr, Anordnung und ästhetisches Profil. Die Herausbildung einer universellen Kulturtechnik in der Frühgeschichte der Schrift, in: *Bildwelten des Wissens. Kunsthistorisches Jahrbuch für Bildkritik* 3.1 (2005), pp. 97–114.

operations can take place. The hanging of the films and the arrangement of cross-sectional images within a film sheet lead the gaze through the image sequences, thereby orienting the mental reconstruction of the body volume. The format of the film and the selection of the image matrix in conjunction with the frame of the lightbox establish a physical and epistemic order for radiological diagnostics.

In addition, radiologists sometimes use magnifying glasses or blinds integrated within the box in order to emphasize certain aspects or limit the illuminated area. In his ethnographic study Barry Saunders describes how radiologists use a few tools such as magnifying glasses or pointers for didactic purposes in particular:

> *Once seated, radiological vision uses few prostheses. Occasionally one sees a reader of mammograms holding a magnifying glass. [...] But diagnostic film viewing, including CT reading, is mostly macroscopic: it employs a 'native' vision, a repertoire of squinting and scanning and gazing, a few feet from the image surface.*[4]

On closer examination, diagnostic practice using light boxes as diagnostic screens for both x-ray and CT visualizations is revealed as a highly orchestrated and instrumented process that requires whole-body involvement – not necessarily the patient's, but rather the radiologist's. Hence, the notion of screen undergoes an almost performative turn to become *screening*. As a "flat surface [...] on which pictures or words are shown", the radiological light box is a screen in the media-technical sense of the term.[5] What is more, a site of image viewing and image operation is established through and before the screen. If the term screen is used in the verb form *to screen*, the epistemic sphere of possibility becomes clear: "to test or examine someone or something to discover if there is anything wrong with the person or thing".[6] The light box opens up, and calls for, both a position and disposition of the viewer that not only screens the visualized body but also draws closer to images or image sequences in order to examine and test them.[7] Here the instrumentation of screening plays an epistemically and aesthetically significant role: image films are weighed up against each other, for example, by comparing them or using instruments such as glasses; the light distribution is limited, and proportions are scaled. In the transition from diagnosis using the lightbox to diagnosis on the computer workstation, which is integrated in a software-based picture archiving and communication system (PACS), interaction and interface designers are faced with the fundamental challenge of establishing new conventions of image viewing by also integrating these hanging and screening routines.[8]

4 Saunders 2008 (as fn. 2), p. 18.

5 Cambridge Dictionaries Online, http://dictionary.cambridge.org/us/dictionary/american-english/screen (accessed February 8, 2018).

6 Ibid.

7 Lisa Cartwright, *Screening the Body. Tracing Medicine's Visual Culture*, Minneapolis: University of Minnesota Press, 1995.

8 A more phenomenological analysis could also be helpful at this point as, more generally, proposed by Introna and Ilharco (2006) to account for the "screenness" of screen dispositives and operations: "The screen is phenomenologically analyzed as the grounding intentional orientation that conditions our engagement with certain surfaces in as much as we comport ourselves towards them as screens [...]. This might be formally indicated as the *screenness* of screen." Lucas D. Introna, Fernando M. Ilharco, On the Meaning of Screens. Towards a Phenomenological Account of Screenness, in: *Human Studies* 29.1 (2006), pp. 57–76, p. 58 [original emphasis].

All Digital – Hanging Protocols and GUIs

The idea of not only digitally generating but also processing and visualizing radiological image data dates back to the end of the 1970s. One of the first subject-specific publications that envisions an "all-digital department" that "includes, besides all-digital diagnostic devices, a complete new digital communication structure and standard" was published by German information science scholar Heinz Lemke and colleagues in 1979.[9] In their paper *Application of Picture Processing, Image Analysis and Computer Graphics Techniques to Cranial CT Scans* they stress the fact that such an integrative system that serves all functions mentioned in the paper's title would need to include digital screen-based workstations to provide "possible working modes in such a system [distributed computing network, KF]".[10] In the filmless era envisioned, the site of radiological diagnosis would shift from the light box to the computer workstation. With the broader realization of digital infrastructures in radiology departments in the 2000s, and more specifically with the implementation of software applications in a PACS, the computer screen and the workstation's Graphical User Interface (GUI) became the primary place of medical image data visualization and examination (fig. 3).[11] The radiologist David Hirschorn notes:

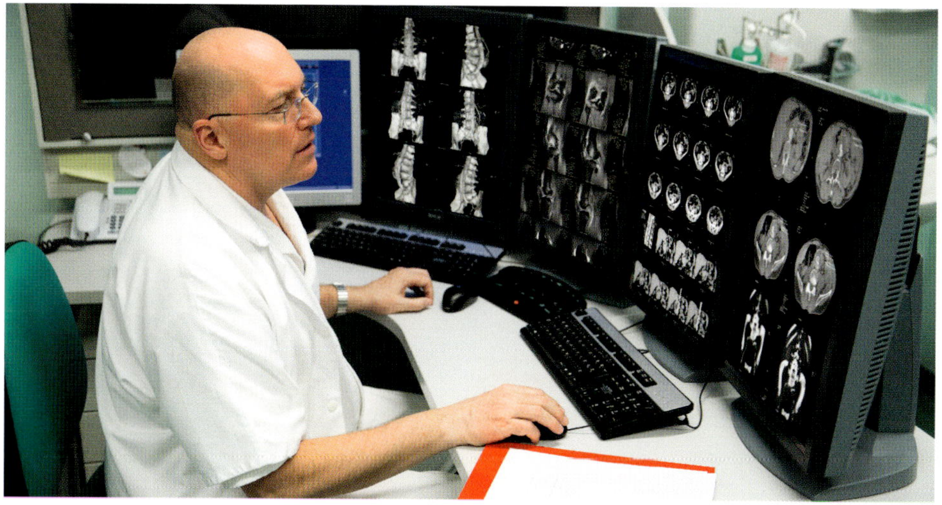

3 Radiological diagnostics at a workstation.

> *CT exams with a thousand images are becoming common and simply cannot be managed effectively on film. PACS viewing software can be used to dissect, analyze, magnify, or reformat image data in an infinite number of ways.*[12]

Hirschorn suggests that digital technologies not only make the dissection, analysis, magnification or reconstruction of image series in real-time possible; all these processes can now take place in an unlimited number of forms and ways.

9 Adrian M. K. Thomas, Arpan K. Banerjee, Uwe Busch (eds.), *Classical Papers in Modern Diagnostic Radiology*, Berlin/Heidelberg: Springer, 2005, p. 332.

10 Heinz U. Lemke, Siegfried Stiehl, Horst Scharnweber, Daniel Jackél, Applications of Picture Processing, Image Analysis and Computer Graphics Techniques to Cranial CT Scans. Proceedings of the Sixth Conference on Computer Applications in Radiology and Computer Aided Analysis of Radiological Images, in: *IEEE Computer Society Press*, 1979, pp. 341–354, p. 341.

11 In a broader perspective, geographer Nigel Thrift identifies screens as a constant place and locus of attention in times of digital processing and visualization: "Screens are one of the constants of everyday life, communicating, informing, entertaining, affecting life, simply being there providing ground." Nigel Thrift, *Knowing Capitalism*, London: SAGE Publications, 2005, p. 234.

12 David S. Hirschorn, Introduction, in: Keith J. Dreyer, David Hirschorn, James H. Thrall, Amit Mehta (eds.), *PACS. A Guide to the Digital Revolution*, New York: Springer, 2006, pp. 3–6, pp. 3–4.

What is proclaimed, in a technological euphoria, as a new, infinite sphere of possibility, made possible by software and GUIs, comes up against radiologists' existing skills and media competences. In the tension between technical possibilities, established ways of image viewing and conventionalized methods of hanging and handling images, the screen and in particular the GUI become enabling yet authoritative interfaces in the act of communication and access. As media theorist Wendy Chun notes:

> *GUIs have been celebrated as enabling user freedom through (perceived) visible and personal control on the screen. This freedom, however, depends on a profound screening: an erasure of the computer's machinations and of the history of interactive operating systems as supplementing – that is, supplanting – human intelligence.*[13]

And in the realm of radiological diagnostics and medical screen operations in general the question of how software structures and screen-based disposition supplement or supplant human intelligence and action is even more pressing as it touches upon responsibilities for making the choices regarding patients' further treatments. Further, which new possibilities for diagnosis does the GUI create as an interactively usable but nevertheless screen-bound interface, compared to film-based image viewing on the light box?

Viewed pragmatically, the GUI forms "a place where individuals and 'communities' meet infrastructures".[14] It thereby establishes its own specific site that concretizes data streams and renders them human amendable. Within the digital infrastructure of radiological imaging software, computer workstations establish a site where radiologists consult their material and gain stationary access to data streams. As in pre-digital times, it is a screen or a battery of screens that frame where and how visualizations are to be viewed. However, the functionality of a workstation is bound *a priori* to the interaction with a number of "interfacial devices", such as the screen, keyboard and mouse.[15] At a workstation, the very external architecture of input and output devices and the involvement of the user reveal that the screen as the image surface is "just a specific sub-interface within a broader human-computer interface".[16] This relativization appears important in order to make clear that, despite the similarity between the light box as a hardware dispositive and the workstation, both the diagnostic and theoretical focus undergoes a fundamental shift. "The screen just reassembles various interfacial processes, translating and returning them as visual representations on a flat visual plane."[17] Therefore, the processes of digital visualization and instrumentation need to be critically untangled to examine which further interfacial processes, such as communication with colleagues or internet searches, are reassembled within the screen to shape the finding of a diagnosis.

Nevertheless, the issue of the ordered hanging and displaying of cross-sectional image series is also relevant in the context of GUIs. Whether the diagnosis is performed in *tile mode* (similar to a light box hanging) or *stack mode*

13 Wendy Chun, *Programmed Visions. Software and Memory*, Cambridge, MA: MIT Press, 2011, p. 59.

14 Adrian Mackenzie, These Things Called Systems. Collective Imaginings and Infrastructural Software, in: *Social Studies of Science* 33.3 (2003), pp. 365–387, p. 366.

15 Marianne Van den Boomen, *Transcoding the Digital. How Metaphors Matter in New Media*, Amsterdam: Institute of Network Cultures, 2014, p. 33.

16 Ibid.

17 Ibid.

determines the possibilities for the diagnostic screening of patients' bodies. In tile mode, individual cross-sectional images in a series can be displayed simultaneously in a number of viewports or tiles, thereby creating a synchronicity similar to a light box hanging that foregrounds comparisons between individual images and, in particular, between different computer graphic representation options (fig. 4).

In order to avoid the need to set up the arrangement layout manually for each study, the hanging protocol function has been integrated in recent software applications, such as Agfa HealthCare's IMPAX software.

> *[T]he purpose of a Hanging Protocol is to present specific types of studies and images in a consistent manner. This can drastically reduce the amount of manual image arrangement and display adjustment required from the radiologist or clinician, thus improving overall operational efficiency.*[18]

The *operational efficiency* of the hanging protocol is intended to ensure that the user does not have to start by virtually hanging images in order to divide the GUI; instead the software automatically assigns specific types of studies (e.g. thorax scans to determine the spread of lesions) to a particular image layout. The software-based shortcut between diagnostic query, visualization modality and operation in the GUI may reduce the amount of time invested, and it may be a response to both collective and individual diagnostic conventions, but it omits the step of getting to grips with the available image material. The comparison of images is immediately delegated to sight with only minimal physical action involved by the co-thinking body as opposed to the hanging practices of the light box.

18 Agfa Healthcare, White Paper. Enhanced Hanging Protocols, 2012, p. 2, http://agfahealthcare.com/global/en/main/resources/white_papers/index.jsp (accessed January 9, 2015).

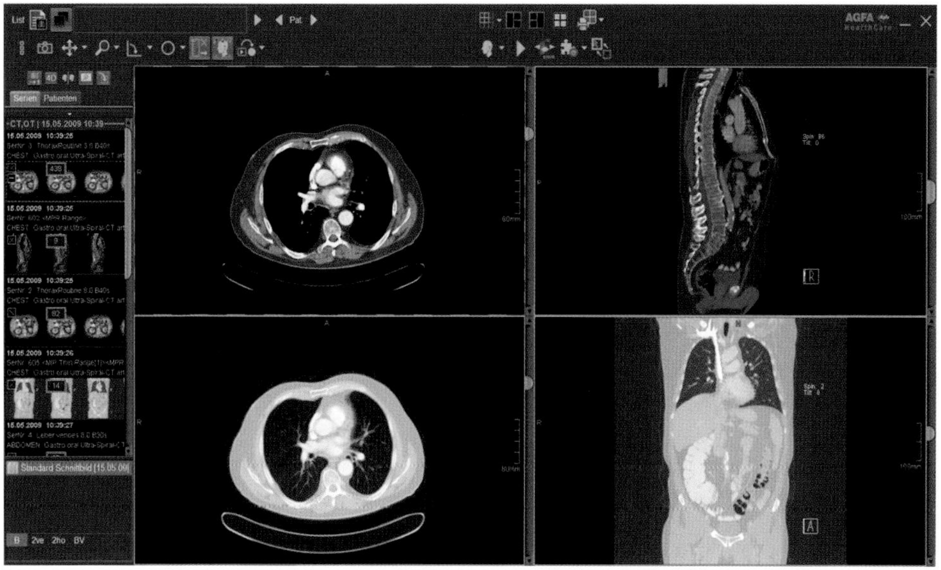

4 Screenshot of Agfa HealthCare IMPAX EE GUI. Display of CT scan in tile mode.

In addition to tile mode, images can also be automatically arranged in stack mode. Whereas tile mode emphasizes an order based on synchronous juxtaposition, the very name stack mode suggests a dimension of depth in this image layout on the screen.

> *Consequently, tile mode is used for cross-sectional imaging only to get the 'gestalt' of one particular series or of the entire examination [...]. In stack mode, images are conceptually placed one on top of each other, like cards in a deck. Only the image at the top of the stack is*

visible. This display mode allows clinicians to create a mental 3D model of the anatomical structure in which they are interested.[19]

Stack mode creates the idea that a reconstruction of cross-sectional images has been piled up on top of each other in a stack, and that the user can work through this stack, thereby reverting back at a conceptual level to the fundamental idea of tomographic imaging: being able to slice a multidimensional space into flat sections to then re-spatialize it virtually through aesthetic and epistemic operations. Stack mode creates the impression of taking a virtual walkthrough of the represented body volume by the operation of scrolling back and forth with the mouse. The key characteristic of the symmetrical and synchronous comparison between different slices, as well as the horizontal comparisons within a series, is not a strictly linear working through of the image series, but rather involves repetitions and loops that constitute a "differential analysis" of images and bodies simultaneously.[20] Even if scrolling is based on a particular individual routine, it is the dynamics of the images that enables a visually guided questioning and searching. The radiological finding crystallizes with each forward and backwards in the stack, with each software-based repetition of a cross-sectional plane. In this respect, the interactive simultaneous interplay of the radiological gaze, visualizations and hand opens up the possibility of an epistemic iteration of diagnostic findings.

19 Adrian Moise, *Designing Better User Interfaces for Radiology Interpretation*, Dissertation, School of Computing Science, Simon Fraser University, 2003, p. 34, http://citeseerx.ist.psu.edu/viewdoc/download?doi=10.1.1.71.8788&rep=rep1&type=pdf (accessed January 9, 2015).

20 Amit Prasad, Making Images/Making Bodies. Visibilizing and Disciplining through Magnetic Resonance Imaging (MRI), in: *Science, Technology and Human Values* 30 (2005), pp. 291–316, p. 292.

Screen and Screening Operations

Screens – light boxes and computer displays – are not only technical *a priori* of diagnostic radiology but also constitute the possibilities for aesthetic and hence epistemic operations. The interplay between screen and screening establishes a collectively embedded and individually embodied practice that is guided by modes and orders of hanging images. By ordering images within the frame of a screen and applying different kinds of instruments and by intersecting these with further media-based infrastructures, the screening of patient's bodies appears to be a routine procedure guided by images. Yet, with the introduction of digital data processing and visualization software the status of screens and images becomes contested. While with film-based screening, images and their order remained relatively static, even the application of digital imaging techniques such as CT, software infrastructures and their GUIs introduced dynamic and instantaneous tools of hanging and handling images. The computer screen also remains a hardware frame that displays radiological visualizations, but now radiologists themselves need to get to know a different layer of screen and screening operations, i.e. the GUI and tools of diagnostic software. While light box viewing established a hierarchy between archive clerks who hung the films according to diagnostic requests and radiologist, with digital technologies radiologists become users. With the available software applications, radiological experts are made responsible for structuring images on screen and having the necessary tools at hand by knowing the possible operations that a certain software application offers. Screen-based actions are now streamlined within a GUI that is part of a software which requires an operational knowledge of its own, even if established analog routines were meant to be predetermined in digital code.

Figures

1 McGovern Historical Center, Gift of Dr. Luther Vaughn, Photo Files, Medical Equipment and Apparatus, Hermann Hospital X-ray equipment, 1953, https://mcgovernhrc.wordpress.com/2014/11/26/hermann-hospital-radiology-department/ (accessed March 29, 2018).

2 Medicshots, Alamy stock photo, Alamy.com (accessed in March 20, 2018).

3 Choia, iStockphoto.com, Photo-ID: 533334108.

4 Reprinted with permission of Agfa HealthCare.

Luci Eldridge

Working on Mars

An Immersive Encounter through the Screen

Lost in the kaleidoscopic colours captured by the Hubble Space Telescope, lured in by the deathly blackness of Comet 67P as represented by Rosetta, and straining to make out Pluto through the first images taken by New Horizons, we are awed by such vastness and intangibility captured within the confines of our screens. Society seems to have a fascination with things beyond the realm of perceptual understanding; presenting us with scenes that we empirically know nothing about, the image of a faraway planet on the NASA website is a mysterious one, far more fascinating than our immediate setting. As viewers of such images we are reliant on space agencies like NASA and ESA to provide us with pictures and information so that we may explore these other worlds from the safety and security of our computer screens, smartphones or tablets. Satisfying the public need for images they reflect humanity's ancient impulse to explore, to discover places with their own eyes, and if not our own, then those of our machines.

This essay is drawn to one particular planet: Mars. More spacecraft, landers and rovers have been sent to Mars than any other planet and as such it is the most imaged (and arguably imagined) otherworldly landscape. The landscape is ostensibly familiar; comparable to the deserts of the American West, the plains of Chile, and the rugged landscapes of Iceland, we have an intuitive understanding as to how it might feel to walk across its surface. The *eyes* of NASA's rovers provide viewpoints through which we regard this alien terrain – windows upon unknown worlds, these images bridge a gap between what is known and unknown, between what is visible and invisible.

This is a planet being explored remotely; data is sent back to Earth, examined and reconstructed into different visualisations, allowing for new commands to be sequenced and uploaded, transforming image into action. Through images, scientists and engineers make daily decisions on how to operate the rovers remotely; data gathered from images is key to constructing visualisations of the terrain. Scientific experiments and rover drive paths are simulated before being acted out by the real rovers on Mars. This is done in a number of ways, and the focus of this essay is one of the most current means scientists are using to explore three-dimensional visualisations of Mars: NASA and Microsoft's OnSight project. Using an augmented reality headset to enable a more immersive encounter with an alien landscape, OnSight is an example of how screens are transmuting from the stationary to the mobile, from the two-dimensional to the spatial and temporal.

A Window onto Mars

NASA's Curiosity is the most recent rover sent to explore Mars (landed in 2012), and there are a whole host of other spacecraft orbiting the planet, with a number of missions planned for the near future. For many of the scientists working on Mars exploration, rovers are the next best thing to being there ourselves; stereo-vision, arms carrying instruments and wheels that enable movement over long distances, the rovers facilitate what computer scientist William J. Clancey has termed "virtual presence".[1] Whilst satellite imagery captures a sense of Mars as a whole, depicting the planet as distant and remote, the rovers, being on the ground, offer visions analogous with perception: *windows* onto a world. As Anne Friedberg points out: "We imagine *perception* to be a kind of photographic view of things, taken from a fixed point by that special apparatus which is called an organ of perception."[2] Although of course we do not see as the camera sees (from a fixed point, through one lens, with a particular focal length, etc.) the notion that a picture is a kind of window dates back to Alberti's first modern treatise on painting, *Della Pittura* (1435). The camera, like Alberti's gridded veil, is a mechanical means of translating subject into image. Discussing the camera, art historian Martin Kemp argues that imaging machines and the way data is presented is always linked to the eye because "it is the human visual system that initiates any kind of photographic activity, [...] the end product is rigged to work within the parameters of our sight, and [...] images are irredeemably subject to our ways and habits of seeing in all their variability."[3] The cameras on Curiosity's mast (the right and left black and white navigation cameras and the right and left colour mast cameras) are located roughly 1.97m from the ground, just above human eye height. The cameras then are the eyes of the rover, and in turn an extension of our vision. Clancey argues that it is through the image that scientists can experience Mars in the *first person*. When looking at a photograph of Spirit's tracks in the sand, scientists liken the marks to the scuffing of their boots: "we have been there and we did this. These are our marks – our boots on the ground of another planet."[4] As "surrogate explorers" these scientists become the rovers, referring to aspects of the landscape as if they had stood there themselves.[5] By looking upon an image that places the human at the centre of the mediated experience, the image makes way for the possibility of a virtual experience.

But we are not there behind the viewfinder to compose the image before releasing the shutter, nor are we able to compare the image with reality once we've tapped the capture button on our smartphone. We are not, therefore, able to verify first-hand the referent these images signify. Jim Bell, the Panoramic Camera's lead investigator for NASA's Spirit and Opportunity rovers, states that "the relation to reality is a particularly strange one" for Mars exploration; the images captured by the rovers are not "abstract" but they do not represent "a reality that any human has quite witnessed yet, either".[6] For Roland Barthes, writing about analogue photography in 1980, the image always contains the referent; the reflected light physically alters the surface

1 William J. Clancey, *Working on Mars. Voyages of Scientific Discovery with the Mars Exploration Rovers*, Cambridge, MA: The MIT Press, 2012, pp. 59–60.
2 Anne Friedberg, *The Virtual Window. From Alberti to Microsoft*, Cambridge, MA: The MIT Press, 2006, p. 142 [original emphasis].
3 Martin Kemp, *Seen Unseen. Art, Science and Intuition from Leonardo to the Hubble Telescope*, Oxford: Oxford University Press, 2006, pp. 268–269.
4 Clancey 2012 (as fn. 1), p. 103.
5 William J. Clancey, Becoming a Rover, in: Sherry Turkle (ed.), *Simulation and its Discontents*, Cambridge, MA: The MIT Press, 2009, pp. 107–127, p. 114. Clancey also explores the use of such phrases as "where are *we* going to go" or "are *we* going to stay here?" to explore how each scientist projects themselves onto and into the body of the rover. Ibid., p. 115.
6 Jim Bell, *Postcards from Mars. The First Photographer on the Red Planet*, New York: Penguin Publishing Group, 2006, p. 3.

of the negative and the image becomes indexically linked to the object it represents, signifying its presence in some past moment.[7] Mars is just over 225 million kilometres away and data – captured digitally – can take between 4 and 24 minutes to reach Earth.[8] Barthes' referent then is completely invisible to the naked eye. The distance between Earth and Mars reduces our experience of the planet to two dimensions, and the screen becomes an impenetrable boundary – both physically and metaphorically – that more immersive visualisation technologies are trying to break. We experience Mars through multiple forms of images; black and white and colour composites standing in for an impossible experience. These images are viewed principally on computer screens, but the image also becomes another kind of screen, a mediator that is imbued with calibration procedures, technological limitations and the science-driven demands of NASA. And yet images are a vital part of the process when it comes to gaining a more immersive understanding of the rover's surroundings, enabling scientists and engineers to make discoveries or decisions on where the rover should drive next. Panoramic visualisations for instance give a 360° view of the terrain, whereby the rover (and thus the human viewer) is placed at the centre of the image. Red/blue anaglyphs are used to get a sense of the topography, and 3D terrain models are constructed from stereoscopic data to simulate drives and build command sequences. The colour of images is manipulated by scientists to draw out geological features and chemical compositions in a practice of working with images to actively reveal what would otherwise remain unseen.[9] Each virtual manifestation of Mars becomes a screen through which a form of invisible vision is enabled. But this is a landscape humans have yet to witness first-hand. Distance, the unknown and the impenetrable lie at the very heart of these images. These predominantly two-dimensional *windows* often fail to offer their viewers a more intuitive grasp of scale, distances and the rover's overall context on the surface. And this is where recent developments in the field of virtual reality are triumphing.

Although scientists at NASA Ames have been working on head-mounted displays (HMDs) since the mid-1980s, the tech world has recently witnessed a surge in developments in this field, particularly in the consumer market. Based on knowledge of stereoscopic vision, the left and right eye views of a virtual environment are projected into either eye to give the illusion that the user is experiencing the image space in real life. Virtual environments may be fully immer-

7 Barthes was referring to un-manipulated analogue photography: "It is often said that it was painters who invented Photography (by bequeathing it their framing, the Albertian perspective, and the optic of the *camera obscura*). I say; no, it was the chemists. For the *noeme* 'That-has-been' was possible only on the day when a scientific circumstance (the discovery that silver halogens were sensitive to light) made it possible to recover and print directly the luminous rays emitted by a variously lighted object. The photograph is literally an emanation of the referent. From a real body, which was there, proceed radiations which ultimately touch me, who am here." Roland Barthes, *Camera Lucida. Reflections on Photography*, London: Vintage, 2000, pp. 80–81.

8 The time it takes signals to reach Mars depends largely on the position of Mars in relation to Earth with the minimum time delay being 4 minutes, the maximum being 24 minutes.

9 These immersive forms of images were explored in my 2017 PhD thesis titled *Mars, Invisible Vision and the Virtual Landscape. Immersive Encounters with Contemporary Rover Images*. The thesis offered a new understanding of human interaction with a landscape only visible through a screen, and explored how contemporary scientific imaging devices aim to collapse the frame and increase a sense of immersion in the image. Arguing that these representations produce inherently virtual experiences, their transportive power was questioned, highlighting the image as reconstructed – through the presence of a glitch, illusion is broken, revealing the image-as-image. The research re-examined scientific forms of images against examples from the history of visual culture (be it art or popular culture) to draw parallels between different ways of seeing, representing and discovering the unknown.

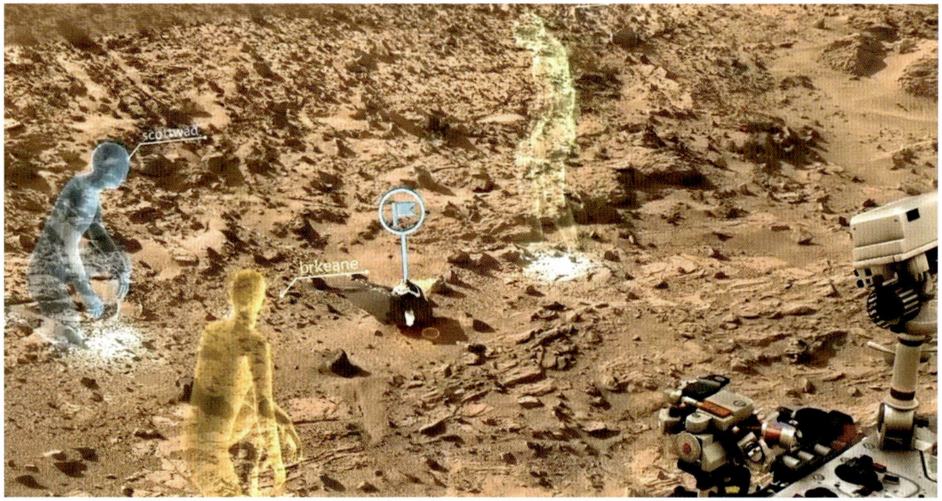

1 Screen view from OnSight, January 21, 2015.

sive (virtual reality) or virtual objects may be overlaid onto real-life surroundings (augmented reality). Head-tracking technologies enable the user to move about in real space and experience environments as if they were really there, placing the viewer at the very centre of the visual experience.

In 2015 NASA and Microsoft launched OnSight, which uses Microsoft's HoloLens (an augmented reality headset) to display a virtual environment constructed from data captured by Curiosity. As a current screen-based medium that synchronises image, action, and space on the spot, OnSight is being used to explore Curiosity's images in more detail, enabling scientists to "work on Mars".[10] As NASA/JPL state: "images, even 3D stereo views, lack a natural sense of depth that human vision employs to understand spatial relationships [...]. [OnSight] provides access to scientists and engineers looking to interact with Mars in a more natural, human way."[11] This technology is being heralded as an immersive means to explore Mars from a scientific perspective, allowing scientists to plan which areas of the landscape they would like to investigate, image and drill, but OnSight is also being used to explore data from previous Martian days (sols) in more detail. In addition, scientists from all over the world can explore the data *together*; each scientist has their own *avatar* within the virtual environment that the other users can see. The avatar's *gaze ray* (a line of coloured light emanating from the avatar's eyes) enables other users to see where they are looking and each user is able to lay flags to pinpoint areas of interest or possible spots for further exploration by the real Curiosity on Mars (fig. 1).

Navigating Mars From the Centre of the Image

The OnSight software constructs a three-dimensional environment from MastCam and NavCam images, together with satellite imagery taken by the HiRISE camera on the Mars Reconnaissance Orbiter.[12] This technology is being used by both NASA scientists and those working with the data at planetary imaging facilities and academic institutions

10 NASA/Jet Propulsion Laboratory, NASA, Microsoft Collaboration Will Allow Scientists to 'Work on Mars,' http://jpl.nasa.gov/news/news.php?-feature=4451 (accessed December 14, 2016).

11 Ibid.

12 Project Manager for OnSight Jeff Norris explains: "The 3D reconstruction is created via a terrain processing pipeline developed by my team that takes as input the stereo images acquired by the rover's cameras. The pipeline extracts range information by using a process called stereo correlation, and then uses that range data to build a 3D model of the shape of the terrain called a 'mesh'. The mesh is then coloured using texture maps that are also derived from the images." Jeffrey S. Norris (Founder and Director of JPL Ops Lab, NASA, California), email message to author, March 26, 2015.

around the world. The OnSight units access the data via the internet by connecting to servers at the Jet Propulsion Laboratory (JPL) in Pasadena, California. When the rover moves to a new location, the scientists are emailed notifications about a new scene, which automatically downloads when the headset is switched on.[13]

The kinds of promotional images that accompanied the announcement of OnSight in January 2015 portrayed an immersive experience whereby the user was able to walk across the surface of Mars (fig. 2). However, the actual experience of OnSight is fairly different; as the software combines NavCam and MastCam images, the virtual environment is a patchwork of colour and black and white. Upon donning the headset and clicking through to the Mars dataset what you see is essentially only a kind of *window* onto Mars; the screen has an aspect ratio of 16:9 and it takes up the centre of the user's vision. Unlike virtual reality whereby you are totally immersed in a simulation, augmented reality overlays the virtual and the real: peripheral vision (anything outside of that screen to the right, left, bottom or top) is taken up by real-life surroundings. Although this might at first be seen as a limitation, Research Associate in the Earth Science and Engineering Department at Imperial College London, Dr. Steven Banham, states that an awareness of the user's real surroundings prevents them from tripping up and counteracts the feeling of nausea so often experienced with fully immersive HMDs.[14] As such, the technology can be used for prolonged periods of engagement. As the user moves their head or rotates on the spot, a three-dimensional rendering using photographic data of the Martian landscape

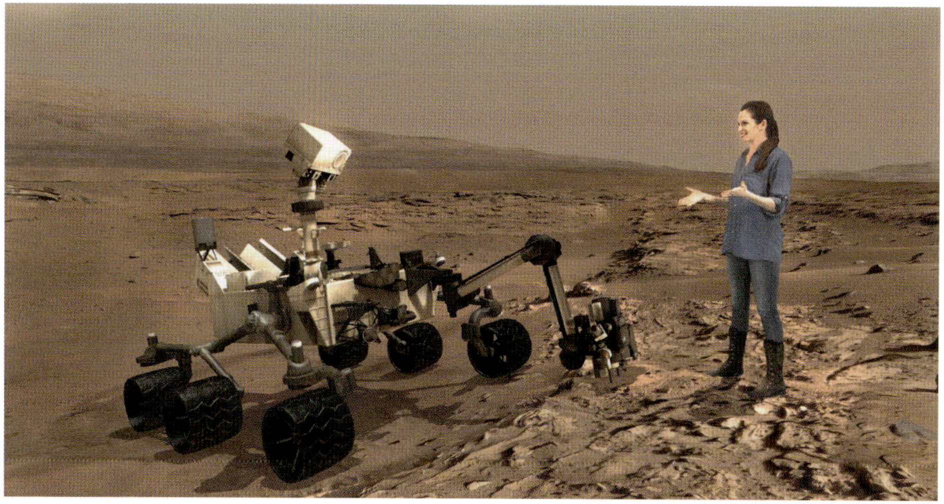

2 Erisa Hines, a driver for the Mars Curiosity rover, based at JPL, talks to participants in Destination: Mars, March 30, 2016.

is revealed behind a window. As the user walks about in real space, so too does the perspective through the window change: through the image-as-screen Mars can be seen from different viewpoints.

Although this tool is not being used by engineers responsible for driving the rovers on Mars, there are similarities between how OnSight constructs a three-dimensional environment and how tools used by JPL engineers model the terrain through stereoscopic data captured by the rover's cameras. Due to the time delay, driving rovers around the surface of Mars in real time is not an option. Instead commands are uploaded on a day-to-day basis. The data from the previous sol is analysed by science and engineering teams and decisions are made on where the rover should drive next and what experiments it should undertake. Commands are written and uploaded at the end of each Earth day and the

13 Dr. Steven Banham (Postdoctoral Research Associate, Earth Science and Engineering Department, Imperial College London), interview by author, London, December 13, 2016.

14 Ibid.

rover carries out these instructions, beaming back its data for the following day, when the whole process starts again. Although scientists return to images taken days, months or even years before, engineering decisions on where to drive the rover are made daily using the most recent data and a different set of tools aid this process.

3D visualisation software is used by rover drivers at JPL to assess the traversability of Curiosity's surroundings from stereoscopic data captured by the navigation cameras, the hazard avoidance cameras and the mast cameras. The multiple programmes used to drive the rover are part of a suite of applications called the Rover Sequencing Visualisation Programme, or RSVP. RSVP takes the stereoscopic image data and automatically produces polygon meshes of the terrain. The programme allows drivers to see the rover's positioning in relation to hazards and holes in image data which are seen as potential obstacles (fig. 3–4). The rover drivers study images and analyse the terrain models, simulate drives over them, and when they feel comfortable with the planned traverse, upload the command sequence to the rover along with instructions from the science team.

These models are used alongside the raw images to provide the rover's up-to-date location (fig. 5); they are *interactable* terrain models that display the rover in the context of its *seeable* surroundings. Such modes of visualisation generate what sociologist of science and technology Janet Vertesi terms as an "immersive view of the Rover's environment", "*draw[ing]* Mars *as* [a] tangible, interactable terrain, and allowing engineers to conjure up the sense of 'being there' virtually".[15] In this case, the practice of working

15 Janet Vertesi, 'Seeing Like a Rover'. Images in Interaction on the Mars Exploration Rover Mission, Dissertation, Cornell University, 2009, http://hdl.handle.net/1813/13524 (accessed March 15, 2014), p. 260, p. 262 [original emphasis].

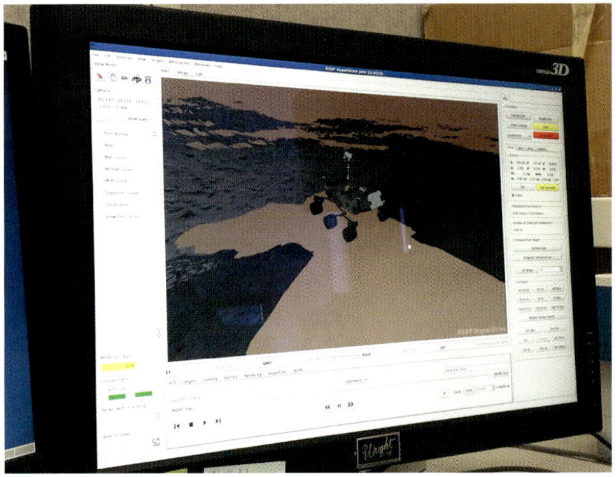

3 Terrain model of Curiosity's surroundings shown in RSVP.

with images and 3D models allows teams at JPL to convert image (simulations of the drive paths) into action (in reality on Mars). In the models, the rover's seeable surroundings are represented by the black and white image data – these images are draped over the underlying polygon mesh, filling in only what can be seen from the rover's central viewpoint. Any holes in image data created by a ridgeline, rocks, or the rover itself are represented in Martian brown – here the polygon mesh is approximated. Rovers are never driven into these spaces because they are unknowns, and for rover drivers these blind spots are as important as the terrain that can be seen. The images then help to define and determine the possibilities and the range of operations. Writing on vision and perception in 1945, Maurice Merleau-Ponty asks how we should experience the existence of absent objects, and how we should experience the nonvisible parts of present objects. "Should we say," he asks "that I *represent* to myself

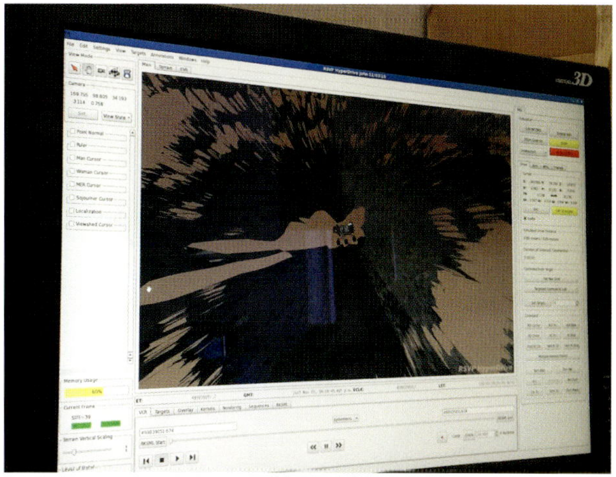

4 Elevated terrain model of Curiosity's surroundings shown in RSVP.

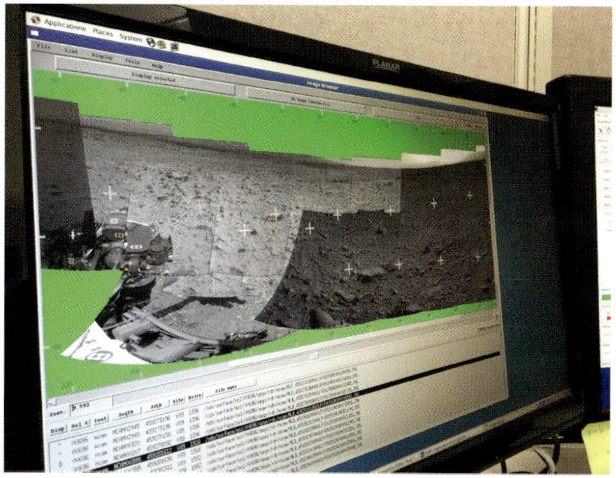

5 Photograph of Curiosity image mosaic shown on screen in RSVP.

the sides of this lamp which are not seen?" The unseen sides of the lamp are anticipated, according to Merleau-Ponty, "as perceptions which would be produced" upon movement.[16] The terrain model in RSVP highlights the gaps in image data, what cannot be seen by the camera's lens. Unlike Merleau-Ponty's lamp, the Martian landscape is not a space we can reach into or around to touch: we cannot physically move through the landscape to reveal the invisible. The visible landscape is framed by the flat colour of the polygon model, showing the "limit of visibility".[17]

In a similar fashion, Paul Virilio writes on shifting one's gaze, one's vision and one's blindness: "Shifting your gaze, whether thanks to the mobility of your head or the mobility of your eyeball, also means effectively shifting your blindness, your own relative blindness."[18] In the RSVP terrain-mapping tool we see according to blindness, distinctly aware of what is missing. Blindness pushes up from underneath the terrain model, giving material form to that which remains invisible. We cannot fully perceive the surrounding landscape, not through Curiosity's eyes or our own. Rover drivers can simulate Curiosity's traverses through the virtual model, but by doing so they cannot reveal any more than what is already there. Curiosity is at the centre of the image. RSVP allows engineers to see what is on the periphery of vision, to rotate, zoom in and out, and generally gain a more encompassing understanding of the terrain they have to navigate, but this vision is dictated by the rover's capabilities, the limits of

16 Maurice Merleau-Ponty, *The Primacy of Perception*, Illinois: Northwestern University Press, 1964, pp. 13–14.

17 Paul Virilio, *A Landscape of Events*, Cambridge, MA: The MIT Press, 2000, p. 38.

18 Ibid., p. 39.

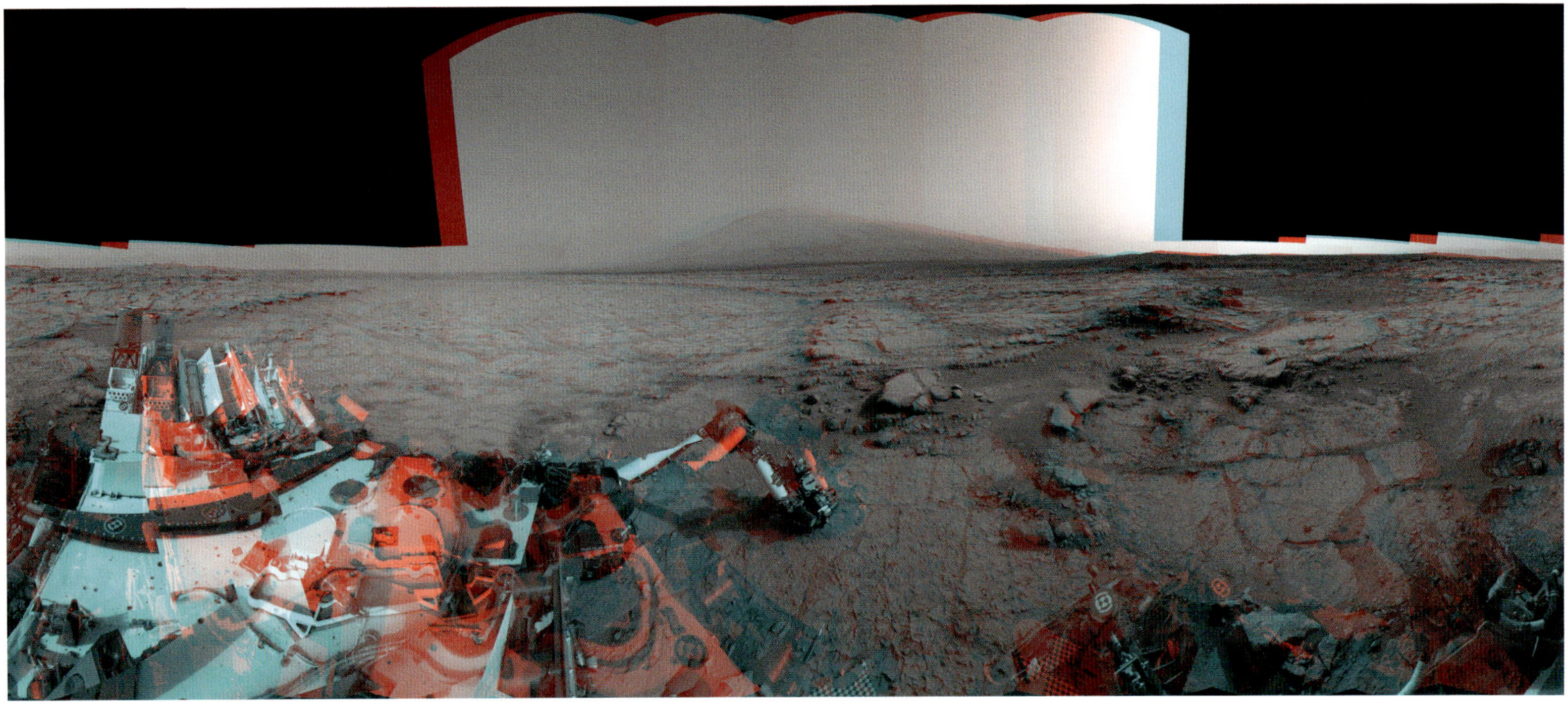

3 *Mars Stereo View from John Klein to Mount Sharp, Raw.* This 360° anaglyph combines dozens of images taken by Curiosity's right and left Navigation Cameras on January 23, 25 and 26, 2013 (sols 166, 168 and 169). Photojournal image addition date: April 23, 2013.

its vision – in a visualisation that is, to quote one of JPL's rover drivers John R. Wright, only "two and a half D".[19]

3D may be achieved through 3D-enabled screens or red/blue anaglyphs which are used to get a more kinaesthetic sense for the terrain. Yet images like *Mars Stereo View from John Klein to Mount Sharp* appear as *windows* (fig. 6); a viewpoint which has been given artificial depth. Such images are shown on computer screens, and OnSight attempts to bridge the gap between an onscreen image and physical experience, allowing a more intuitive exploration of landscape. Scientists spend prolonged periods of time looking at image data and editing it to produce certain results. Before OnSight, what was lacking was the ability to physically move about within the landscape, to inspect a particular rock, or to walk about and get a *feel* for the terrain surrounding the rover.

19 John R. Wright (Data Visualisation Developer IV at Jet Propulsion Laboratory, California), interview by author, Pasadena, CA, November 3, 2015.

But there are still limitations, primarily relating to the fact that the environment can only be constructed through data captured by the rover.

Donning the OnSight headset and *airtapping* through the drop down menus to load the scene, the virtual environment appears through a window. This window is overlaid onto real-life surroundings and shifts with the motion of the user's head, always directly in front of their eyes. From its vantage point Curiosity can image its near surroundings in high resolution, but as with RSVP, *seeing* behind objects is not a simple case of walking into the landscape and looking from a different perspective; the virtual environment is dictated by the rover and its cameras' stereoscopic reach. Unlike RSVP, which represents the unseen sides of the landscape in a different colour, OnSight's objective is to increase levels of immersion, so unseen sides of rocks and terrain features are estimated, rather than being left blank. The further away from the rover the user gets, the more the software has to fill in. As a result, these features appear slightly distorted as the photographic data is stretched over the underlying polygon mesh. As long as the user remains close to the rover, the distortion does not impede a great deal on the level of immersion.

Nonetheless, OnSight enables its user to gain greater situational understanding for the terrain around the rover, just as they might if they were *there on Mars*. Project manager for OnSight Jeff Norris elaborates on this:

> *OnSight tries to engage many of the same senses that a geologist would have when exploring a location on Earth. A very important sense that OnSight engages but a traditional computer monitor does not is* proprioception, *the body's sense of its own position and movement. Because we rapidly and accurately track the position of the scientist's head as they move around in their office, we can show them the views of Mars that they would have if they were moving in the same way on Mars. This is what creates [a] 'first-person perspective'.*[20]

The importance of experiencing the world from within, with the body as the locus of perception was set out by Merleau-Ponty: "I do not see [a space] according to its exterior envelope; I live in it from the inside; I am immersed in it. After all, the world is all around me, not in front of me."[21] The image is a space in which the objects do not surround us and immersive technologies attempt to deceive us otherwise by enveloping us in the image. With OnSight the body is quite literally placed at the centre of the experience; the user must move his/her body to reveal more of the landscape. Artist and critic Brian O'Doherty describes a similar experience: "[T]he Eye urges the body around to provide it with information – the body becomes a data-gatherer."[22]

With OnSight, however, our vision and body become oddly detached; reaching our hand out in front of us – as if to point towards something through the window – it disappears, existing behind the screen within physical and not virtual space. Despite our body being integral to *how* the illusion is revealed, the eye is isolated, being the only entity present in the image of Mars. To this end, Merleau-Ponty discusses the idea of the seer and the visible:

> *[...] without even entering into the implications proper to the seer and the visible, we know that, since vision is*

20 Norris 2015 (as fn. 12).
21 Merleau-Ponty 1964 (as fn. 16), p. 178.
22 Brian O'Doherty, *Inside the White Cube. The Ideology of the Gallery Space*, Berkeley: The Lapis Press, 1976, p. 52.

> *a palpation with the look, it must also be inscribed in the order of being that it discloses to us;* he who looks must not himself be foreign to the world that he looks at. *As soon as I see, it is necessary that the vision (as is so well indicated by the double meaning of the word) be doubled with a complementary vision or with another vision: myself seen from without, such as another would see me, installed in the midst of the visible, occupied in considering it from a certain point [...]* he who sees cannot possess the visible unless he is possessed by it, unless he is of it.[23]

To appropriate another O'Doherty quote, with OnSight "the eye is abstracted" from a mobile body "and projected as a miniature proxy into the picture to inhabit and test the articulations of its space".[24]

It is important to note that although OnSight is perhaps the most immersive means of experiencing images of Mars in a screen-based technology that synchronises image, action and movement, the tool is not without its limitations. Being an augmented reality headset, the window onto Mars appears against our real-life surroundings and as such it is not fully immersive. OnSight's window onto Mars only allows the user to *glimpse* a virtual image of Mars. Floating about, occupying a strange space between the user and their real surroundings, the screen becomes the frame that withholds the image of Mars just beyond our grasp. The level of immersion for OnSight then is not the technology's ability to give a full 360° view of an environment (for Mars to invade all areas of vision) but is in the act of movement to reveal the depth of the virtual image.

With OnSight there is the definite wow-factor and seductive novelty of new illusions; like the Victorian stereoscope or 3D TV. With new technologies appearing all the time, perhaps there is something in the ephemeral nature of technologies in re-presenting images of Mars that reflects our human desire to see ever more clearly and in a more immersive manner, to get closer and closer to a feeling of touching and being in the landscape, if only – for now at least – on the level of vision. In a sense then, OnSight reveals a deeper, more insatiable desire that lies at the heart of all types of imaging; to re-live, re-construct or imagine something that is unseen because of its distance from us in time and space.

An Immersive Encounter

This essay concludes with a subjective encounter of OnSight which took place on December 13, 2016 at Imperial College London, and was kindly facilitated by Dr. Steven Banham. The virtual environment encompassed datasets from sols 1526 – 1547 (November 22 –December 13, 2016). During this time the rover had been parked for a few days whilst engineers ran diagnostics on the drilling mechanism; as such the rover was able to image its immediate surrounds in high-resolution detail.

23 Maurice Merleau-Ponty, *The Visible and the Invisible, followed by Working Notes*, Evanston: Northwestern University Press, 1968, p. 135 [original emphasis].

24 O'Doherty 1976 (as fn. 22), p. 18. Original quote: "One 'steps' firmly into such a picture or glides effortlessly, depending on its tonality and colour. The greater the illusion, the greater the invitation to the spectator's eye; the eye is abstracted from an anchored body and projected as a miniature proxy into the picture to inhabit and test the articulations of its space. For this process, the stability of the frame is as necessary as an oxygen tank is to a diver. Its limiting security completely defines the experience within."

Lowering the headset over my eyes and adjusting the headband I looked through tinted glasses at the office surroundings of the Royal School of Mines at Imperial College London. A window slotted down into view. With an almost opaque but luminous translucency this window was hard edged and glowing against the dull grey of the real office carpet and surrounding white walls. But unlike Alberti's fixed veil this window was mobile, almost fragile. Floating and glimmering the window followed the motion of my head, persistently present within my direct field of vision, in front of and against, yet within the office interior. A screen which was simultaneously a window, appearing only for me. A personal portal out onto Mars.

The screen flickered and the laying out of a polygon mesh announced the forthcoming emergence of landscape. The terrain began to materialise, expanding outwards rapidly from my immediate surroundings and into the distance, a patchwork of greys and Martian browns in high and low resolution. Revolving on the spot I looked out towards the mountainous rim of Gale Crater; a dusty grey in the distance, offset against a shimmering soft pink sky.

As I knelt down to examine a portion of the ground, the window shrank in size. Zooming in physically and virtually I saw cracks and crevices in the rocks, the strata in the bedrock, granules of sand and tiny pebbles. As I reached out to touch and feel the surface under my gaze my hand evaporated, my body belonging to a space exterior to my vision.

As I stepped back Curiosity flickered into view. The large immobile body of the rover was coated in a thin film of dust, trapped here, in the virtual image of Mars. As I advanced forth in an attempt to inspect its wheels, Curiosity vanished. In an instant I became the rover, seeing the surrounding terrain from its vantage point, its body merged with mine.

As I walked backwards once more I looked out towards the distant horizon. The environment appeared perversely trapped within a pixel-thin layer, a three-dimensional image held somehow within a two-dimensional display. This was a virtual opening that did not require a click or swipe of the finger to reveal what lay beyond the borders. Here I was present virtually in the image, a presence that relied on my own physicality; the position of my head in relation to my body. A three-dimensional image of Mars that I was in control of revealing.

Revealing. The act of revealing coincided with the act of concealing. Movement enabled me to penetrate the environment contained within the image, but did not allow me to bypass the screen. Movement revealed depth but concealed width. The window could not be enlarged, the frame could not be collapsed, the image-as-screen could not be stepped into.

I looked upon markers in the landscape, upright poles that marked where Curiosity had been and for how long. My gaze lingered and the rover's path became illuminated, snaking through the landscape from one point to the next. A glowing path into the past of a landscape it would not see again. Upon walking towards this point in the landscape I revealed the depth of image, a depth of space, a depth of time.

And yet I was not limited to walking alone, nor to the four walls of the office. I could reach the outer edges of Curiosity's vision through teleportation. Speeding back through time, through space, and into the reconstruction of Mars-as-image.

Figures

1 NASA/JPL-Caltech, http://jpl.nasa.gov/news/news.php?feature=4451 (accessed December 14, 2016).

2 NASA/JPL-Caltech/Microsoft, http://jpl.nasa.gov/news/news.php?feature=6220 (accessed December 14, 2016).

3–5 JPL-Caltech, Photo: Luci Eldridge.

6 NASA/JPL-Caltech, http://photojournal.jpl.nasa.gov/catalog/PIA16847 (accessed March 15, 2014).

Nina Franz and Moritz Queisner

The Actors Are Leaving the Control Station

The Crisis of Cooperation in Image-guided Drone Warfare

The Ground Control Station in Remote Warfare[1]

Under the condition of extreme remoteness, as is facilitated by computer-aided and robotic weapon technologies, military interventions, both geographically and in relation to the laws of war, have become increasingly "limitless".[2] Paradoxically, remote military interventions are, by their very nature, restricted to the *closed system* of computerised, heavily mediated environments of control. The potential for crisis in the cooperation of human and non-human actors is nowhere more apparent than in the ground control stations (GCS) in remotely-controlled drone warfare. A GCS provides image-guided control over the deployment of so-called *unmanned* weapon systems. It forms the central operative unit for decision-making and action in remote warfare by linking human perception to the sensor technology of the drone. It is, thus, a crucial component of a control setting, where questions of agency culminate in a distinctly political operation – the remote execution of a military command to kill.

Due to their remote setting, drone operations structurally rely on visibility and controllability via a complex system of sensors, control instruments, software interfaces, and transmission technologies. In such cases, knowledge of a situation in an area of operation, the *situational awareness*, is based mainly on operative images in the form of visualised sensor data, but even where images form the primary, often sole, basis for action, the analysis must not be narrowed down to the level of depiction. The situation in which the images are being applied must also be considered. Against the background of an expanded understanding of what constitutes a military operation, we will discuss the ground control station as a site of image-operations in contemporary warfare.

Contact Zones of Sensing

The terminology of *unmanned aerial vehicles* is misleading in the context of remote-controlled military interventions, in that both the infrastructures for guiding and controlling drones at a distance and the decisions underlying the actions are the result of intricate cooperation between a large

1 This text first appeared in German in: Johannes Bennke, Johanna Seifert, Martin Siegler, Christina Terberl (eds.), *Prekäre Koexistenz*, Paderborn: Fink, 2018. The authors would like to thank Timothy Cullen for the opportunity for critical discussion, Deborah Curtis for the translation, and Habib William Kherbek for editing.

2 Derek Gregory, The Everywhere War, in: *The Geographical Journal*, 177, 3 (2011), pp. 238–250; Caroline Holmqvist-Jonsäter, War as Perpetual Policing, in: Caroline Holmqvist-Jonsäter, Christopher Coker (eds.), *The Character of War in the 21st Century*, London/New York, 2010, pp. 103–118.

number of people in a situation of shared responsibility,[3] as well as the complex technological processes that could be said to facilitate those actions. While traditional aviation, even if cockpits have become computerised and automatised, still demands a comparatively high degree of independence on the part of pilots, the operational structure for the deployment of drones is essentially based on the communication between very different actors, such as cameras, relay stations, pilots, ground troops, military lawyers, data analysts and imaging specialists.[4]

Marie-Luise Angerer has highlighted the fact that the "the zones of contact between the interior and exterior domains of sensing can be understood to be simultaneous processes of connection, disruption, and translation", which generally proceed with a high degree of friction and conflict and are ultimately shaped by the "radical in-translatability (of 'sensing' and 'sense-ability')".[5] Accordingly, the contact zones between the human sensorium and machine-based sensing can be investigated as spaces in which some of these sources of friction and conflicts come to the surface while others disappear entirely behind the surfaces of the interface design and are rendered imperceptible to the operators. As a particular contact zone of sensing, the CGS points to the question of human-technological *co-existence* as a fundamental problem of military intervention and, arguably, as one of the most extreme means of political agency.

Analysing the use of the GCS, as illustrated in the available literature, and based on conversations with US Air Force pilots, we want to highlight the ways in which the conditions for human-machine cooperation refer to the circumstances of their design, and, thereby, to the tightly interwoven political, scientific and economic relations that extend far beyond the dedicated technological processes of control. In doing this, we seek to reach beyond a description of "distributed agency" and "chains of operations" to put the emphasis on the question of how the distribution of control is implemented and negotiated, i. e., on which parameters it is based, and where specifically *human* agency is distributed in these human-machine configurations.[6] In focusing on the moments of crisis within this cooperative process, we moreover seek to shed light on the apparent methodological reduction that comes with the terminological *symmetrisation* of human-technological co-existence which pervades not only the discourse in current cognitive science and media theory, but, also, military policy and action. Thus, we hope to return the focus to the politically pertinent – and explicitly human – interests invested in these processes.

3 Up to 200 people are required to run an aerial patrol with Reaper or Predator drones. In addition to the crew and technical personnel, this includes data and image analysts. See also Derek Gregory, From a View to a Kill. Drones and Late Modern War, in: *Theory, Culture & Society* 28.7–8 (2012), pp. 188–215, p. 195. See also M. C. Elish, Remote Split. A History of US Drone Operations and the Distributed Labor of War, in: *Science, Technology & Human Values* 42.6 (2017), pp. 1100–1131.

4 More detail in ibid., pp. 195–197.

5 Marie-Luise Angerer, *Ecology of Affect. Intensive Milieus and Contingent Encounters*, translated from German by Gerrit Jackson, Lüneburg, 2017, p. 46.

6 As Lucy Suchman and Jutta Weber have argued in their theorisation of the use of drones in a military setting the position that in a situation of shared "human-machine autonomy", agency should not be thought of as something pertaining specifically to humans or machines, but should be treated as a conglomeration of "effects of specific human-machine configurations", the analysis of which requires the careful setting of a framework, a "cut in the network", that allows a certain unit in the expansive system of agency to be rendered comprehensible, and its specific entanglement to be identified. Lucy Suchman, Jutta Weber, Human-Machine Autonomies, in: Nehal Bhuta, Susanne Beck, Robin Geiß, Hin-Yan Liu, Claus Kreß (eds.), *Autonomous Weapons Systems. Law, Ethics, Policy*, Cambridge, 2016, pp. 75–102, p. 91.

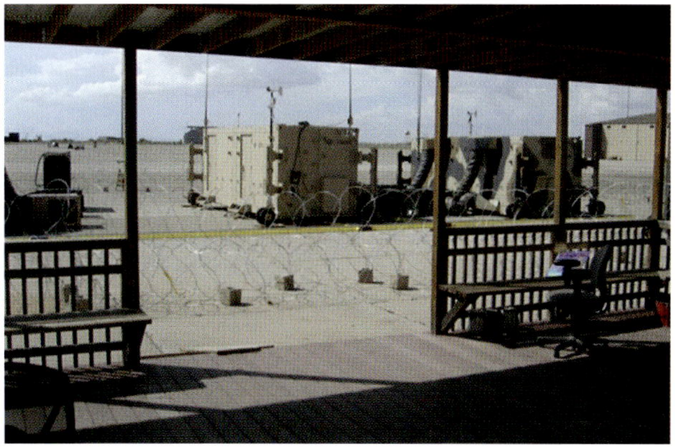

1 Ground Control Station, external view, Holloman Air Force Base.

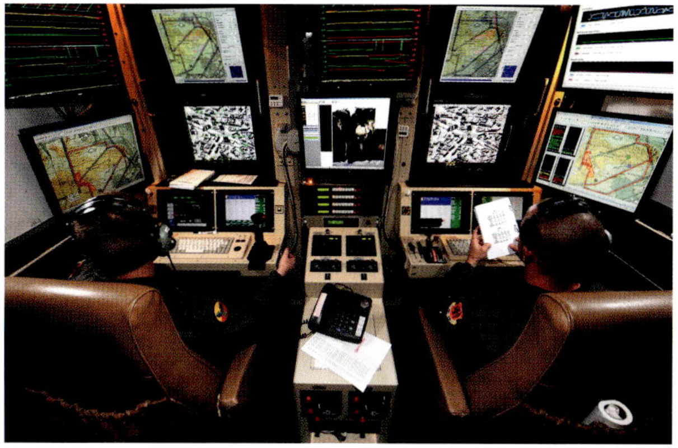

2 Pilot and sensor operator in a Ground Control Station, Holloman AFB.

The Ubiquity of Operations

The GCS is composed of multiple workstations that are usually installed in a mobile shipping container (fig. 1). The configuration of hardware and software in the central control unit defines the options of visual access to the combat zone. The members of the crew are referred to generally as *operators,* and are then subdivided into pilots and sensor operators on the one hand, and a mission intelligence coordinator overseeing the procedure on the other. Pilots and sensor operators sit in front of multiple vertically and horizontally arranged monitors (fig. 2). These provide a range of options for visualising the sensor data from the multi-spectral targeting system[7], an array of different sensors that is attached to the body of the drone, also called the sensor ball. Furthermore, the screens show air traffic and cartographic information, as well as software applications for chat and email, in addition to other mission data, such as maps, command and control options, and the warning system.

The pilots guide the aircraft and control the weapons system while the sensor operators control the sensor system and are responsible for targeting. The mission intelligence coordinator prepares the deployment of the aircraft and provides support in the analysis and interpretation of the incoming sensor data, their comparison with external sources, and coordination with ground forces, lawyers, and superior officers. The information exchange between individual actors happens through chat clients and audio link, which connect the crew to the other participants in the so-called *remote split operations.*[8] Military research on the control and navigation of unmanned aircraft con-

7 The Multi-Spectral Targeting System is composed of an infrared sensor, an image amplifier, a daylight camera, a laser marker, and a laser illuminator.

8 Mark C. Elish, Remote Split. A History of US Drone Operations and the Distributed Labor of War, in: *Science, Technology & Human Values* 42 (2017), pp. 1100–1131.

ceives the relationship between human and machine as an immersive synthesis, in which processes relating to action, decision-making, and perception are realized in cooperation with humans. Correspondingly, military operations are increasingly understood as a convergence between human users and technical processes at the interface between senses, sensors and computational processes. In cultural and media theory this form of entanglement of the human and non-human is commonly framed by terms such as *cultural technique*[9], *operational chain*[10], *a priori*[11] and *hybrid*[12]. The debate about human machine cooperation has recently been expanded by the inclusion of older approaches from various disciplines, in which the symmetrical relation between humans and machines is taken as a given.[13] The way in which these approaches contribute to clarifying the precarious constellation of the actions of humans and machines in the context that concerns us here can only be elucidated based on the concrete observation of such operations through relevant practices with attention paid to the terminology of *operation* in a military context as well as in current media theory.

The US military regards any military action that serves the purpose of achieving a planned objective or military mission as an operation.[14] Military operations are increasingly understood as cooperation between automated and (partially) autonomous technologies and the human *operating crew*, whose agency and decision-making abilities are placed in a precarious relationship with the efficiency of technical systems. Somewhat oblivious to current media-theoretical debates on the recognition of non-human actors, in the military and technological discourse, *agency* is readily understood as *co-agency*, and the human actor is regarded as an "element"[15] or "component"[16] of an operative system. Against this background of the military understanding of operation, we will discuss the ground control station as a site of image-operations in contemporary warfare. To

9 See Bernhard Siegert, Cultural Techniques. Or the End of the Intellectual Postwar Era in German Media Theory, in: *Theory, Culture & Society* 30 (2013), pp. 48–65; Thomas Macho, Christian Kassung (eds.), *Kulturtechniken der Synchronisation*, Munich, 2013, pp. 16–18; Erhard Schüttpelz, Die medienanthropologische Kehre der Kulturtechniken, in: Lorenz Engell, Bernhard Siegert, Joseph Vogl (eds.), *Archiv für Mediengeschichte 6. Kulturgeschichte als Mediengeschichte (oder vice versa?)*, Weimar, 2006, pp. 87–110.

10 See André Leroi-Gourhan, *Gesture and Speech*, translated from French by Anna Bostock Berger, Cambridge, MA, 1993, pp. 219–235; as well as the texts referred to in fn. 3. For a critique of the concept of operativity in current German media theory, see Dieter Mersch, Kritik der Operativität. Bemerkungen zu einem technologischen Imperativ, in: Dieter Mersch, Michael Mayer (eds.), *Techne/Mechane. Internationales Jahrbuch für Medienphilosophie 2* (2016), pp. 31–52.

11 See Friedrich Kittler, Manfred Schneider, Editorial, in: Friedrich Kittler, Manfred Schneider (eds.), *Diskursanalysen 2. Institution Universität*, Opladen, 1987, pp. 7–11; Lorenz Engell, Joseph Vogl (eds.), *Archiv für Mediengeschichte 1: Mediale Historiographien*, Weimar, 2001; Erich Hörl (ed.), *Die technologische Bedingung. Beiträge zur Beschreibung der technischen Welt*, Frankfurt/M., 2011.

12 See Bruno Latour, *We Have Never Been Modern*, translated from French by Catherine Porter, Cambridge, MA, 1993; see also Gustav Roßler, Kleine Galerie neuer Dingbegriffe. Hybriden, Quasi-Objekte, Grenzobjekte, epistemische Dinge, in: Georg Kneer, Markus Schroer, Erhard Schüttpelz (eds.), *Bruno Latours Kollektive. Kontroversen zur Entgrenzung des Sozialen*, Frankfurt/M., 2008, pp. 76–107, pp. 79–82.

13 Among them Human Factor Studies, Work Place Studies or Science and Technology Studies. See: Erhard Schüttpelz, Sebastian Gießmann, Medien der Kooperation. Überlegungen zum Forschungsstand, in: AG Medien der Kooperation (eds.), *Navigationen. Zeitschrift für Medien- und Kulturwissenschaften* 15.1 (2015), pp. 7–54.

14 See Charles Messenger, *Dictionary of Military Terms*, published by the US Department of Defense, Greenhill, London, Stackpole, 1995, p. 274; see also Jimena Canales, Operational Art, in: Niels van Tomme (ed.), *Visibility Machines. Harun Farocki and Trevor Paglen*, Baltimore, 2015, pp. 37–54, p. 37.

15 Assistant Secretary of Defense for Research and Engineering, Human Performance Training and Biosystems Directorate: Human Systems, http://acq.osd.mil/rd/hptb/programs/human_systems (accessed January 8, 2017).

16 Ibid.

date, there is a notable lack of detailed descriptions of the interactions between drone crews and the systems of control, communication, and the sensors on which they rely,[17] despite armed drones having been deployed by the US Air Force since the Yugoslav Wars in the mid-1990s, and by now have become a cornerstone of US-military strategy.[18]

Cooperation and the Eroding Boundaries of Agency

The English term operation is ubiquitous in contemporary military discourse. The *US Department for Defense Dictionary of Military Terms* defines an operation as "[a] military action or the carrying out of a strategic, tactical, service, training, or administrative military mission; the process of carrying on combat, including movement, supply, attack, defense and maneuvers needed to gain the objectives of any battle or campaign."[19] In other words, the term operation describes *all* actions required to achieve a military objective, independent of their hierarchical level or the means to carry them out.

Jimena Canales, a historian of science, emphasises that the term operation, already very broad in its scope and usage within military strategy, has undergone a further expansion in recent times through the development of the related term of *operational art*. In the *Field Manual of US Military Doctrine*, published in 1999, the term operational art, previously used to refer exclusively to military interventions, was used for actions of war, as well as for all "operations outside warfare", including "diplomatic activity, economics and information", as well as "political and other non-military factors".[20] In our view, this expansion of the doctrinal meaning of military operations must be taken seriously, as it refers to a change in the understanding of military actions and of the actors in the military itself.

A tendency to group together different military personnel (e.g. pilots, operators of weapon and sensor systems) under the term *operator*, previously only used for certain participants in secret service missions becomes apparent. In the light of the numerous other meanings of the word *operator*, such as *machine operator*, *user*, or *supervisor*, military actions are correlated to a greater extent with technological equipment, without which modern warfare is inconceivable. The interplay between the human operator and the increasingly complex and networked technological infrastructure is described as a state of "cooperation" and "collaboration"[21] in which human and non-human actors participate on an equal level, leading to an apparent symmetrisation between human and non-human actors. However, under the condition of automated technological systems, the military discourse characterises this cooperation primarily

17 Notable exceptions are Peter M. Asaro, The Labor of Surveillance and Bureaucratized Killing. New Subjectivities of Military Drone Operators, in: *Social Semiotics* 23.2 (2013), pp. 1–29 and David J. Blair, Nick Helms, The Swarm, the Cloud, and the Importance of Getting There First. What's at Stake in the Remote Aviation Culture Debate, in: *Air & Space Power Journal* (2014), pp. 33–52.

18 By now, the number of pilots for remotely piloted aircraft (RPA) exceeds the number of those operating piloted aircrafts within the US Air Force. See Oriana Pawlyk, Drone Milestone: More RPA Jobs Than Any Other Pilot Position, March 8, 2017, http://military.com/daily-news/2017/03/08/drone-milestone-more-rpa-jobs-any-other-pilot-position.html (accessed August 17, 2017).

19 Messenger 1995 (as fn. 14), p. 274. See also Canales 2015 (as fn. 14), p. 37.

20 Ibid., p. 37.

21 James A. Winnefeld, Frank Kendall, *Unmanned Systems Integrated Roadmap FY2013-2038*, 2013; Robotics Collaborative Technology Alliance / Army Research Laboratory, *FY 2012 Annual Program Plan*, 2011, https://arl.army.mil/www/pages/392/rcta.fy11.ann.prog.plan.pdf (accessed December 7, 2017).

as a technological process of action and decision-making, which the human operator approaches in an increasingly passive manner,[22] namely, as a supervisor who monitors autonomous actors. In a different disciplinary context, but in a somewhat parallel way, the term "operative image" has recently attracted much attention in contemporary media studies and cultural theory. In this context the term links the use of an image to an "object-constituting", "generative" function of "manageability and explorability" that is understood as "operativity".[23] Particularly in the context of automated image processing those practices constitute a new type of image that is defined by the exclusion of the human actor.[24] From this point of view, images become *actors* that not only facilitate human actions within chains of operations via the screen, as in the case of camera-aided remote control, or in the context of graphical user interfaces, but they also are attributed to act with a degree of autonomy.

Distributed Cognition

Thanks to the study *The MQ-9 Reaper Remotely Piloted Aircraft: Humans and Machines in Action*[25], carried out by Timothy Cullen, a Lieutenant Colonel in the US Air Force, between 2009 and 2011 at the Engineering Systems Division of the Massachusetts Institute of Technology (MIT), we can base our considerations on a relatively detailed – albeit also heavily redacted and undoubtedly partisan – description of the theatre of operations of the *MQ-9 Reaper* drone. We can add, for consideration, the perspective of the drone crews and trainers working for the US Air Force, with whom we held discussions in the jointly organised workshop on the topic *Technology and Expertise in Remote Warfare* at Maxwell Air Force Base in Montgomery, Alabama, in February 2017.

Cullen's study, carried out using methods from Science and Technology Studies, is a significant document, in that a member of the military who is familiar with the internal culture of knowledge within the U.S. Air Force provides information on the setting of remote warfare. The study not only reveals some of the less known operative processes of modern warfare, but it also, inevitably, provides information on the way in which members of the "RPA community"

22 Grundel et al. provide the following definition for "cooperative systems", which also include weapons systems: "They have some common elements: 1) more than one entity, 2) the entities have behaviors that influence the decision space, 3) entities share at least one common objective, and 4) entities share information whether actively or passively." The objective of "cooperative technical systems" in these contexts is no longer solely the revaluation of non-human actors, but the minimisation of human participation: cooperative systems "capitalize on the availability of various interconnected resources and on the sharing of key information among the networked entities with minimal involvement of the operating crew". Don Grundel, Robert Murphey, Panos M. Pardalos, Oleg A. Prokopyev (eds.), *Cooperative Systems. Control and Optimization*, Berlin/Heidelberg, 2007, preface, without page numbering.

23 Sybille Krämer, Operative Bildlichkeit. Von der ‚Grammatologie' zu einer ‚Diagrammatologie'? Reflexionen über erkennendes ‚Sehen', in: Martina Hessler, Dieter Mersch (eds.), *Logik des Bildlichen. Zur Kritik der ikonischen Vernunft*, Bielefeld, 2009, pp. 94–123, p. 98.

24 Harun Farocki, on whose investigation of automated image-based navigation techniques the term is based, for example in remote-controlled rockets (*Auge/Maschine I–III*, 2001–2003 and *Erkennen und Verfolgen*, 2003), defined "operative images" as those images that "do not represent an object, but are rather part of an operation." Harun Farocki, Phantom Images, in: *Public* 29 (2004), pp. 12–24, p. 17. Volker Pantenburg explains further that this new type of image is "in no way any longer a 'separate entity' and located opposite a potential observer, but becomes fully integrated into an electronic technical operation". Volker Pantenburg, *Film als Theorie. Bildforschung bei Harun Farocki und Jean-Luc Godard*, Bielefeld, 2006, pp. 189–234. Trevor Paglen calls operational images "images made by machines for other machines". Trevor Paglen, Operational Images, in: *e-flux Journal* 59 (2014), http://e-flux.com/journal/59/61130/operational-images (accessed May 7, 2017).

25 Timothy Cullen, *The MQ-9 Reaper Remotely Piloted Aircraft. Humans and Machines in Action*, Cambridge, MA, 2011.

wish to be perceived, how they reflect on their own position within the scope of military interventions, and what role they claim for themselves in this process.

Cullen describes his work in *Human and Machines in Action* as an illustration of "how social, technical, and cognitive factors mutually constitute remote air operations in war".[26] From a methodological perspective, the study is guided both by Bruno Latour's *Science in Action*[27] and by Edwin Hutchins' 1995 study, *Cognition in the Wild*[28], with its concept of situated and socially distributed cognition. The latter study develops this concept based on the example of pre-modern navigation practices in Micronesia, Hutchins' personal observations gleaned from his time serving on the bridge of a US Navy ship, as well as on Lucy Suchman's descriptions from the 1980s of the "use, combination and re-representation" of information in so-called "intelligent machines", for which she introduced the term "situated cognition".[29]

Hutchins' approach lends itself to the description of the military control environment, as humans and things are described as participants in one and the same "system" in both contexts, namely, as participants in "a distributed process composed of emergent interactions among people and tools".[30] At this point, the common epistemic roots of both the US military discourse on technology and the discourse of cognitive science and media anthropology become apparent, each of which, in their own way, can be traced back to their origin in the systems thinking of cybernetics.[31]

Similar to Hutchins, Cullen follows the navigational control sequences as examples of "socially distributed cognition"[32] through describing and mapping the actions of a crew in the cockpit. He finds the concept of computation useful, without which a horizontal description (in contrast to the hierarchical description of a human as the sole actor) would not be possible. According to Hutchins, navigation takes place based on a sequence of activities, "in which representations of the spatial relationship of the ship to known landmarks are created, transformed, and combined in such a way that the solution to the problem of position fixing is transparent".[33] In his view, this results in a generalised definition of computation, a very broadly defined concept in cognitive science, as "the propagation of representational states across a series of representational media".[34] This definition is striking in its rejection of any clear distinction between the media of the representation, whether these are internal images produced by human imagination, a diagrammatic sketch, a map, or a computer-aided model that is shown on a screen. Defining the process of computation as such an act of constituted translation allows the nominal reduction of the friction between senses and sensors, algorithms and human cognition, decision-making, and programming.

26 Ibid., p. 37

27 Bruno Latour, *Science in Action. How to Follow Scientists and Engineers through Society*, Cambridge, MA, 1987.

28 Edwin Hutchins, *Cognition in the Wild*, Cambridge, MA, 1995; Edwin Hutchins, Understanding Micronesian Navigation, in: Dedre Genter, Albert L. Stevens (eds.), *Mental Models*, Hillsdale, NJ, 1983, pp. 191–225.

29 Lucy A. Suchman, *Plans and Situated Actions. The Problem of Human-Machine Communication*, New York, 1987, as cited in Cullen 2011 (as fn. 25), p. 29.

30 Cullen 2011 (as fn. 25), p. 29.

31 See Paul N. Edwards, *The Closed World. Computers and the Politics of Discourse in Cold War America*, Cambridge, MA, 1996. On the cybernetic origins of the cognitive sciences, see Jean-Pierre Dupuy, *On the Origins of Cognitive Science. The Mechanization of the Mind*, Cambridge, MA, 2000. On the history of cybernetics in US military science, see for example Antoine Bousquet, Cyberneticizing the American War Machine. Science and Computers in the Cold War, in: *Cold War History* 8.1 (2008), pp. 77–102.

32 Hutchins 1995 (as fn. 28), p. xii, xiii and 129, see also Cullen 2011 (as fn. 25), p. 30.

33 Hutchins 1995 (as fn. 28), p. 117.

34 Ibid., p. 117.

Building on this idea, Cullen also recognises the drone cockpit as a system in which both humans and automated tools are participating: "[P]ilot, sensor operator, automated tools, and other elements of Reaper were part of a larger computational system that performed in ways specific to the environment and circumstances of operation."[35] However, his characterisation of the operating crew can be seen as an attempt to separate human action from the automated and technologically defined sequences of action.

The processes that are very broadly defined as *computation* by Cullen and Hutchins can be clarified based on the operations that are essential to remote-controlled warfare. Central to operating the GCS are voice, images, and computation, these serve to visualise and synthesise complex relationships, and render them legible on screen – a procedure that Cullen refers to as "building a picture".[36]

Vision at a Distance

The deployment of drones for targeted killing, missions in warfare, and surveillance presents a paradigm for a type of military intervention that is defined and organised by imaging, sensor and network technologies. It is based on a configuration of humans and machines that not only permits seeing without being seen, but also killing in real time without being physically present. On the one hand, this is made possible by the spatial mobility of sensor technologies that are ever more independent from human presence, and, thus, become the preconditions for human decision making. On the other hand, it is based on the almost immediate temporal availability of data provided by transmission and visualisation technologies, through which operations at a distance are moved into the sphere of real time.

Imaging technologies, such as thermographic or electromagnetic measuring techniques, as well as light- or sound-based methods, allow visual and operative access to a situation in the conduct of war. While the pilots of manned aircraft are generally neither in the position to observe a target over the longer term, nor of making it visible at a small distance or in high resolution, the visual practices of drone crews are based on a continuous video stream, which constitutes an important criterion for distinction according to Peter Asaro:

> *In most manned combat missions, the target is simply a set of geographic coordinates that were obtained from another source, such as soldiers on the ground, an aircraft or satellite up above, or the outcome of the analysis of multiple intelligence sources. They also rarely remain close to a target to observe the consequences of their attack, a task called 'battle damage assessment' that is often given to unarmed surveillance aircraft or soldiers on the ground.*[37]

In contrast, drone operations present a visual practice in which vision becomes a cooperative process. Based on the example of the merging of the visualisation process and the ability of human vision, Cullen shows how the awareness of the boundary between human and machine is strategically eliminated in military training and practice. According to Cullen, the operation of the sensor system by the sensor operator will only function successfully when operators dissociate the ability to see from the presence of their own

35 Cullen 2011 (as fn. 25), p. 32.
36 Ibid., p. 117.
37 Asaro 2013 (as fn. 17), p. 14.

bodies: "Instructor sensor operators taught their students to visualize themselves being on the Reaper aircraft, floating above the ground and looking down at their quarry from the belly of the aircraft"[38]. The eye takes the place of the sensor, negating machine action, but also acting as a sensor itself: "A sensor operator's close relationship with the sensor ball helped them to do their jobs well. Experienced sensor operators who 'flew' the sensor ball from an 18-inch monitor became the machine. They became the eye in the sky".[39]

This demonstrates the extent to which the work of the crews is dependent upon the production of visibility. Wherever imagery intervenes between soldiers and the battlefield, the interplay of structures and processes, behind and in front of the screen, are crucial in order to understand how operators act via sensor and imaging technologies. Published video feeds that are based on the sensor data have contributed to making the practice of drone warfare more visible.[40] However, the existence of such material hardly reveals *what* the crews themselves see, and, above all, *how* they saw a given situation. The actors' remote interaction, their diverging perspectives, and the underlying workflows can only be vaguely surmised, if at all. Even so, the essential arguments that are linked to vision at a distance can be revealed based on the debate about these documents.

Advocates of drone use emphasise the aspects related to safety and the minimisation of risk, in particular, as vision and action at a distance do not necessarily require the presence of human actors.[41] Further, they note that this distance implies, above all, the option of the "projection of agency without vulnerability".[42] Soldiers' lives are not at risk during an operation. Additionally, from a military technological perspective, a positive cost-benefit ratio is attributed to surveillance at a distance in comparison to the options for observation by ground troops or manned aircraft.[43] To this end, it is predominantly techniques for visualisation that are listed: the methods for obtaining ISR (intelligence, surveillance, reconnaissance), the argument proposes, produce a continuous and ubiquitous visibility that forms the basis for a clean, almost surgical conduct of a war.[44]

In contrast, critics regard drone technology as no less than "the technical – and technological – solution *par excellence* for the political problem of imperial overreach".[45] They argue that "death of distance enables death from a distance,"

38 Cullen 2011 (as fn. 25), p. 166.

39 Ibid.

40 See, for example, projects such as Forensic Architecture (http://forensic-architecture.org), Dronestagram (http://dronestagram.tumblr.com) or Airwars (https://airwars.org), which visualize the locations and consequences of drone attacks (accessed March 1, 2018).

41 See Peter W. Singer, *Wired for War. The Robotics Revolution and Conflict in the Twenty-First Century*, New York, 2009.

42 David Deptula, *The Use of Drones in Afghanistan*, CNN *Amanpour*, November 24, 2009, as cited in Grégoire Chamayou, *A Theory of the Drone*, New York, 2015, p. 12.

43 See Department of Defense, *Unmanned Systems Integrated Roadmap, FY2011-2036*, 2011, http://acq.osd.mil/sts/docs/Unmanned%20Systems%20Integrated%20Roadmap%20FY2011-2036.pdf (accessed January 8, 2017).

44 See Conor Friedersdorf, Calling U.S. Drone Strikes 'Surgical' is Orwellian propaganda, in: *The Atlantic* 27 (2012), http://theatlantic.com/politics/archive/2012/09/calling-us-drone-strikes-surgical-is- orwellian-propaganda/26292 (accessed January 8, 2017); John O. Brennan, The Ethics and Efficacy of the President's Counterterrorism Strategy, Lecture at the Woodrow Wilson International Center for Scholars, Washington, DC, April 30, 2012, http://cfr.org/counterterrorism/brennans-speech-counterterrorism-april 2012/p28100 (accessed January 8, 2017); Harold Koh, The Obama Administration and International Law, in: *Annual Meeting of the American Society of International Law*, Washington DC, March 25, 2010, https://geneva.usmission.gov/2010/04/01/obama-administration-international-law/ (accessed July 27, 2017).

45 Laleh Khalili, Fighting over Drones, in: *Middle East Report*, No. 264; Pivot, Rebalance, Retrench. The US Posture in the Middle East (2012), pp. 18–21, p. 21.

and that this "replaces one tyranny of geography with another".[46] One of the ramifications of this that is often criticised is that the visual presentation and simulation of the events implies an emotional distance, which leads participants to dissociate from the consequences of their action.[47] For example, the aesthetics, as well as the software interfaces and control instruments of the GCS, are often compared to those of video games; the term, "push-button-war" is frequently used to characterise this phenomenon, in which it is no longer possible to distinguish between simulation and reality.[48] Conversely, arguments are promulgated stating that the spatial proximity and chronological continuity created by the real time video feeds moves the participants far *closer* to events, and, also, renders the consequences of actors' actions visible:

> *You're 8,000 miles away. [...] But it's not really 8,000 miles away, it's 18 inches away. [...] We're closer in a majority of ways than we've ever been as a service. [...] There's no detachment. [...] Those employing the system are very involved at a personal level in combat. You hear the AK-47 going off, the intensity of the voice on the radio calling for help. You're looking at him, 18 inches away from him, trying everything in your capability to get that person out of trouble.*[49]

Peter Asaro suggests that most of these arguments miss the point in relation to the actual complexity of drone technologies, as they only view the agency of the technology in the sense of a simple reaction without considering how the use of the technology also changes human agency.[50] Derek Gregory has pointed out that the visually-conditioned spheres of action of drone operations are not technical, but techno-cultural phenomena, the problems of which can only be revealed through their use.[51] In order to understand the actions and decisions relating to the deployment of drones, it is, therefore, not only their workflows that must be documented, and their consequences that must be made visible, but, also, the pragmatic conditions of the intervention. In this sense, it appears necessary to question the interplay between human and technical actors in relation to how this dynamic engenders or prevents specific forms of vision and visibility; this is the key practice of remote-controlled warfare.

Crafting a View

Based on the example of the live video feed, Cullen describes how the cooperative production of the image in real time becomes a source of identity for the different actors distributed in space: "The feed is distributed and networked. It is the *product* of the aircrew. The crew ties their identity and their worth to this feed."[52] He emphasises the importance of

46 Gregory 2012 (as fn. 3), p. 192.

47 James Der Derian, *Virtuous War. Mapping the Military-Industrial-Media-Entertainment Network*, Boulder, 2001, pp. 9–10.

48 Rachel Plotink, Predicting Push-Button Warfare. US Print Media and Conflict from a Distance, 1945–2010, in: *Media, Culture & Society* 34.6 (2012), pp. 655–672.

49 Col. Pete Gersten, commander of the 432nd Air Expeditionary Wing at Creech Air Force Base, in Megan McCloskey, The War Room. Daily Transition between Battle, Home takes a Toll on Drone Operators, 2009, https://stripes.com/news/the-war-room-daily-transition-between-battle-home-takes-a-toll-on-drone-operators-1.95949#.WYSF0dPyjmE (accessed December 8, 2017).

50 See Asaro 2013 (as fn. 17), p. 5.

51 Gregory 2012 (as fn. 3), p. 190.

52 Timothy Cullen, "MQ-9 Reaper Operations and the Evolution of Remote Warfare", Presentation during the workshop *Technology and Expertise in Remote Warfare* February 1, 2017, Air University, Maxwell Air Force Base, transcript by Nina Franz.

the chat room, in which members of the crew communicate on the content of the feed, in which events are described and interpreted. Vision becomes a cooperative process through this interplay between language and image. The production of the video feed by the crew, as highlighted by Cullen, implies the involvement of non-human actors, because in order to recognise anything on the screen at all, the human observers require context. Context is not provided primarily by the written communication in the chat rooms alone, but "is facilitated with the help of tactical displays, moving maps [...] and chat rooms".[53]

The cooperative production of the feed as an actual task of the crew represents a stark contrast to the role of fighter jet pilots, from whom the US Air Force drone pilots distinguish themselves, sometimes in a polemic form, in their self-image as "desk workers in flying uniform".[54] While aircraft pilots profess to act on an individual, and accordingly autonomous basis, the operators of the GCS are co-authors[55] in an interactive software environment. However, Cullen simultaneously positions this creation of visibility as predominantly a "product of the crew".[56] Contrary to the strategic elimination of human-machine differences that he observes in the training of Reaper crews, human actors remain superior to technology for Cullen: they not only recognise the errors and weaknesses of the technology, but also anticipate these by using their knowledge-based experience, and they develop work-arounds to exploit the technology in individual situations.

This reveals Cullen's own perspective as an erstwhile Air Force pilot, a tradition into which he attempts to enlist the role of drone pilot. Cullen argues in favour of understanding drone crews as *autonomous*, and, thus, decidedly human decision-makers: "They struggled to be human", he states, and, thereby, locates the problem in the transformation "from automatons and technicians into military professionals who viewed the interpretation and manipulation of the virtual world they created as matters of life and death".[57] According to Cullen, this battle for the identity of an acting human subject, who is not simply an element in the technical process, takes place at the level of interfaces that are designed by engineers without adequate input from the users. The design of the control surfaces is revealed to be a point of contention, based on statements by members of the crew:

> *General Atomics engineers initially designed the aircraft to fly autonomously for the bulk of a mission, but pilots modified the ground control station and their procedures to share aircraft control with the autopilot*

53 Ibid.

54 Workshop participants at Maxwell Air Force Base pointed out that the requirement to wear a flying suit in the GCS appeared to them to be slightly absurd. On the modification of the use of the classification of "pilots", and the term Air Force Operator, Cullen 2011 (as fn. 25), p. 20.

55 In this, the situation in the GCS as opposed to the cockpit shares similarities with the set-up that Timothy Lenoir and Sha Xin Wei have described as the "operative theatre" of computer-mediated surgery that is marked by the "necessary cooperation between human and machine". The surgical, like the remote military intervention, is today mediated by a technological infrastructure, in which individual operators are "replaced by software-mediated, machine-human collectivities". Accordingly, the unified authorship of the operator/agent is transformed into "co-authorship" within "interactive 3-D simulations". Timothy Lenoir, Sha Xin Wei, Authorship and Surgery. The Shifting Ontology of the Virtual Surgeon, in: Bruce Clark, Linda Henderson (eds.), *From Energy to Information*, Stanford, 2002, pp. 283–308, pp. 284–285.

56 Cullen, *MQ-9 Reaper Operations and the Evolution of Remote Warfare*, 2017, handout on the lecture with the same title given in the workshop *Technology and Expertise in Remote Warfare* February 1, 2017, Air University, Maxwell Air Force Base, p. 13.

57 Cullen 2011 (as fn. 25), p. 119.

> *in order to maneuver more quickly and destroy a target at a specific time. […] To make the system work for them, Reaper operators determined the time and place to use automated tools; avoided modes of operation known to trigger failure; adjusted to the erroneous behavior of subsystems and other operators; and translated data into formats other humans and machines could receive, interpret, and evaluate […].*[58]

Here, Cullen emphasises the self-reliance of the prospective drone pilots in the 29th Attack Squadron of Holloman Air Force Base and highlights their options for intervention and sovereignty of action in comparison to weapons systems that are designed for "autonomy". According to Cullen, their own expertise in the handling of the control panel allows the members of the crew to intervene in the automated processes and "share" control of the aircraft with the autopilot, such that targets that are being aimed at can be destroyed "more effectively". What is notable here is the fact that it is not those qualities that enable a secure and reliable assessment that are emphasised (for example, to identify a legitimate military target under the laws of war), but those that have the objective of ensuring speed of decision. Likewise, "failure" and "erroneous behaviour" of subsystems and other operators that are designed to prevent intervention do not primarily refer to causing collateral damage or civilian victims.

The drone crew's practice of "translation" of the data produced by computer systems into "formats others could receive, interpret and evaluate"[59] that Cullen observes, follows the definition of computation introduced by Hutchins. While these forms of human *translation* were apparently possible in the Reaper cockpit described by Cullen, it is improbable that this still applies in current and future control stations in the same way. This is because modern computer systems can collect and analyse ever greater quantities of data practically in real time and visualise these data on the screen at an increasingly uncircumventable level of sensory perception; users can only accept or reject in the specific application contexts, but are now hardly capable of comprehending the weight of such data.[60]

Interpretation and Decision

The high resolution, real time video feed that forms the basis for action and decisions taken by drone crews amounts to a use of images that is fundamentally different from previous military practices in planning, surveillance, reconnaissance, and intervention. The circumstance that both control and navigation, as well as the visual access to the area of operations, are guided by images, implies a structurally different image practice to that suggested by the visual methods and competencies of traditional military reconnaissance.[61]

When visualization practices mediate between drone crews and the area of operations, this not only demands an examination of what can be seen and recognised in and on

58 Ibid., p. 119.
59 Ibid., p. 43.

60 Mark Hansen, Feed-Forward, in: Robin Mackay (ed.), *Simulation, Exercise, Operations*, Falmouth, 2015, pp. 57–61, p. 57.
61 On this point, see Antoine Bousquet, *The Martial Gaze. The Logistics of Military Perception in the Age of Global Targeting*, Minneapolis [publication in 2018], therein Chapter 3 Imaging [We would like to thank Antoine Bousquet for sharing an unpublished version of the manuscript]. On the civilian use of image-guided navigation techniques, see Manovich, *The Language of New Media*, Cambridge, MA, 2001, therein The Poetics of Navigation, pp. 259–268; Tristan Thielmann, The ETAK Navigator. Tour de Latour durch die Mediengeschichte der Autonavigationssysteme, in: Georg Kneer, Markus Schroer, Erhard Schüttpelz (eds.), *Bruno Latours Kollektive. Kontroversen zur Entgrenzung des Sozialen*, Frankfurt/M., 2008, pp. 180–218.

these images. How images facilitate, complicate, or even prevent vision and action in an operative process is also a subject for negotiation. In this case, the visualisation of sensor data not only creates static situations for a temporally and spatially subordinate reception, but also essentially guides and controls the actions and decisions of the crews. Drone operations, therefore, correspond to a new type of intervention in which military action is directly guided, or misguided, by what and how images depict or obscure.

The interpretation of sensor data in real time now represents a significant number of decisions taken during military interventions. While the techniques and methods of obtaining information, surveillance, and reconnaissance that are required for this were separated from the use of weapons in conventional warfare, the selection and observation of the target and the decision to kill are now part of the remit of the crew. This confronts soldiers with complex cognitive demands: decisions taken by drone crews are based on a practice that Asaro defines as a "fast-paced multimedia and social media environment of intelligence gathering and killing".[62] The practices of seeing and visualising for the Reaper crews are based on a view according to which humans and machines must enter into a cooperative "partnership"; they must become a "functional system"[63], as outlined by Hutchins.

Even if the active design, manipulation and intervention in the feed, that Cullen describes as "building a picture"[64] and as "growing a video track"[65], still forms part of the core competencies and learning objectives of drone crews, it may become progressively less of a requirement in future operative scenarios where drones are deployed. Given the newer sensor systems, the commonly cited "view through a soda straw"[66], which continues to be opposed to the popular concept of total visibility through omnipotent *seeing* sensor systems, now hardly seems applicable to the visual practices of drone crews. Based on more recent studies, human-based visual scanning of the surfaces of the combat zone will hardly form part of the core competencies of drone crews in the future.[67] Target acquisition and observation is, indeed, still carried out based on restricted fields of view, zooms, and camera perspectives. However, sensor systems, such as the so-called Argus-IS (Autonomous real time ground ubiquitous surveillance Imaging System), now record moving images of areas on the scale of entire cities.[68]

There has been a corresponding rise in the requirement for personnel tasked with data analysis and interpretation. In distributed common ground systems, infrastructures for the processing and analysis of the data from different sensor systems that are firmly established as weapons systems; screeners are employed to monitor incoming data signals and video streams – an activity that is also increasingly being delegated to private companies (fig. 3).[69] In this case,

62 Asaro 2013 (as fn. 17), p. 13.
63 Hutchins 1995 (as fn. 28), p. 170, see Cullen 2011 (as fn. 25), p. 198.
64 Cullen 2011 (as fn. 25), p. 117.
65 Cullen 2017 (as fn. 56).

66 Ibid., p. 130: "The ability of Reaper pilots, sensor operators, and mission coordinators to communicate, develop, and execute a plan to 'look through a soda straw' was a critical skill necessary to defend ground forces and to obtain the best video possible."
67 See Valerie J. Gawron, Keven Gambold, Scott Scheff, Jay Shively, Ground Control Systems, in: Nancy J. Coke, Leah J. Rowe, Winston Bennett Jr, DeForest Q. Joralmon (eds.), *Remotely Piloted Aircraft Systems. A Human Systems Integration Perspective*, Chichester, 2017, pp. 63–109.
68 Defense Science Board, Office of the Secretary of Defense, *Autonomy*, 2016, p. 50.
69 Abigail Fielding-Smith, Crofton Black, Revealed. The Private Firms Tracking Terror Targets at Heart of US Drone Wars, in: *The Bureau of Investigative Journalism*, July 30, 2015, https://thebureauinvestigates.com/stories/2015-07-30/revealed-the-private-firms-tracking-terror-targets-at-heart-of-us-drone-wars (accessed December 8, 2017).

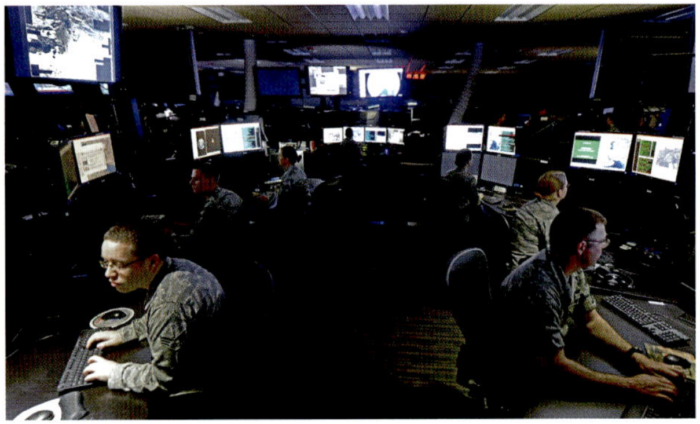

3 Distributed common ground system, US Air Force.

the forensic monitoring is still based on culturally acquired knowledge, as noted by Gregory.[70] However, due to the enormous quantity of video material, the observation of images within the scope of anthropological vision is increasingly impossible as "no human eye is capable of analysing such images with a volume of several terabytes per minute, which is why movement profiles of humans and vehicles are pre-sorted through automated pattern recognition"[71], as Michael Andreas points out. Using a "Global Information Grid, i.e. a communication network or raster that is interconnected with the military databases [...], the interconnection of surveillance drones and precision weapons thereby renews the fiction of military real time [...], the phantasms of which have penetrated through to the acronyms used in military terminology"[72], Andreas argues.

The use of automated software systems for image analysis, for example, to filter out abnormal activities, or to pursue moving targets, demonstrates how seeing and visualising in the deployment of drones cannot be understood as solely the distributed activity of human actors, for example, as a result of the collaborative observation of the screen by a crew. This not only changes "the ideas about who the agent of image production is in situations of war"[73], but the situations in which data is depicted visually at all. Through the establishment of forms of automated recognition and selection, data visualisation may only be required if human actors need to verify or falsify pre-filtered results of calculated decisions, i.e. when they need to comply to juridical routines or military workflows. In other words: the exclusion of the human observer goes so far that action and decision-making do not require images at all.

However, it is not only the visualisation, but also the operationalisation of the combat zone that is increasingly system-controlled. The drone cockpit of the future intends to fuse the division of labour between the sensor operator and pilot – which has existed to date – into labour carried out by only one person.[74] Given the increasing automation of the controlling of an ever greater number of aircraft by ever fewer operators[75], and the division and partition of combat zones into geometric decision spaces, so-called kill boxes, this appears to herald the start of a new, worrying para-

70 Gregory 2012 (as fn. 3), p. 195.

71 Michael Andreas, Flächen/Rastern. Zur Bildlichkeit der Drohne, in: *Behemoth. A Journal on Civilization* 8.2 (2015), pp. 108–127 [translated by the authors].

72 Ibid., p. 114.

73 Carolin Höfler, Eyes in the Sky. Körper, Raum und Sicht im bildgeführten Krieg, in: Martin Scholz, Friedrich Weltzien (eds.), *Design und Krieg*, Berlin, 2015, pp. 13–34, p. 31 [translated by the authors].

74 Maia B. Cook, Harvey S. Smallman, Human-Centered Command and Control of Future Autonomous System, Power Point presentation at the *18th International Command and Control Research & Technology Symposium* Track C2 in Underdeveloped, Degraded, and Denied Operational Environments, June 21, 2013, Alexandria, https://dodccrp.org/events/18th_iccrts_2013/post_conference/presentations/090.pdf (accessed May 7, 2017).

75 Gawron et al. 2017 (as fn. 67), p. 99, p. 102.

digm of control and operation in which the the human factor becomes a precarious element which is only loosely attached to increasingly autonomous processes of computation.

Crisis of Cooperation

Weapons manufacturers, with the Californian company General Atomics leading the way, are still designing GCS control and communication interfaces based on the classic cockpit architecture of manned aircraft, in which a crew acts and observes within a specific aerial space. However, recent research in Human Factor Studies is resulting in the design of the GCS more like platforms, where operators monitor automated processes. Gawron et al. view operators as no longer being capable of translating the increasing quantities of data into actions:

> *[M]onitoring a systems status is burdensome and requires continuous effortful filtering of relevant versus irrelevant information, but emerging technologies can make this a supervisory task by presenting operators only with those alerts that require operator attention, in turn freeing up operator resources for other tasks or even making some monitoring tasks obsolete.*[76]

Cook and Smallman, on the other hand, regard the demand for a new operative paradigm in the design of GCS as justified by the fact that future crews will coordinate numerous activities carried out by different "autonomous platforms and agents"[77] in parallel, instead of tasks being allocated to a single aircraft. In such cases, operators function as supervisory decision-makers,[78] instead of as observers.

In conversation with the US Air Force drone crews on the Maxwell Air Force Base, participants openly discussed their frustration with the fact that more and more responsibility and power of decision-making is being transferred from the operators of the weapons systems to the engineers, i.e. from members of the military to actors who are pursuing fundamentally different, primarily market-driven, interests and have no military responsibility. Criticism was aimed, in particular, at the armaments group General Atomics, which not only manufactures the most frequently used armed drone systems, Predator and Reaper, but which is also responsible for the design of the control stations that are currently used. Complaints focused mainly on the growing *rigidity* of the interfaces and the lack of scope for influence on the system by military personnel, especially when defining the requirements of the GCS.

David Blair and Nick Helms, both US Air Force drone pilots, contrast what they refer to as the *capability*-oriented view of the military users with a *cybernetics*-oriented view taken by developers and manufacturers:

> *From a capabilities view, crew members—in partnership with a fleet of maintainers and support personnel – take 'their' aircraft into the fight to hunt down threats. Conversely, a cybernetics view uses a crew to supply a set of inputs that in turn produces x number of hours of intelligence, surveillance, and reconnaissance.*[79]

76 Ibid., p. 93.
77 Cook, Smallman 2013 (as fn. 74), p. 2.
78 See Gawron et. al. 2017 (as fn. 67), p. 93.
79 Blair, Helms 2014 (as fn. 17), pp. 40–41.

Their "capability perspective" can be assigned to the heroising tradition of classic pilots, who are first and foremost self-reliant and use technology as "amplifiers of human will".[80] Similar to Cullen, Blair and Helms accordingly view the problems of a cybernetics perspective mainly within the context of a diffusion of agency, through which crews become "subsystems within larger sociomechanical constructs", which locks them into closed "control loops"[81] that regulate the systemic variables of specified parameters.

In contrast, a "capabilities perspective", such as is also advocated by Cullen, highlights the "technical" aspect of the skills required for control. In order to "tease out" details from an image, parameters such as "gain, level and focus" must be manipulated manually, or different imaging modalities must be organised in various ways and overlaid.[82] According to this description, that is in contrast to more critical perspectives on what constitutes these *image operations*,[83] what becomes visible is based primarily on the competence of the operator.

Even so, the crew members are convinced that the sphere of influence they are granted is significantly affected by software and hardware engineers and developers, and that these actors are simultaneously determining, to an ever greater extent, what is visualised and how, i. e. what can actually become the focus of attention. The selection process that precedes the workflow in the cockpit is not transparent to the operators, who, as the human actors, bear the responsibility for the decisions taken based on data visualisations, and it is hardly recognisable as such, or even comprehensible. This circumstance is further complicated by applications based on artificial intelligence or machine learning; this is especially true in the case of automation of the data analysis[84] in surveillance missions and for target recognition, wherein the identification of a legitimate target is precisely the critical function, a function that then sets a precedent.

A new Defence Advanced Research Projects Agency (DARPA) programme called Explainable AI (XAI) demonstrates that the US Air Force is aware of the problems posed by increasing automation. The head of the programme, David Gunning, was also responsible for the DARPA programme CALO (Cognitive Assistant that Learns and Organises), whose most prominent spin-off is Siri, Apple's language recognition software. In a public statement on XAI, Gunning explains the objectives of the new programme, the focus of which, as cited in the mission statement, "is the development of a model that will enable human users to understand, appropriately trust, and effectively manage the emerging generation of artificially intelligent partners".[85]

The anthropomorphism that pervades military references to *cooperation*, *partners* and *human-machine teams*, expresses a new turn in the rhetoric of Explainable AI, where not only agency, but also the ability to think is attributed to the machines:

80 Ibid., p. 40.
81 Ibid.
82 Cullen 2011 (as fn. 25), pp. 165–167.
83 See Kathrin Friedrich, Moritz Queisner, Anna Roethe (eds.), *Image Guidance. Bedingungen bildgeführter Operation*. Berlin, 2016; Jens Eder, Charlotte Klonk (eds.), *Image Operations. Visual Media and Political Conflict*, Manchester, 2016.
84 An example of this is provided by the identification of military targets through the analysis of behavioural patterns, see Patrick Tucker, A New AI Learns Through Observation Alone. What That Means for Drone Surveillance, in: *Defense One*, September 6, 2016, http://defenseone.com/technology/2016/09/new-ai-learns-through-observation-alone-what-means-drone-surveillance/131322/?oref=d-channeltop (accessed September 30, 2017); see also Nina Franz, Targeted Killing and Pattern-of-life Analysis. Weaponised Media, in: *Media, Culture & Society* 30.1 (2017), pp. 111–121.
85 Explainable AI, https://darpa.mil/program/explainable-artificial-intelligence (accessed September 30, 2017).

> *Continued advances promise to produce autonomous systems that will perceive, learn, decide, and act on their own. However, the effectiveness of these systems will be limited by the machine's inability to explain its* thoughts and actions *to human users. Explainable AI will be essential, if users are to understand, trust, and effectively manage this emerging generation of artificially intelligent partners.*[86]

The step from the controlling human to "manager", intelligent "partner" or decision-maker, therefore, appears completed, at least rhetorically. It is perhaps no coincidence that Timothy Cullen was asked to act as a consultant on the programme, given his intensive investigation of the practitioner's perspective with reference to the user interface in the increasingly automated ground control stations. He was invited to participate in a working group involving behavioural psychologists from the field of naturalistic decision making, the objective of which is the development of an explanatory model for decision-making processes. In turn, this model is to be used by computer scientists for the development of a "system" to explain the performance of "other systems", such as so-called "deep neural networks".[87]

In this case, *cooperation* remains primarily a matter of technical requirements at the level of model development and the provision of trustworthy technological *partners* that also supply the comprehensibility of their own decisions remains a desideratum. It does not seem unlikely that this initiative is more of a symptom anticipating an apparent crisis of the *human* operator rather than a realistic perspective for opening the black box of highly complex neuronal networks, as Explainable AI seems to do little more than adding further, deeper levels to the operating interfaces providing instructions, and increase the epistemic distance between an automated decision and the human executing the decision.

While the symmetrisation of agency appears to be a progressive notion for thinking the complex relationships between humans and machines in media theoretical reflection, especially in the wake of Latour, the tendency to put human and machines on an equal plain in the context of military operative discourse is recognisable as a rhetoric that nominally reduces the confrontational nature of this co-operation and obscures the influence of powerful actors that are not necessarily part of the command chain. A report by the Bureau of Investigative Journalism from the year 2015 warned that service providers of powerful defense suppliers like General Atomics are increasingly taking over responsibilities that are defined as "inherently governmental functions".[88] This is true for instance, when surveillance missions are outsourced to contractors, as one commentator notes:

> *The Pentagon may not have plans to allow contractors to fire missiles off drones. But allowing them to feed targeting data to the uniformed trigger-puller takes the world one step closer in that direction.*[89]

86 David Gunning, Explainable Artificial Intelligence (XAI), DARPA/I20, Distribution Statement A, https://cc.gatech.edu/~alanwags/DLAI2016/(Gunning)%20IJCAI-16%20DLAI%20WS.pdf (accessed July 30, 2017) [emphasis by authors]; Explainable Artificial Intelligence (XAI), [official website], https://darpa.mil/program/explainable-artificial-intelligence (accessed July 30, 2017). A later version of the statement on the website for the programme replaces "thoughts and actions" with "decisions". Explainable Artificial Intelligence (XAI), [official website], https://darpa.mil/program/explainable-artificial-intelligence (accessed September 30, 2017).

87 Timothy Cullen, email correspondence with the authors on July 22, 2017 and September 10, 2017.

88 Fielding-Smith, Black 2015 (as fn. 69).

89 Laura Dickinson, Drones and Contract Mission Creep, in: *Just Security*, August 5, 2015, https://justsecurity.org/25223/drones-contractors-mission-creep (accessed September 15, 2015).

Adding to that, over the course of the development of increasingly automated control interfaces that form the eye of the needle for a technological authority, and through the introduction of the newest generation of AI technologies, which diverge from the path of strictly rule-based cybernetic models, the Command and Control functions are increasingly disappearing behind a rhetoric of cooperation that no longer only places objects, humans, and algorithms on the same cognitive plane, but essentially removes the agency from the human element within the control environment. This discourse, which attributes the ability to think and act to things and which obliges humans to primarily believe in the explanations provided by technology, pays no attention to the fact that the real decisions have been taken by engineers during the design process: The actors are leaving the control station.

Figures

1 US Air Force, Timothy Cullen, 2010.

2 US Air Force, Christopher Flahive, https://media.defense.gov/2009/Apr/20/2000593687/-1/-1/0/090409-F-0502F-001.jpg (accessed January 2, 2018).

3 US Air Force, 2014, http://af.mil/About-Us/Fact-Sheets/Display/Article/104525/air-force-distributed-common-ground-system (accessed January 3, 2018).

Shannon Benna

Systems and Practices to Produce Stereoscopic Space on Screen

When we as filmmakers, photographers and artists try to create an image, we are bound by the technical constraints of the machines or mediums we use. Through increasingly complex mechanisms or processes, we have been able to reproduce the images we perceive with our eyes more and more realistically. We have also been able to manipulate images to express an imagined aesthetic, intended to convey an artistic vision, an emotion, or a forced perception designed to inspire sympathy/empathy for an unusual perspective. Whether we are discussing black and white, color, still, moving, realistic, fantastical, 2D, or stereoscopic 3D images, we must not only consider how the image is technically created, but also how it is delivered to the viewer. An image viewed on a small, personal screen (like a tablet or cell phone) will create a different relationship with the viewer than the same image viewed on a theatrical screen. Likewise, a shared viewing experience in public will create a different response than one experienced via a virtual reality (VR) headset.

As a filmmaker who has worked in both 2D and 3D media, I find that although 3D imaging provides an avenue to recreating the most realistic version of the world that we see every day, it is not always presented in a way that gives the viewer a real-to-life experience. Additionally, stereoscopic 3D can provide a way for an artist to create a fantastical world-view and present an alternate version of reality to the viewer. I think this is more often the way stereoscopic 3D has been used, however, I don't think it is always because it was the intent of the creator, but due to the way the content is delivered to the viewer. For example, if you create a 3D image that is true-to-life in depth and scale, accurately recreating the way the average person would view another person, or an orthostereo image, you must also present it to the viewer in such a way that they are viewing the image from the same position that the camera was capturing it, and without their being aware the image is being projected. Then the viewer would be unaware they were experiencing an optical illusion. Whereas, if an orthoscopic image is projected onto a screen in a theater, the viewer experiences the image as part of an audience, as opposed to as an one-on-one experience, and the image is now larger than life in scale, rendering it no longer realistic, but fantastical. With the advent of modern immersive media we are given the opportunity to place viewers inside of the world of our choosing in a whole new way. Although VR can give us a complete sphere to visit, only with stereoscopic imaging can the environment become truly immersive in the sense of granting the individual a real-life perception opportunity. Via this presentation method we can engage a viewer in an orthoscopic version of an alternate reality, whether it is realism or fantastical, allowing for a whole new way for artists to connect with their audience.

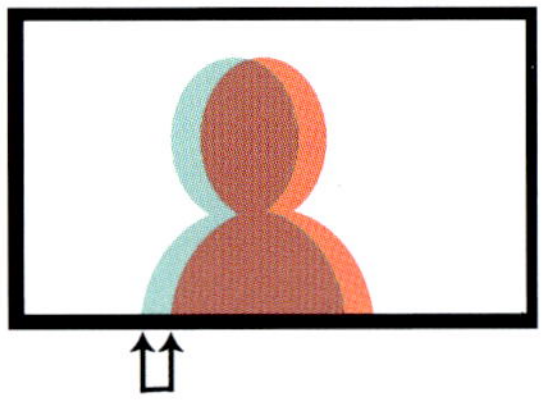

1 Parallax, R/C (all images marked R/C can be viewed stereoscopically with red/cyan anaglyph glasses).

Therefore, to utilize stereoscopic imaging to its fullest capability, we must understand its technical constraints and requirements and we must make a conscious decision about the intended purpose of the images from the beginning. We must not only go through a procedure that ensures the images we create will be viewable as 3D images, and not cause pain or nausea to our viewer, but also choose the right set of tools to capture and deliver our images, accurately expressing our intentions towards reality or fantasy and immersion vs. presentation. Then, once we have chosen our toolset, the subtle adjustments we make to the settings of our tools will convey the emotional and physical sensations we intend, as well as the perspective we present, including the immersive, intimate, spectator and/or voyeuristic quality of the images.

As both a practitioner and an educator of stereoscopic 3D imaging and media production, I believe the key to this and any discipline is communication. Herein lies a bit of a problem, as our terminology is not truly standardized. There are multiple ways to refer to the same effect, error, process, or action, which are not only diversified between production teams or academic institutions, but also between stages of production. The way one speaks of 3D in pre-visualization, development and appreciation is often different than the terminology used in production, or in post-production. There isn't even consistency between editorial, finishing, and conversion, which are all post-production practices. To aid us in this publication, any **terms** will be typeset in a visually different manner for ease of recognition. There is a glossary of terms and accompanying illustrations to facilitate comprehension as well.

Rules for Creating and Manipulating Stereoscopic Space

To begin, one should understand the basic premise behind stereoscopic imaging. It requires two images, a left and right, representative of and ultimately delivered to each of the viewer's eyes, discretely. As we all see in three dimensions everyday (unless we are part of the 6–12% of the population that suffers **stereo blindness**, or an inability to see in depth) we are pre-equipped with the firmware in our brains to take in these two separate right and left eye images and **fuse** them into an all-encompassing perception of space, complete with volumetric objects living at some distance from us physically. This process is called **stereopsis**, and those who can perceive depth possess **binocular vision**.

The amount of separation between the eyes is known as **inter-ocular (IO)** separation, while the distance between the imaging sensors is known as the **inter-axial (IA)** distance. These terms are often used interchangeably and are measured in a variety of different units: inches, centimeters, percentage of screen size, and pixels. Although it may seem extreme, we *are* concerned with pixel accurate measurements, as this can mean the difference between an image that is comfortable to view, and *too deep* for the average person to fuse. Through stereoscopic imaging we are able to *trompe l'œil* or fool the eye into believing these two images, captured from two slightly different perspectives are actually a three dimensional object, landscape or world. The wider the separation between the same object in each of the images when overlaid, or **parallax** (fig. 1), determines the amount of depth in the image. Too much parallax can render the image unresolvable, as the eyes would be forced to **diverge** or strain towards the periphery in order to attempt to fuse the stereo pair into a single image.

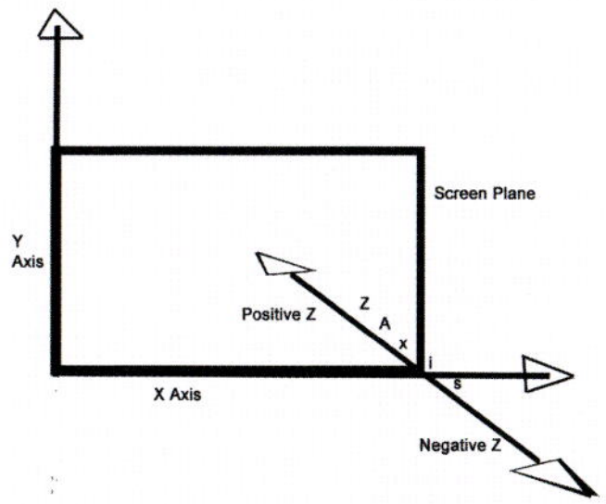

2 Z-Space.

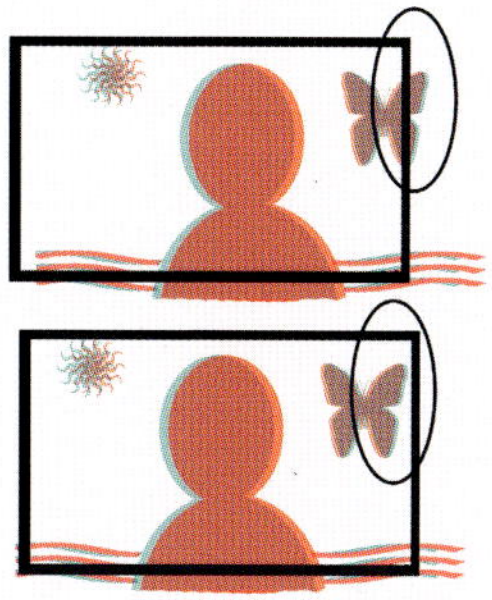

3 Edge violation above, correct framing below, R/C.

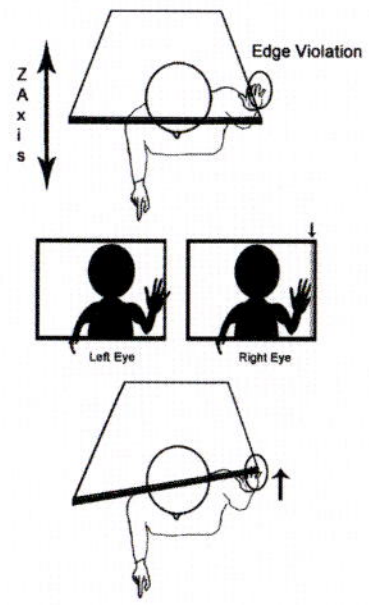

4 Floating window correction.

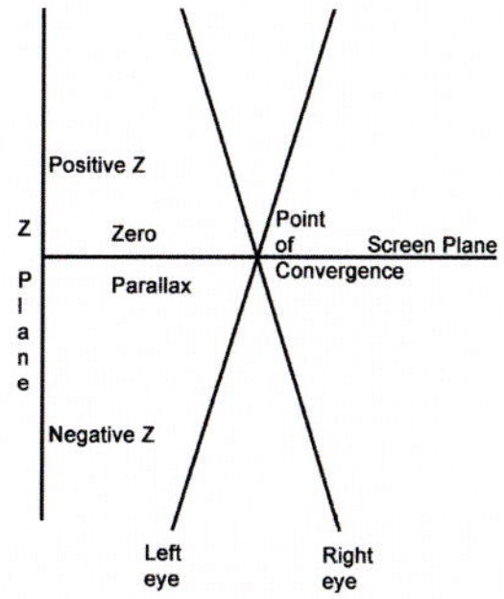

5 Convergence in stereo-space.

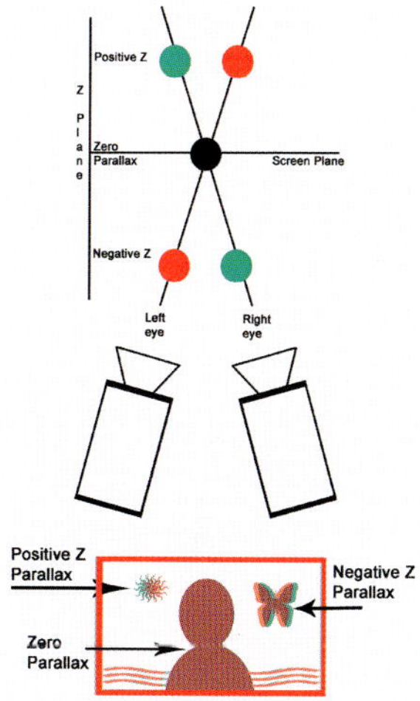

6 Parallax seperation in 3D image, R/C.

Along with being able to control the amount of depth in a stereoscopic image, we are also able to choose where objects are placed in relation to the screen plane, or where they are placed along the **Z axis** (fig. 2). Objects behind the screen plane are in **positive Z space**, while objects that are in front of the screen, or float into the audience's space are in **negative Z space**. It is often difficult to perceive the actual screen plane, and in some cases, especially VR presentation, the screen plane does not truly exist. We do use the concept of the screen plane to describe **zero parallax**, for ease of discussion, especially while creating images and using traditional on-set monitors.

If an object is intended to be positioned in negative Z space, it must not be clipped off in any unnatural fashion but be imaged as a whole. If part of the object exists off screen it must also be in positive Z space. If the object/subject is both partially off screen *and* in negative Z space, the effect is called an **edge violation** (fig. 3) and causes confusion in the brain. A person or thing *cannot* be simultaneously outside of a window and inside at the same time without being angled. We can angle the screen plane to accommodate this creative choice by employing a **floating window** (fig. 4), or a black mask that is applied to the side of one eye to bevel that side of the screen plane backwards in space. This technique is used exclusively in a traditional presentation method, in VR there is no *off-screen*, and thus no **stereo window** to consider.

Objects that appear at the screen plane are at zero parallax and are considered to be the point of **convergence**. The point of convergence is the point at which our eyes cross when we naturally observe the world (fig. 5), or where the two imaging devices' lines of sight cross when capturing images. If we choose to set convergence during the initial imaging stage, we are said to be shooting **converged** and we set a degree of rotation that the cameras will turn in towards each other for each shot to decide where objects/subjects will be positioned along the Z axis (fig. 6).

It is not necessary to converge the lines of sight when initially imaging. If we do not, it will give the effect that all of the image appears in negative Z space and is referred to as shooting **parallel** (fig. 7). Using this method, all of the image

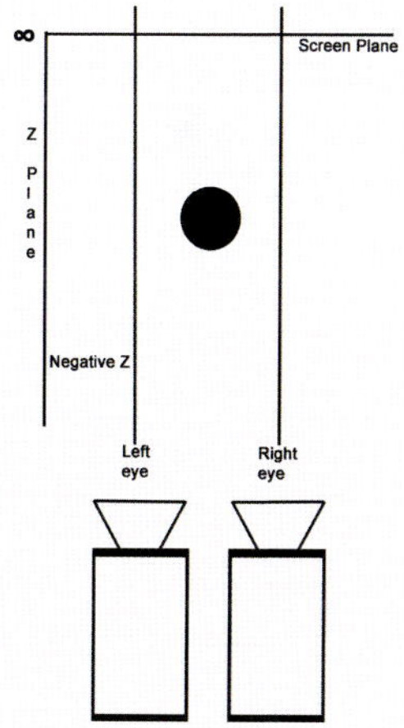

7 Shooting Parallel.

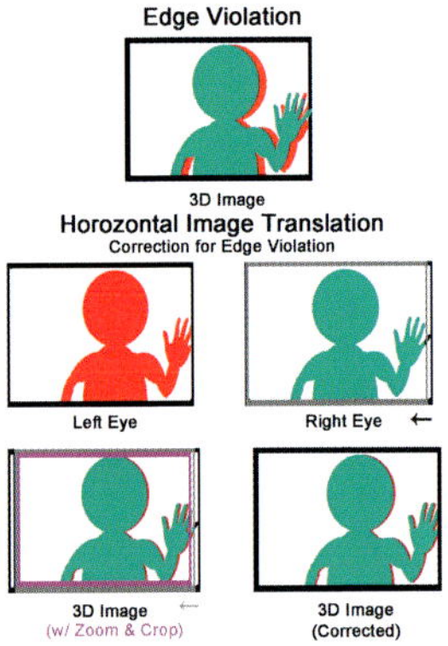

8 Horizontal Image Translation correction for edge violation, R/C.

9 Stereo Pair, R/C.

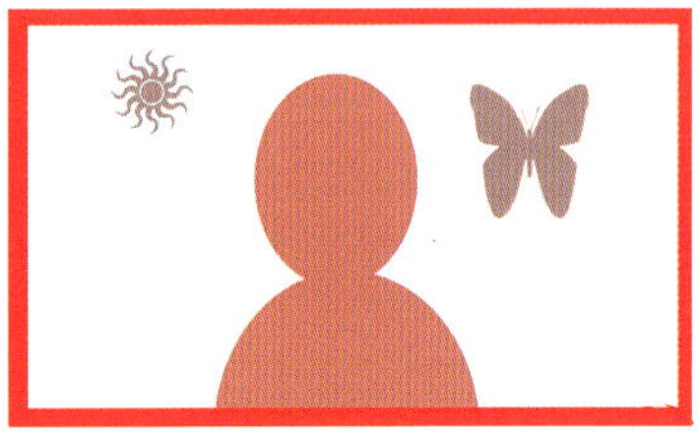

10 Monoscopic / 2D Image, R/C.

can be pushed out into the audience's space, all the time, like in an IMAX 3D film. When shooting parallel, the point of convergence can also be chosen during the post-production process. However, there will be a loss of resolution and an alteration of framing to compensate for the change in overlap of the images when zooming in and cropping the image to eliminate superfluous edges. An adjustment to where the images overlap and where the subsequent point of convergence exists, is called a **Horizontal Image Translation, or H.I.T.** (fig. 8).

Stereoscopic Imaging Toolsets

Below are brief descriptions of five ways to capture stereoscopic images, using either one or two cameras. The choice of toolset will ultimately be decided by physical restrictions, as well as by the desired effect of the images to be captured.

Any of these toolsets can be used in a parallel or converged configuration, dependent upon the intended immersive or invasive quality of the content, and the chosen delivery format(s). Projects shot and delivered in parallel will only be viewable on a Giant Screen, where the viewer is relieved from looking at all of the screen at the same time and can choose to fuse smaller portions of the image selectively. If the content is also going to be available on home theater or TV screens, the images will all have to be adjusted to push some of the image back into positive z space, so there is not too much parallax separation present in all parts of the screen to be fused at one time.

Single Camera Methods

2D–3D conversion: Images captured with a single camera then post-production processed to create second eye view. This process can be very useful when **native capture** (stereoscopic imaging as produced using a naturally occurring **stereo pair** (fig. 9) or left & right eye images) is not practical, i.e., when physical size, scale diversity, or camera/rig mounting and/or operating space is very restricted. Although conversion can be expensive and time consuming, it can also be very helpful when creating fantastical scenarios, as the creation of a second eye image from the existing **monoscopic** (2D), or single eye image (fig. 10) can utilize more than one inter-axial setting at a time, creating more complex geometry than is achievable when using physical cameras in a physical environment.

Stop motion/stepper motor: Two images captured by a single camera/lens, in succession, but after shifting perspective between the first and second images. The amount of shift between images is equivalent to the inter-axial distance. The subject must remain static during the acquisition of both images intended for a single stereo pair, and the camera must only shift along the X axis. The stereo pairs are then strung together in an image sequence and played back at speed, each stereo pair equal to one frame in *frames per second*.

Dual Lens/Camera Methods

Single body/dual-lens camera systems: There are a variety of stereoscopic 3D video/digital cinema camera options raging in size, weight, image compression method and quality, but with fixed inter-axial distance and permanent lens choices. These devices allow for less setup and no **alignment** necessary or possible (some aberrance in hardware may require post-production geometry and/or color correction). Dependent upon the inter-axial distance, the cameras innately restrict the amount of depth an image can have, as the distance from subject and/or the furthest background point is limited to avoid excessive parallax. They do grant extreme freedom and mobility for the content creator, although they limit creative flexibility. These solutions are ideal for ENG (Electronic News Gathering), Reality and documentary situations, where time and crew/equipment space is limited. One must camera test to assure the fixed inter-axial does not prohibit the necessary distance from the subject to create both ideal framing and intimacy/immersive effect, as well as viewable 3D images.

If the fixed inter-axial distance is too wide, the camera will not be able to approach near enough to the subject to get a close-up shot and will leave the audience feeling an emotional separation from the topic. Also, it could create too large of a parallax separation in the background elements, causing a divergence among viewers. This effect becomes exaggerated in an immersive delivery method, such as the cinema or a VR set-up.
Light levels that will be standard for the subject matter must also be considered as some camera systems handle low light better than others. If a lack of light is innate in the subject, one should be aware that a greater than usual amount of light is recommended for stereoscopic image viewing, as there is light loss when projecting through **polarizing** or color filters, and the viewer is wearing 3D glasses. Dim images can cause eye-strain, and low lumen projection in theaters has been the cause of much 3D media criticism and distaste. Additionally, lighting techniques for 3D production are more demanding, as accenting individual areas along the Z axis create **depth cues**, or planes of light for the viewer to distinguish, creating a richer depth environment.

Side-by-side rig (fig. 11): Dual camera configuration utilizing a simple bar with both cameras oriented next to one another. This is convenient for shooting larger, spectacle style subjects with an increased potential inter-axial distance but limits cameras from getting any closer to one another than the physical body and or lenses will allow.

Shooting with an exaggerated inter-axial setting can create an effect called **hyperstereo** (fig. 12), which allows us to perceive depth in environments or objects that are very large or further away than we are able to observe depth in naturally. Due to the average inter-ocular separation of human eyes (about 6.3–7.6 cm or 2.5–3 inches) we are restricted from observing depth in anything more distant than 18.3–30.5 m or 60–100 feet away.

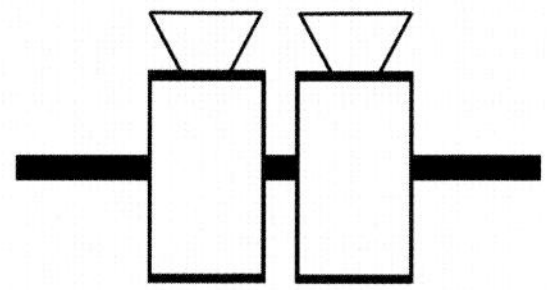

11 Side-by-side rig.

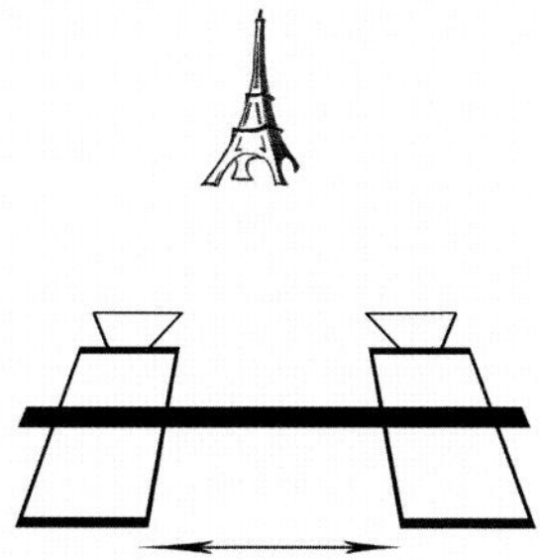

12 Hyperstereo.

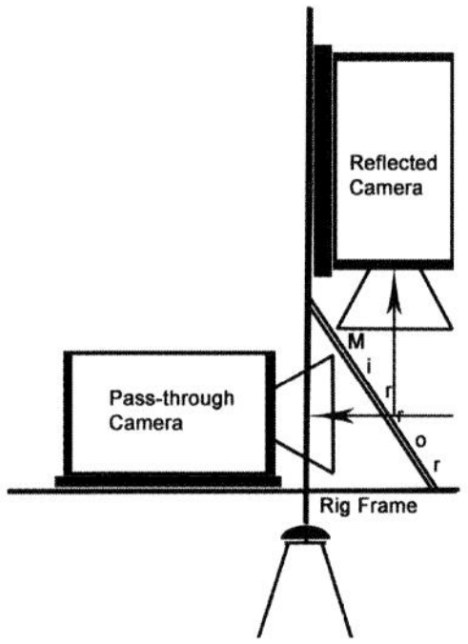

13 Beamsplitter rig.

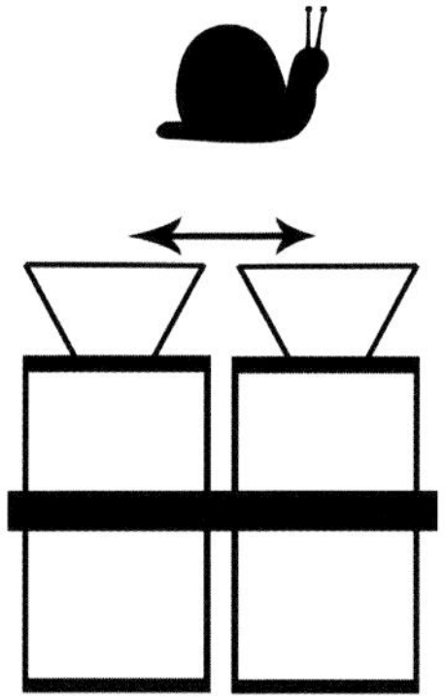

14 Hypostereo.

While this process is ideal to allow the viewer to see depth in grand vistas, large objects, architectural structures or natural wonders, when used to perceive depth in normal sized objects that are a great distance from the camera, an effect called **miniaturization** can occur. This effect makes objects and people seem very small, like toys, or as if the viewer were a giant. This can be used creatively to simulate a fantastical scenario, rather than to observe sporting events etc. from a distance, as miniaturization will pull the viewer out of the experience and remind them they are watching something artificially produced.

Beamsplitter rig (fig. 13): Dual camera configuration utilizing a mounting bracket with mirror to reflect the desired image into one camera lens (**reflected camera**), while the second camera looks through the mirror to capture (**pass-through camera**), allowing for a far narrower inter-axial separation than the physical cameras would be able to achieve if mounted next to each other. There is significant light loss due to the mirror in the rig, about 1.5 to 2 stops decrease, so lighting conditions must be of significant consideration when choosing to shoot with a beamsplitter.

Shooting with a narrower inter-axial separation than our average inter-ocular separation of 2.5 inches will allow for use of extreme/close-up shots, which are a standard part of cinematic storytelling when we want to convey focus or attention to a specific object or to intensify the emotional connection with the subject. It is not a realistic situation, as we rarely put our faces very close to an object or person when we are paying attention to a particular thing but is useful as a cue to the audience that something important is happening.

The narrowness of the inter-axial setting must be used with care, because although it can help us to get closer to a subject, an effect called **hypostereo** (fig. 14) can result, in which one can perceive the world as if one were very small. This effect can be used creatively to immerse the viewer into the world on a microscale, like documenting insects, surgical procedures, electronics construction or granting the viewer a first-person POV as the inhabitant of a fantasy world of tiny people like in Jonathan Swift's *Gulliver's Travels* (Benjamin Motte, 1726).

Once we are able to communicate verbally about our intentions and desires concerning the stereoscopic images we wish to produce, we can begin to choose the toolset, pipeline and workflow that will help us to achieve these aims.

Stereoscopic Imaging Systems Set-Up and Variables

No matter what camera or rig is chosen for the job, there are a consistent set of parameters that must be adhered to in order to ensure the two images that are produced are useable. The two images must be exactly matched in every way, except for their slight difference in vantage point along the X axis. Any discrepancy between the images is called **retinal rivalry** (fig. 15), as it can cause eye strain, because the viewer will be seeing something in one eye but not seeing it in the other. Each of the following items listed must be precisely matched and aligned to achieve a perfect stereo pair. It is useful to label each piece from the cameras, lenses, and media cards to cables card to cables etc. as *Left* and *Right*. Further, one should maintain use of each item in the same position consistently as it alleviates confusion and allows for easier troubleshooting if a problem arises. One basic guideline is to eliminate potential variables at every opportunity.

Cameras should be the same make, brand, and have the same firmware version. They should ultimately be able to genlock. The frame rate and shutter speed/shutter angle settings should be the same on both cameras. Exposure and color temperature should also match as closely as possible, but note that the mirror will cause a color and light shift in one camera. Final exposure and color balancing will need to be done in post-production.

Lenses should be a matched pair (as close as possible), lens mapping helps with and aberrations in the physical elements, and using wider lenses is preferred to long lenses as zooming compresses space, which manifests in 3D images as **cardboarding**, or the flattening of objects so they appear like cardboard cutouts. Prime lenses (lenses with only 1 focal length) are most preferable, although for speed in production, wider zoom lenses (lenses with multiple focal lengths, but not to exceed 50 mm ideally) can be used as vari(able)-prime lenses, allowing one to have multiple focal length choices without having to change the lenses on the rig and re-align every time. In order to perform the initial alignment, zoom lenses must be set to the same actual focal length. Although the lens marking may denote a particular focal length, the optical reality of 50 mm being exactly the same on different lenses is very improbable. A visual confirmation of zoom scale must be checked.

The back of the lenses should be the same distance from the sensor when attaching them to their respective cameras. Not all lenses are adjustable, but most professional digital cinema cameras do have the option to adjust back focal distance. Also be sure to focus both cameras to properly align them.

FIZ motor calibration is essential if we desire to have dynamic framing or to zoom in or out during a scene in 3D imaging. These motor systems are used in 2D production to control Focus, Iris, and Zoom of the lenses remotely. The same motor systems can be used during stereoscopic production, but two sets will be needed (one for each camera). They will need to be calibrated to work in unison if they are

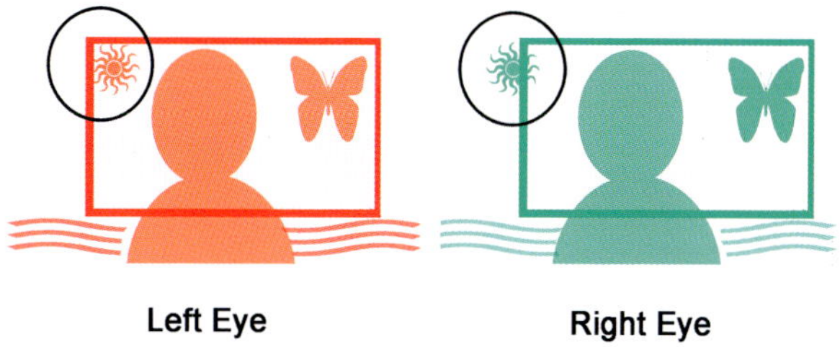

15 Retinal rivalry, R/C.

to match the images in a stereo pair. A third FIZ control can be used on some beamsplitter rigs to motorize the adjustments for inter-axial and convergence.

Filters may be used but must be matched in density and orientation. The most commonly used filter is a quarter wave retarder, placed in front of the mirror in the beamsplitter. This protects against polarization issues, which result from shooting reflective surfaces from slightly different angles. As the reflection will look different in the left and right eye, when viewed as a stereo image, strobing will occur and take the viewer out of the moment or cause eye-strain. This effect can be used creatively but should be intentional.

Timecode is important for both imaging and post-production practices. Some cameras generate their own timecode, while others must be fed timecode. This provides not only a point-of-reference for post-production to find specific events within footage, but also allows for multiple camera shoots to have a common standard. It also provides a regulated cadence by which the cameras may be synced.

Genlock is a process by which two cameras may be linked together. Regardless of the shutter speed chosen, the cameras must open and close shutter at the exact same time. Genlock guarantees the two cameras will maintain

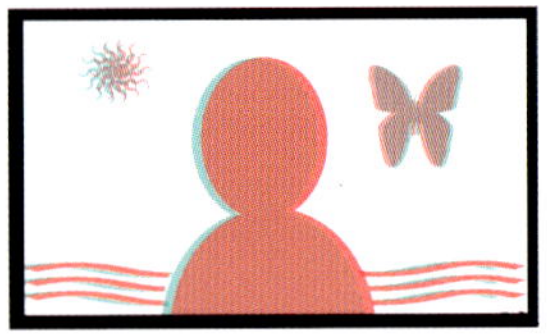

16 Anaglyph 3D Image. R/C.

their sync with each other, so that the left and right cameras capture the exact same frame without any discrepancies being introduced by the shutter angle. If not there will be temporal differences and the images will be unusable. Also, if a camera system is not genlocked but both cameras are started at the same time, the shuttering tends to drift out of sync over time.

There are a few configurations one might apply to the cameras being linked, but in stereoscopic imaging, we tend to use the *master/slave* configuration. This allows changes on one camera (master) to automatically be applied to the other camera (slave), so we are certain the settings are always the same on both cameras and eliminate user error.

Keeping cameras in sync is not just concerned with shuttering, but also with scanning. Film cameras capture images by the opening and closing the physical aperture, allowing light to pass through the iris and strike the film, leaving an impression. Digital cameras have two types of shutters available: A **global shutter**, which behaves like a film camera and takes one massive impression each time the camera shutters, or a **rolling shutter**, which basically acts like a scanner or copy machine, scans the lines of resolution in succession starting at the upper left hand corner and ending with the lower right corner. When using cameras with a rolling shutter, the scan progression must be in sync as well. This is what we are referring to when we describe the cameras as being in phase. There are some additional issues when using rolling shutter cameras including dealing with fast motion. One arises when the first part of the image is scanned before the last part of the image, thus causing motion artifacts. Additionally, one or both cameras on the rig are sometimes mounted in inverted positions (because of mounting necessity), so the cameras may actually be scanning in the opposite direction.

When an inversion of image occurs, whether it is because one of the cameras on a beamsplitter is recording a reflected image and it appears backwards, or because the camera is mounted upside down, that image must be flipped and/or flopped to allow for alignment and monitoring during production. **Flip/flop** can be done in camera (if the option is available on the camera) or at the monitoring level. It does not matter which stage it is done at, but it must be done at some point, either during initial image capture, or during the post-production process.

Some specialty monitors exist for onset 3D monitoring and make the process much easier. They usually have multiple modes that are helpful during alignment, and when setting inter-axial & convergence. They range in size from 5 inches (which would be used for onboard camera monitoring for the operator) to large TV size, which would be used by the stereographer when setting the depth in each scene or by the director to view and compose shots and on set action or playback and review. They can be fully polarized stereoscopic monitors (that use polarized lens glasses) or an **anaglyph** (fig. 16) display (which utilize the red/cyan glasses). Polarizing filters orient the left and right eye images' light waves to only pass through their correlating glasses' eye. This same technology is used in theaters and on 3D TVs.

There are also some pass-through devices that will output a 3D image in a variety of formats, although anaglyph images are able to be viewed on any standard monitor or computer screen. Some form of 3D monitoring is essential for at least the alignment stage, and is highly recommended during production, although with the use of complex calculations, inter-axial and convergence can be set without it. This will definitely increase on-set production time and will often increase the required amount of post-production corrections.

Alignment Settings on Beamsplitter Rigs and in Post-Production Correction

Alignment is one of the most important aspects of 3D imaging, as it can be the difference between communicating an emotional message via an image on screen and having a meaningful experience lost in translation due to a technical aberration that prevents it being viewed comfortably. As there are so many variables in these imaging systems, and the adjustments that effect the outcome are so minute, the following procedure is very helpful in establishing sound geometry for image capture. Because we are working in three-dimensional space, any defects will create alignment issues, and render the images un-viewable. Although some corrections can be made during the post process, some aspects are permanent and thus must be matched as closely as possible before imaging.

The following are the basic points of alignment and/or geometry correction:

1. Mounting: When mounting the cameras onto the rig, they must be square and level, meaning the camera must be exactly perpendicular to the mirror at a 90 degree angle and when a bubble level is placed on top of the camera rig it should be level to the ground.
2. Horizontal: During the alignment, set the horizontal separation between the cameras to zero (zero rig); this will produce the exact same image in both eyes, essentially creating a 2D image. Once the images are exactly the same, as you proceed through the process of alignment, your images will be geometrically matched, and the rig will be properly aligned for stereoscopic imaging.
3. Scale: The images must be at the same scale in order to match geometrically. Therefore, when mounting the cameras on the beamsplitter rig, the lenses should be equidistant from the mirror and zoomed to the same focal length (optically). Also, the images output from the cameras should be in focus, as it is nearly impossible to ascertain pixel accurate alignment in an out of focus image.
4. Vertical alignment: This setting is sometimes controlled by adjusting the height of the camera. If this is the case, the potential for pitch and roll issues arises. To combat this, many rig designers place the vertical alignment adjustment at the mirror.
5. Pitch: This is the measurement of the height of the front or back of the physical camera body as mounted to the rig. As some internal sensors are not always physically squarely seated in the camera, adjustment of pitch can help to correct for this geometrical issue. However, it is very easy to use this to compensate for misalignments created via another variable and thus make the geometry incorrect in either the background or foreground. Ideally one should be able to zoom from one end of the lens to the other without loosing alignment, but as lenses have physical weight, they can shift center of gravity when on one end or the other, so one must align for the object of interest on a shot by shot basis.
6. Roll: This is a measure of the camera body's flush mounting to the rig as well but is concerned with the tilt left or right.
7. Rotation: As one camera either hangs down or points upward into the mirror, the camera can have a tendency to pivot on the mounting plate.

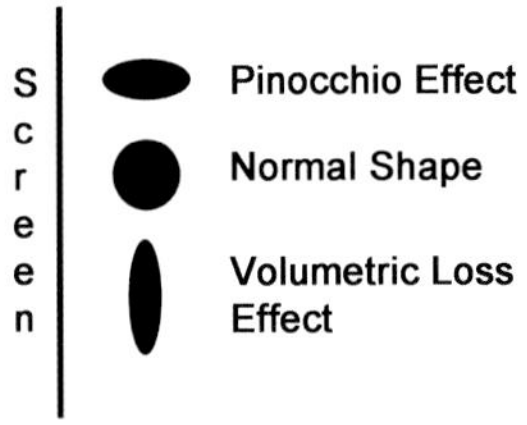

17 Distortion effects.

Process of Formulating Depth in Images on Screen

After properly aligning the beamsplitter rig, we are ready to begin capturing stereoscopic images. We need to make a series of creative choices to decide how to set the inter-axial and convergence (if we are shooting converged). There are many methods that involve complex processes of measuring and mathematical computation. However, a simple method for creating natural and comfortable 3D is called the Natural Depth Method, developed by stereoscopic filmmakers Alan & Josephine Derobe. This process creates stereoscopic images by choosing a natural convergence point for the viewer to focus on, then balancing the amount of parallax separation between the negative and positive Z spaces. The procedure is as follows:

1. Set IA to roughly 2.5 inches.
2. Set the convergence on the object/subject you want to be at the screen plane.
3. Check the parallax of the nearest object/subject, at the closest point it will pass in front of the rig, to make sure it is not too diverged to be fused.
4. Check the parallax of the furthest object or background for excessive disparity. (If the background is out of focus, there can be more forgiveness, as the audience will not be attempting to fuse a blurry image.)
5. Decrease the IA to adjust for acceptable amount of parallax in both.
6. Re-converge.
7. Repeat until entire image is comfortable to view.
8. Finally, when parallax is within acceptable **depth budget** (allowed amount of separation, usually measured in percentage of screen size, as dictated by creative design or distribution channel):
9. Check for distortion within the subject/object of interest.
10. If there is too much IA + convergence, objects will distort, elongating along Z axis, creating a *Pinocchio effect* (fig. 17).
11. If there is too little IA + convergence, object will distort elongating along X axis, either flattening the image, or rendering a lack of volume or roundness (fig. 17).

Controlling Parallax for various Delivery Formats

A larger size of screen proportionally increases the amount of depth a given parallax separation will display. There are often several versions of a 3D product that have been re-converged for each distribution channel. For example, media intended to be viewed on a cell phone, personal tablet or small monitor will have greater parallax separation than the version distributed for movie theater projection. Conversely, projects designed for Giant Screen Cinema (IMAX) will be unviewable on a smaller screen. Most feature movies/documentaries are routinely reconverged before being released for home theater consumption.

Creative Design in Producing Stereoscopic Images

When designing a piece of stereoscopic media one must take into consideration the demands of production, as per the physical elements (scale of subject, size of environment, size of budget, desired pace of production, un/predictive nature of subject matter), but also the way the eye receives information and the time it takes the brain to process the stereoscopic images into a three-dimensional environment. Often to ease this process, we will divorce from traditional cinematic practice and embrace a cinematic theatre approach. This means longer takes of action in which we move within space as if the camera is a character/anthropomorphized, rather than executing the traditional set-ups (establishing shot, two-shot, over the shoulder, coverage, etc.). This allows for a more natural viewing experience, as if we were there, as opposed to being reminded that we are viewing a captured image. This serves us well in both orthoscopic imaging as well as fantastical settings and helps the viewer to stay immersed in the scene.

If we do adhere to traditional set ups, it is advisable to use longer shots and cut less often when editing, as it takes the brain several seconds to fuse a new shot. One should also attempt to transition from one shot to the next while having the point of interest at the same position along the Z axis, at least at the end of one shot and the beginning of the next. The convergence point, or placement along the Z axis can be dynamic, softly easing the viewer from one shot to the next. Once the images have been fused, the subject can visually move freely within the stereo space, and in and/or out of the **stereo window** (fig. 18), into positive or negative Z space.

The use of more or less depth as well as placement of objects/subjects along the Z axis can render many diverse psychological effects on the viewer. There is no set of defined rules by which one can construct specific reactions. On the production side we discover them as the stereoscopic medium is explored. Dynamic inter-axial and/or convergence can be used to draw a viewer into the image or to push objects away from the viewer, creating a sense of dramatic/emotional tension, or loss. Objects/subjects that are pushed into negative Z space can be used as a gimmick, to throw something into the viewer's face or to create a sense of intrusion into the audience's space, while placement of an objects/subject in positive Z space can create a sense of voyeurism. Increased depth in a scene, with the subject in the distance, can make a viewer feel as though the space is vast and provide a sense of occhiolism, while increased depth with the subject and environment filling up the scene can create a sense of claustrophobia. A diminished depth when observing a subject or their environment can provide a sense that something is lacking in that person's life, while realistic volume and depth can create a sense of well being and richness in the subject's life. The freedom of the camera to move within the three-dimensional space can create a sense of weightlessness or buoyancy, while a static or locked-off shot can provide the physical sensation of restriction or captivity. The only true way to learn these tricks is to watch as much stereoscopic content as possible, both with glasses on and off. When an emotional reaction is noted, look at the image without glasses to ascertain the parallax and convergence setting. Also, there is nothing like practicing and experimenting to help one grasp the subtle art that is **stereography** or designing depth for expression on screen.

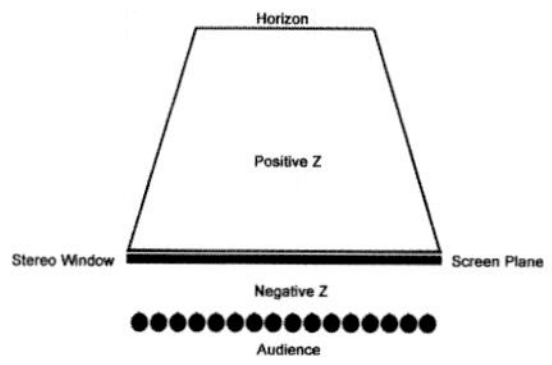

18 Stereo window in display environment.

Figures

1–18: Shannon Benna.

Glossary of Stereoscopic 3D Terms

2D-3D conversion Postproduction process by which a second eye view is generated from a single eye image.

3D image A volumetric image consisting of a left and right eye view, delivered discretely to each of the viewer's eyes, and perceived to have depth. Also called **stereoscopic 3D**.

Alignment The matching of geometric elements/configuration when assembling a stereoscopic imaging system, or the postproduction process of correcting geometric mismatches in image pairs.

Anaglyph A method of displaying stereo images that utilizes contrasting color images & filtering glasses to discretely deliver left and right images to their respective eyes.

Beamsplitter rig Dual camera configuration utilizing a mounting bracket with mirror to reflect the desired image into one camera lens (reflected camera), while the second camera looks through the mirror to capture (pass-through camera).

Binocular vision The ability to perceive depth while viewing with two eyes.

Cardboarding An effect in stereoscopic imaging that comes from shooting an object/subject at great distance while zoomed in using telephoto lenses.

Converged a method of stereoscopic imaging where the cameras will turn in towards each other a specified degree of rotation for each shot. This decides where objects/subjects live along the Z axis.

Convergence The point at which our eyes cross when we naturally observe the world, or where two imaging devices' lines of sight cross.

Depth cues Points of interest along the Z axis, accenting layers of depth in space.

Depth budget The sanctioned amount of overall parallax separation in a 3D image, as dictated by creative design or the distribution channel.

Diverge The effort to fuse two images by straining towards the periphery.

Edge violation An object/subject that is both partially off screen and in negative Z space.

FIZ Motorization & controls for the focus, iris, and zoom of a lens.

Flip/flop The vertical/horizontal re-orientation of an image. (There is not a standardized direction = term.)

Floating window A black mask that is applied to the side of one eye to bevel that side of the screen plane backwards in space.

Fuse The physical process of combination of separate images in the brain, creating a single image, complete with volumetric objects living in space at some distance from us physically.

Global shutter Digital imaging shutter that captures all of the image at once, like a traditional film aperture.

Horizontal Image Translation (H.I.T.) Post-production process where the convergence point is changed in a shot, by shifting one or both of the images left or right in relation to one another.

Hyperstereo A method of stereoscopic imaging using exaggerated inter-axial settings, intended to give depth to objects at a great distance.

Hypostereo A method of stereoscopic imaging using very narrow inter-axial settings, allowing for extreme close-up or macro images of very small objects/subjects.

Inter-Axial (IA) The distance between the nodal centers of acquisition point on digital imaging sensors.

Inter-Ocular (IO) The amount of separation between the pupils of the eyes.

Miniaturization A side effect of shooting hyperstereo, where objects appear to be miniaturized.

Monoscopic A single eye or 2D/flat image.

Native capture Stereoscopic imaging using 2 cameras or lenses to create a left and right eye image.

Negative Z space In front of the screen plane/stereo window.

Orthostereo A realistic depth image, reproducing the natural scale and volume of an object/subject.

Parallax The separation between the same object in the left and right images of a stereo pair when the images are overlaid.

Parallel A method of stereoscopic imaging where the cameras do not tow in towards one another, placing the screen plane at infinity.

Pass-through camera The camera that looks through the mirror of a beamsplitter to capture an image. Also called the Direct Camera.

Polarized A method of displaying stereo images that utilizes special filters to orient the left and right eye images' light waves. Electrons from the left eye light waves are spun to the left, while electrons from the right eye light waves are spun to the right. Left and right polarizing filters are worn in front of the viewer's eyes, which only allow the correlating light waves to pass through.

Positive Z space Behind the screen plane/stereo window.

Reflected camera The camera that captures the image reflected in the mirror of a beamsplitter.

Retinal rivalry When all or part of an object appears in one eye of a stereo pair, but not in the other.

Rolling shutter A digital imaging shutter that scans lines of resolution starting with the upper right corner and finishing in the lower left.

Side-by-side rig Dual camera configuration utilizing a simple bar with both cameras oriented next to one another.

Stereo blindness An inability to perceive depth, when viewing with two eyes.

Stereo pair A left and right eye image that when viewed create a 3D image.

Stereo window The screen plane boundary between the negative and positive portions of the Z axis.

Stereography The creative art and technical science of creating stereoscopic 3D images.

Stereopsis The perception of depth by fusing two images into one volumetric image.

Z axis Towards or away from the audience.

Zero parallax At the screen plane.

Jens Schröter

Viewing Zone

The Volumetric Image, Spatial Knowledge and Collaborative Practice

Volumetric Images

James Cameron's *Avatar* (2009) is undoubtedly not the most original contribution to science fiction cinema – think of the stereotypical the-good-indians-against-the-evil-capitalist-plot. Nevertheless it was hailed as a major contribution to cinematic techniques because of its use of stereoscopy. While the use of stereoscopy might be quite conventional in Avatar, there is a sequence, which is quite interesting and convincing (fig. 1a–c).

The sequence shows a kind of control room, a "center of calculation"[1], as Bruno Latour would have pointed out, or a "center of coordination" as Lucy Suchman notes:

> *Centers of coordination are characterizable in terms of participants' ongoing orientation to problems of space and time, involving the deployment of people and equipment across distances, according to a canonical timetable or the emergent requirements of rapid response to a time-critical situation.*[2]

Information from the outside is collected and processed to generate strategies for action – in the case of Avatar: the *evil capitalists* trying to convince the native species on the planet to leave their village, which is located in a very big tree, because of the deposit of a demanded material with the funny name *Unobtainium* in the ground. The scene in fig. 1a–c shows a visualization of a landscape based on fictional (or nearly fictional, I will come back to that) display technologies, which render an image space even more convincingly by the stereoscopic presentation of the film. However, the shown fictional images are not stereoscopic: they are *volumetric*, no glasses are needed to see a spatial image, which can be seen from all sides, is colorful and half-transparent. It has several properties:

1. It represents space not only like an image based on linear perspective, it is itself spatial. That means it is a post-perspectival image: it does not need to represent the scenery with the aid of perspective because it does not render the scene on a plane. It is a *transplane image* (this is similar to holography, but volumetric displays are normally not holographic).[3] Thereby it avoids the spatial problems of perspectival representation: By foreshortening, perspective

1 Bruno Latour, *Science in Action. How to Follow Scientists and Engineers through Society*, Cambridge, MA: Harvard University Press, 1987, Chap. 6.

2 Lucy Suchman, Centers of Coordination. A Case and Some Themes, http://citeseerx.ist.psu.edu/viewdoc/download?doi=10.1.1.627.4590&rep=rep1&type=pdf (accessed September 19, 2017).

3 See Jens Schröter, *3D. History, Theory and Aesthetics of the Transplane Image*, New York, NY: Bloomsbury Academic, 2014.

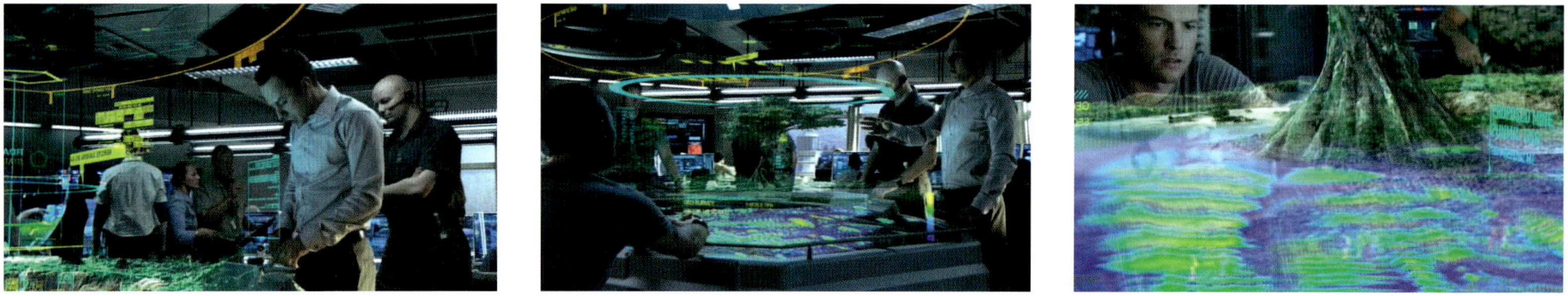

1a–c Screenshots of Avatar, TC: 00:47:49, 00:48:03, 00:48:10.

changes the relative length of all lines to each other and it changes all angles – and in this sense a perspectival image is not a very reliable representation of space (except under very controlled conditions). That's why technical and architectural drawings made for the construction of objects (and this is still true for the images in the instruction manuals which accompany IKEA furniture) are not in linear perspective but in different forms of *parallel perspective*.[4]

Another way to avoid the distortions and misrepresentations of perspective is to make the image itself spatial, which of course already begins with sculpture or scaled models made from wood or clay used by engineers and architects for example. The fictional volumetric computer display in *Avatar*, designed by the special effects company Prime Focus, stands in a long tradition of images which avoid perspective to represent a situation undistorted and therefore in a more efficient and operational way.[5] It is not surprising that we see military personnel in the scene. Military decision-making requires representations, which allow immediate understanding of spatial situations and structures.

2. This shows another important aspect: Such a transplane image is ideally suited for collaborative work in which decisions often have to be made. People stand around the representation, see the scene from different angles, point to specific aspects and discuss what to do. Of course this can be done (and is done) with flat images as well, and normally these are preferred for the simple reason that they are more available, but the technical arrangement (in *Avatar*) is especially helpful in situations where the structure of space plays a central role. In this sense the display shown is in the tradition of, what in German is so beautifully called, a *Lagebesprechungstisch* or a table for discussing the situation. There is for instance a long tradition of planning and education in the military done with sand tables (fig. 2–3).[6]

One important point that connects with the aforementioned aspect is that a spatial representation of this kind does not align viewers in the same way a screen (or a linear-perspectival representation on a screen) does. Linear-

4 See Jeffrey Z. Booker, *A History of Engineering* Drawing, London: Chatto and Windus, 1963.

5 Prime Focus World, http://primefocusworld.com/projects/ (accessed September 30, 2017).

6 See Hans Hemmler, *Die Arbeit am Sandkasten*, Aarau: Sauerländer, 1942.

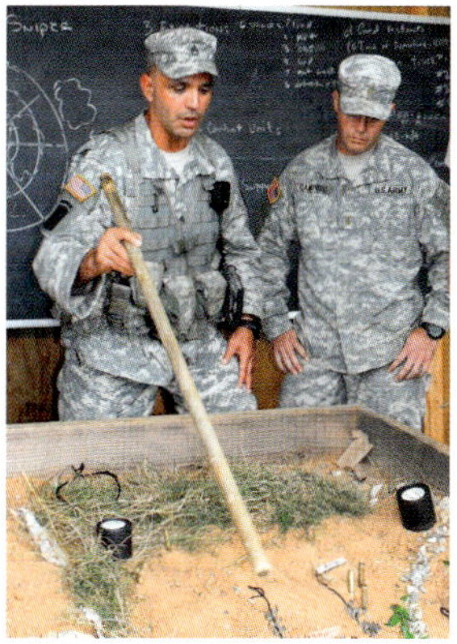

2 Sand table (conventional).

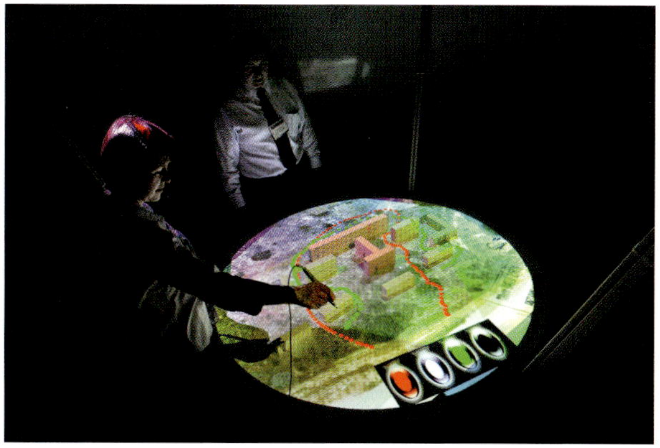

3 Sand table (virtual).

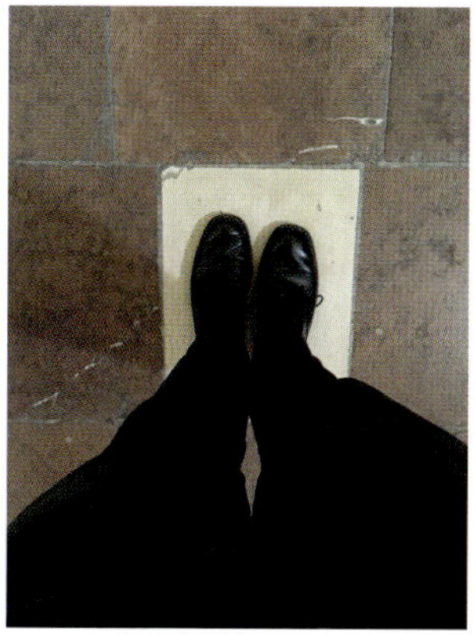

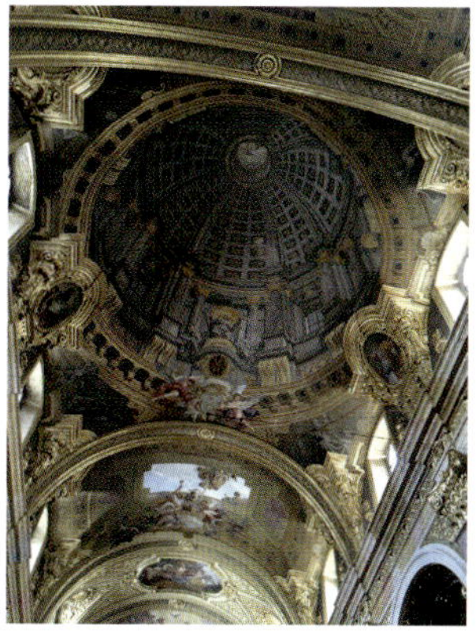

4–5 Jesuit church in Vienna.

and even more central perspectives, with one vantage point, position the viewer at the eye point – at least in principle: Extreme cases are, for example, the remarkable trompe l'œil dome in the Jesuitenkirche in Vienna, executed by Andrea Pozzo in 1703, which looks correct only from one standpoint. And that point is explicitly marked on the floor (fig. 4–5).

Screens usually direct the attention in the direction where the screen is. Spatial displays like a sand table or the fictional strategic volumetric display in *Avatar* do not prescribe any specific positions for the viewer (except of course that you have to look towards the display). The representation can be seen from different perspectives – in the literal, and that's the point, also in the metaphorical sense. It does not privilege any position but rather opens up a viewing zone (the direct positioning of the viewer is also avoided in parallel perspective, which is used in, e.g., simulation games, like Sim City, and is often called god's eye view, because it prescribes no position. The gaze comes from no specific place at all).[7] The representation in *Avatar* can be scaled up and down, it is enriched with further information, and it can be scrolled and rotated and therefore allows differentiated analysis. As it was put in one review of *Avatar*:

> *The Holotable was a lovely way of displaying an interactive map. As opposed to being displayed on a screen, the 3D hologram allows people to view the map from all angles and have a better sense of scale. It's also much better than a physical model, as it allows users to see the internal structure of the terrain and to point inside the model, it's also more movable, updatable, can be endless and takes up less room. There's countless ways this can be used to display information.*[8]

7 On different forms of parallel perspective see Benjamin Beil, Jens Schröter, Die Parallelperspektive im digitalen Bild, in: *Zeitschrift für Medienwissenschaft* 4 (2011), pp. 127–137.

8 HUDS+GUIS, The Design of Avatar UI, https://hudsandguis.com/home/2011/01/16/the-design-of-avatar (accessed September 20, 2017).

The display in *Avatar* is purely fictional, but that makes it no less interesting: The fictional representation in *Avatar* is a projection of a futuristic technological practice intertwined with real developments. Often there is an immediate connection, like when technology developers become film consultants. Kirby uses the notion of the "diegetic prototype" to designate fictional technologies that operate "normally" in the diegetic world of a film.[9] Some people explicitly connect the displays in *Avatar* to real developments: As a technician discussing the possibilities of realizing such a "holo table" puts it in a blog post, beautifully titled "holo-tables-avatar-style-are-cool": "The solution, both hardware, software and computer power, seems pretty expensive. An elliptic 3D display used horizontally as a 'holo table', but price aside, it seems doable with today's technology unless I'm missing something."[10] And then he adds: "Zebra Imaging, a long-time producer of 3D holographic prints has been awarded a contract by DARPA back in 2005 to develop a real-time interactive holographic display map. The Urban Photoinc Sandtable Display (UPSD) is the result of that. It supports up to 20 participants, 360 degree view points, 12 inch depth and displays that scale up to 6 feet in length, enabling full Parallax without requiring special glasses or goggles."[11]

Sheila Jasanoff argues: "Science Fiction [...] is a repository of sociotechnical imaginaries, visions that integrate futures of growing knowledge and technological mastery with normative assessments of what such futures could and should mean for present-days societies."[12] Often movies show futuristic technologies that become central and driving metaphors for a certain line of technological development – as became the fictional "Holodeck" from Star Trek for the discourses in the development of virtual reality in the nineties.[13] This is an example for how an imaginary technology directly becomes part of a development process.

Fictional three-dimensional displays exist in popular cinema as well as in technical papers. But in both of these discursive fields they have different functions. In popular cinema the representation of future technology can work – as I analyzed in my PhD and as was discussed in the already mentioned text by David Kirby – as a means to produce the desire for potential new technologies in mass audiences, normalize them as parts of a potential future and in this way help developers to receive funding, which historically *has happened*.[14] In that sense Kittler once described popular cinema as an instruction manual for new media.[15]

In technological papers such fictional entities help to orient researchers towards a common goal and help to get funding too, think of the role of projected technological artifacts in patents. In this sense the fictional presentation of future technological practices is by no means external or secondary but rather an integral part of the development of

9 David Kirby, The Future Is Now: Diegetic Prototypes and The Role of Popular Films in Generating Real-World Technological Development, in: *Social Studies of Science* 40 (2010), pp. 41–70. See Jens Schröter, Das Holodeck als Leitbild, in: *Bildwelten des Wissens* 14 (2018), pp. 90–99.

10 Arie Tal, Augmented Reality Science-Fiction vs. Science-Fact: Are We There Yet?, http://augmentech.blogspot.de/2012/12/holo-tables-avatar-style-are-cool-and.html (accessed September 20, 2017). The author alludes to https://zebraimaging.com (accessed September 20, 2017).

11 Ibid.

12 Sheila Jasanoff, Imagined and Invented Worlds, in: Sheila Jasanoff, Sang-Hyun Kim (eds.), *Dreamscapes of Modernity. Sociotechnical Imaginaries and The Fabrication of Power*, Chicago/London: The University of Chicago Press, 2015, pp. 321–341, p. 337.

13 See Jens Schröter, *Das Netz und die virtuelle Realität. Zur Selbstprogrammierung der Gesellschaft durch die universelle Maschine*, Bielefeld: transcript, 2004.

14 Ibid.

15 See Friedrich Kittler, Synergie von Mensch und Maschine, in: Florian Rötzer, Sarah Rogenhofer (eds.), *Kunst Machen. Gespräche über die Produktion von Bildern*, Leipzig: Reclam, 1993, pp. 83–102, p. 101.

technology. In the next part I will analyze the steps in the actual development process of volumetric display technologies and how fictional representations of potential future practices are to be found in developer's accounts and other forms of *diegetic prototypes*.

History of Volumetric Displays

In 1948 a paper called *Three-Dimensional Cathode-Ray Tube Displays* by Parker and Wallis appeared, where they state:

> *Since the screen of a c. r. tube [= cathode ray tube] is only two-dimensional, only two coordinates of the object's position can be thus directly displayed. This has until relatively recently been adequate, the radar set being called upon to scan in only a single angular coordinate, usually with a 'fan beam', but the modern set may scan in two angular co-ordinates with a 'pencil beam'. It is with these volume-scanning radar sets, where the object's position in three coordinates is derivable, that we are concerned here.*[16]

Obviously, the concern here is to represent spatial information in a three-dimensional way and – since we are dealing with radar – to achieve this as fast as possible in critical situations where decisions have to be made quickly. "When a human operator is involved in the loop, however, all the n channels have to pass simultaneously through the bottleneck of his senses, consciousness and movements."[17] The slow *human operator* thus has to get optimal information on space. This can also be seen in a paper published in 1963, regarded as an important early text: "A real need exists for a three-dimensional display in almost any spatial navigation problem, whether it is through water, air, or outer space. Faster and faster vehicle velocities have outmoded visual navigation, even when direct visual observations are possible. [...] The navigator's ability to react should not be limited by his position display."[18]

The solution to the problem of the ineffectiveness of the human operator could be to develop a real three-dimensional display: "A truly three-dimensional display is one in which the echoes appear as bright spots in an actual volume of light, at points representing the spatial positions of the corresponding objects."[19] This is the decisive point in volumetric displays. The image is not being created on a plane, nor on two, as in stereoscopy; it is created in a volume. As a result the image is perceived as spatial. How can this be done? According to the authors,

> *[t]he echoes are displayed in the volume of light as bright spots, by an intensity modulation of the c. r. t. spot. The deflections must be suitably synchronized with the scan of the aerial beam, in order that the echoes may appear consistently at points representing the objects' spatial positions. The deflection produced mechanically can be either 'real' or 'apparent'. An example of the former would be obtained if the c. r. tube itself were moved axially. This is, for mechanical reasons, undesirable. A similar effect can be obtained, however, by projecting the c. r. t. picture on to a moving*

16 E. Parker, P. R. Wallis, Three-Dimensional Cathode-Ray Tube Displays, in: *Journal of the IEEE* 95 (1948), pp. 371–390, p. 371.
17 Ibid., p. 379.
18 R. D. Ketchpel, Direct-View Three-Dimensional Display Tube, in: *IEEE Transactions on Electron Devices* 10 (1963), pp. 324–328, p. 324.
19 Parker, Wallis 1948 (as fn. 16), p. 372.

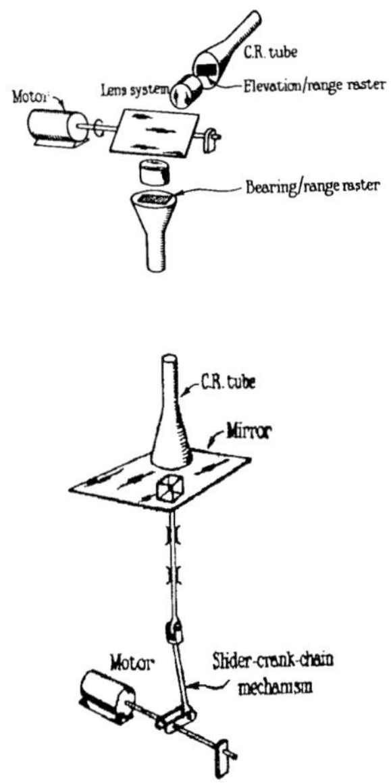

6 Early Diagrams for a *Moving Screen* and a *Moving Mirror* Display. Two Fundamental Forms Of Volumetric Display Of The *Swept Volume*-Type.

> *screen. An 'apparent' deflection can be obtained, for example, by observing the c.r.t. picture in a mirror which is moved in a suitable manner.*[20] (*fig. 6*)

Here, Parker and Wallis describe two fundamental types of the class of volumetric displays that create the volume of the image with movable parts ("swept volume"). In the first case the screen is rotating and the light-points are projected onto it. In the second case the plane is multiplied into a volume through a translational moving mirror. This means that similar to film, volumetric displays function on the basis of the series of physiological optics with the addition of the third dimension. Human perception visualizes a three-dimensional image produced by the fast succession of projections onto the rapidly moving planes. It can (in principle) be viewed from all sides without additional glasses. Contrasted to geometrical optics, this plane is being moved, thereby becoming transplane. The image then appears in the volume, described by many authors as image-space or image-volume. Obviously, these are very primitive concepts to realize a volumetric image – at least compared with the presumably computer-generated smooth image in *Avatar*. *Avatar* shows that technological progress is, of course, unavoidable – and so we are in the midst of the discursive level of these imaging technologies.

Even though (or maybe because) volumetric imaging technologies require extensive funding, fictional, phantasmagorical ideas are surrounding this topic even in research projects. In a text on volumetric displays from 2004 Rieko Otsuka and others state:

> *The motivation for this work is the dream of realizing real stereovision images in space. Most of us remember the scene in the 1977 movie 'STAR WARS' in which the robot R2-D2 projects a three-dimensional image of Princess Leia, who begs Obi-Wan Kenobi for help. Besides 'STAR WARS,' there have been many movies that contain scenes in which holograms appear [...]. These films indicate a desire or a premonition in many of us to see this kind of technology brought to life.*[21]

Even though the "desire for 3D" does not have to be hyped into an anthropological constant, it seems that it is an actor in this matter.[22] Here we can already glimpse that fictional representations do play a role in orienting research towards certain goals to be achieved. Fig. 7 shows a (fictional) representation of a volumetric display from a text published in 1989. This sort of centralized traffic control room is remarkably similar to representations of futuristic displays in *Star Wars Episode III – Revenge of the Sith* (USA 2005, George Lucas) (fig.8).

Specific institutions subject a distant place to analysis and bring it under control with the help of volumetric displays. Bruno Latour has argued that the "simple drift from watching confusing three-dimensional objects, to inspecting two-dimensional images which have been made less confusing" is a central technique of producing knowledge.[23]

20 Ibid.

21 Rieko Otsuka, Takeshi Hoshino, Youichi Horry, Transpost. All-Around Display System for 3D Solid Image, in: *Proceedings of the ACM Symposium on Virtual Reality Software and Technology*, Hong Kong, 2004, pp. 187–194, p. 187.

22 Meaning the thesis, that there is a naturally given "desire for illusionistic images", which sometimes seems implied in positions like André Bazin's The Myth of Total Cinema, in: idem., *What is Cinema?*, Berkeley: University of California Press, 1967, pp. 23–27.

23 Bruno Latour, Visualisation and Cognition. Drawing Things Together, in: *Knowledge and Society. Studies in the Sociology of Culture and Present* 6 (1986), pp. 1–40, p. 15.

"No matter what they [the scientists, but one could also say: the military, J. S.] talk about, they start talking with some degree of confidence and being believed by colleagues, only once they point at simple geometrized two-dimensional shapes."[24] However, as the above quoted discussions on volumetric displays have shown, it is at least problematic if, for locating something within an image, collaborative viewing as well as the discussions and the decision-making connected with it are perhaps more successful with three-dimensional representations. Through these display technologies space or spatial constellations themselves become "immutable mobiles", in the sense of Latour.[25] In this way a spatial situation is opened up to discussions and control.

The literature on volumetric displays goes beyond our current discussion, but in these texts one can repeatedly find commentaries on the viability and necessity of volumetric display technologies:

> *With vendors lowering the barrier to adoption by providing compatibility with new and legacy applications, volumetric displays are poised to assume a commanding role in fields as diverse as medical imaging, mechanical computer-aided design, and military visualization.*[26]

Military and medical visualizations are the most mentioned fields of applications. Most often the goal is to control spaces filled with people or to control the human body itself. According to Blundell and Schwarz the means of control is usually a god's eye view in which the user either observes

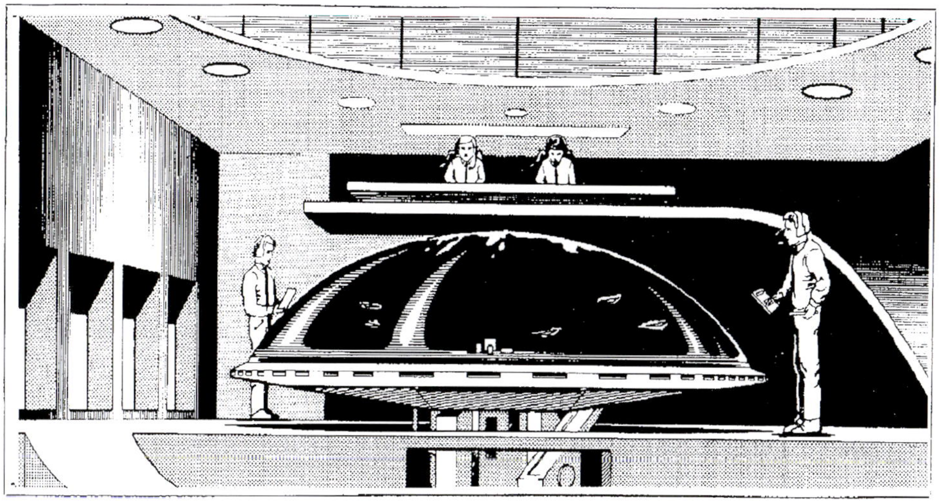

7 Fictional volumetric display used in a sort of traffic control.

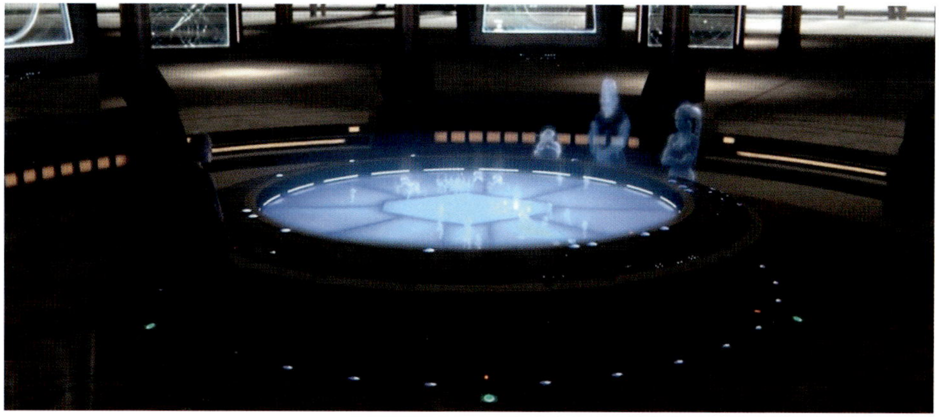

8 Fictional volumetric display from Star Wars Episode III – Revenge of the Sith.

24 Ibid., p. 15–16.
25 See Ibid.
26 Gregg E. Favalora, Volumetric 3D Displays and Application Infrastructure, in: *Computer* 38 (2005), pp. 37–44, p. 37.

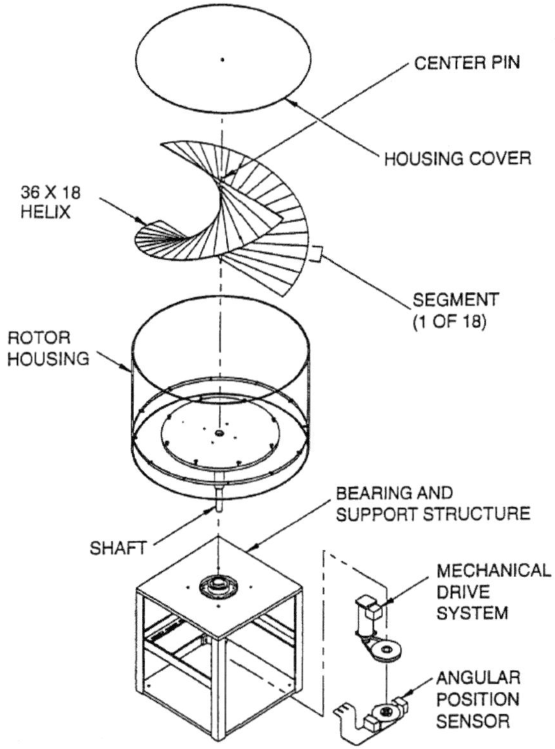

9 Volumetric Display in which a transplane image is projected onto a rotating plane in form of a helix.

a space from outside or is able to effortlessly penetrate the body: "In the case of all volumetric display systems known to the authors, the generation of images occurs within a containing vessel from which the [observer] is excluded. Volumetric systems therefore provide a 'God's-eye' view of any image scene."[27]

To conclude I will discuss a paper on a military development of a volumetric display, which is rich in fictional projection. Fig. 9 shows a volumetric display in which a transplane image is projected onto a rotating plane in form of a helix.

In the paper the authors discuss possible usages of this display: "A logical application for the 3-D volumetric display is for control and management of air traffic in a volume of aerospace for the FAA, Air Force, or Navy."[28] It should be noted that in this military setting that only men are watching the display and thereby direct their controlling gaze on the targets (be they hostile or friendly), although women are not excluded from the military in the USA (fig. 10). The corresponding text explains:

> *The Department of Defense Science and Technology Initiative identifies seven thrust areas. One of these is Global Surveillance and Communications, a capability that can focus on a trouble spot and be responsive to the needs of the commander. A three-dimensional display of the battle area – such as the LaserBased 3-D Volumetric Display System – will greatly facilitate this capability. Tactical data collected for command review can be translated and displayed as 3-D images. The perspective gained will contribute to quicker and more accurate decision-making regarding deployment and management of battle resources.*[29]

27 Barry Blundell, Adam Schwarz, *Volumetric Three Dimensional Display Systems*, New York, NY: John Wiley & Sons Inc, 2000, p. 4.

28 Parviz Soltan et al., Laser-Based 3D Volumetric Display System, in: Richard M. Satava et al. (eds.), *Interactive Technology and the New Paradigm for Healthcare*, Amsterdam et al.: IOS Press, 1995, pp. 349–358, p. 356.

29 Parviz Soltan et al., Laser-Based 3-D Volumetric Display System (The Improved Second Generation), 1996, http://handle.dtic.mil/100.2/ADA

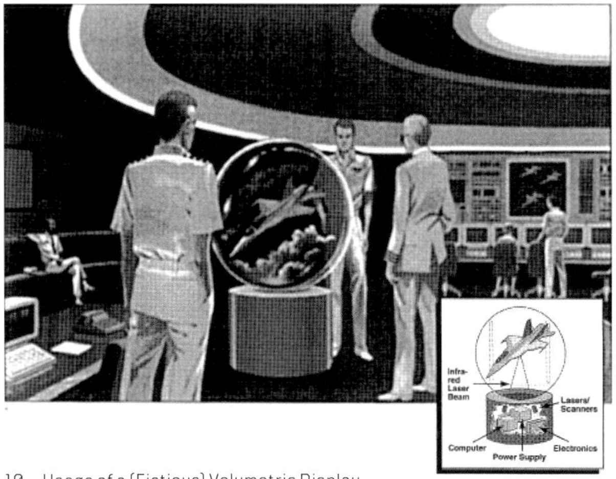

10 Usage of a (Fictious) Volumetric Display for a Command-and-Control Situation.

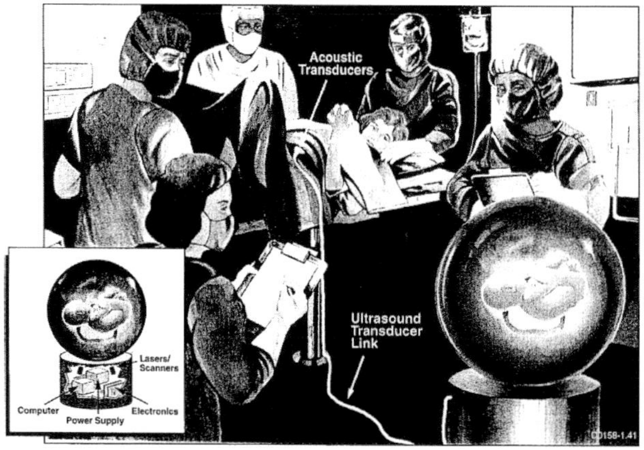

11 Usage of a (Fictious) Volumetric Display for the Control of a Female Body during Birth.

This is quite explicit.

In fig. 11 it is a woman who is at the center – but in this case she is the object of the medical gaze via the volumetric display. It seems as if there is a gendered bias of the space-controlling gaze.

This is only a very small fraction of the rich literature on volumetric display technologies, but some elements are clearly visible: These displays aim to produce truly three-dimensional and therefore post-perspectival images. These images render scenes and situations, in which the spatial structure is of paramount importance, without the distortions of perspective. Because the images do not prescribe and situate viewers in any strict sense, they are ideally suited to be seen, discussed and used by a group of people in collaborative work.[30] And the discourses on these images are full of fictionalized projections about their possible future uses especially in military and medical practices, centered on control and surveillance. These potential image technologies are understood as assembling and situating people in a situation room or a *center of calculation* to control a situation.

Such projections of futuristic medial practices are of course more revealing about the present than about any future to come. Obviously, all the different extrapolations on volumetric display technologies converge in that they are mainly used for strategic planning, control, surveillance,

306215 (accessed August 30, 2007), p. 17.

30 Of course: In difference to flat images, where everyone sees the same image (even if not standing on the position prescribed by the perspectival construction), while watching volumetric images everyone in a group of observers sees a different image – as would be the case with real three-dimensional objects. That the image does not prescribe a certain viewpoint opens up the space for several viewpoints, which can be brought into discussion.

analysis – in *Avatar* to plan the exploitation of Pandora by the military-industrial complex. *Avatar* is explicitly about anthropocracy, about humans trying to establish their control over the resources of a foreign world by, amongst other things, three-dimensional displays. The following part will analyze the cultural imaginary of anthropocratic power as sedimented in the discursive history of volumetric displays – be they real or fictional.

Volumetric Display as Symbolic Form

Erwin Panofsky's famous paper *Perspective as Symbolic Form* ends with the sentence: "It is thus no accident if this perspectival view of space has already succeeded twice in the course of the evolution of art: the first time as the sign of an ending, when antique theocracy crumbled; the second time as the sign of a beginning, when modern 'anthropocracy' first reared itself."[31] Here anthropocracy is directly connected with (central-)perspectival representation. Panofsky's argument is, put simply, that perspective shows the world *as seen* or at least: *approximately as seen by a human observer*, man becomes the center of the shown world: "This view of space [...] is the same view that will later be rationalized by Cartesianism and formalized by Kantianism"[32] – and, as we all know, in Descartes the cogito is the only secure knowledge and in Kant the world appears only according to the transcendental structures of consciousness. Man is in the center. Panofsky argued that perspectival representation is an expression and/or one performative realization of an anthropocentric worldview.

First there is the obvious problem that it might be too big an argument to correlate a certain form of representation with a certain anthropocentric episteme. Might not the anthropocentrism of perspective be more gradual and depend on different practices with perspectival images? These were arguments already made to criticize so-called *apparatus theory.*[33]

Closely connected with this is, secondly, a more fundamental problem with Panofsky's argument. Perspective was invented in the renaissance. But it was only decades later, namely in the 19th century, that industrial modernity had its big takeoff, an anthropocracy if you will, resulting nowadays in ecological disaster. And here is the central point: Very important forms of representations to be used in this upheaval for the constructing of technologies, for increasing the effectiveness and speed of individuals in decision making etc. were not at all structured by linear or central perspective, but were, as I said, parallel-perspectival or material 3D models or maps – e.g. in engineering drawing, meaning the transfer of technological knowledge or architecture. Although parallel perspective and maps on the one hand and real volumetric representations on the other are of course different in that the former two are still forms of projection, where the latter is not (it is more a scaling and filtering), they are similar in that they do not imply a positioned viewing subject.[34] Does that mean, when we follow Panofsky's argument, that they are not anthropocentric because they do not imply a viewing (although one-eyed) body? Yes, perhaps that is what it means – but in a very special sense: The forms of power relevant during the renaissance

31 Erwin Panofsky, *Perspective as Symbolic Form*, trans. by Christopher S. Wood, New York, NY: Zone Books, 1991, p. 72.

32 Ibid., p. 66.

33 See Hartmut Winkler, *Der filmische Raum und die Zuschauer. 'Apparatus' – Semantik – 'Ideology'*, Heidelberg: Carl Winter, 1992.

34 An exception may be the implied position of potential observers in sculpture.

were mostly feudal forms, implying personalized forms of power, e.g. peasants in serfdom to their landlords or power concentrated in the two bodies of the king, according to Kantorowicz.[35] Panofsky's perspectival anthropocracy is a personalized form of power, the world made to conform to the gaze of an idealized person. One shouldn't forget that in some types of baroque theater architecture the emperor or another type of king had the only place from which the perspectival scenery on the stage was completely coherent – here the body of the emperor and the eye point of central perspective were *literally* matched.[36]

But in modernity, as has often been noted, personalized power disappears and new objectified, anonymous forms of power took its place; a power we describe as *Sachzwänge*, factual constraints, the subject of deep analysis by Marx, Weber, Schelsky and others.

A question becomes unavoidable: Can we formulate the speculative thesis that the view from nowhere – in parallel perspective, material models and volumetric display technology – is the view of objectified power, which no one in particular possesses? Which is only to be found in abstract structures – as Marx, Luhmann and Foucault, for example, have shown in very different ways?

In parallel perspective it is more important that the relative length of the lines and the angles are preserved, that you can measure it: It is a representation of the object in itself and not as someone sees it, or as architectural theorist Robin Evans put it:

> *In orthographic projection the projectors do not all converge to a point, but remain parallel. Because this is not the way we see things, orthographic drawing seems less easy to place. It does not correspond to any aspect of our perception of the real world. It is a more abstract and more axiomatic system. [...] The advantage of orthographic projection is that it preserves more of the shape and size of what is drawn than perspective does. It is easier to make things from than to see things with.*[37]

These abstract, measurable representations – think of how in *Avatar* the three-dimensional representation is enhanced by information – are the expression of modern power, which has always already transcended human standards and scale.[38] The combination of spatiality, the enhancement of collaborative work and the saturation of images with information is characteristic for a wide field of display technologies. Such display technologies are less the expression than the performative realization of modern power. The view is not a god's eye view – but it is the view of the successor of god, what was already precisely described by Benjamin in his beautiful fragment *Kapitalismus als Religion* – it is the view of Capital (or of capital and the military as one of its executive forces – as is shown in *Avatar*).[39] It is a view or a

35 See Ernst H. Kantorowicz, *The King's Two Bodies. A Study in Medieval Political Theology*, Princeton, NJ: Princeton University Press, 1957.

36 See John Peacock, Inigo Jones's Stage Architecture and Its Sources, in: *The Art Bulletin* 64 (1982), pp. 195–216.

37 Robin Evans, Architectural Projection, in: Eve Blau, Edward Kaufman (eds.), *Architecture and Its Image. Four Centuries of Architectural Representation. Works from the Collection of the Canadian Centre for Architecture, Montreal*, Cambridge, MA: The MIT Press, 1989, pp. 18–35, p. 20.

38 See Jens Schröter, Tristan Thielmann (eds.), Display I: Analog, in: *Navigationen* 6.2 (2006) and Tristan Thielmann, Jens Schröter (eds.), Display II: Digital, in: *Navigationen* 7.2 (2007).

39 See Walter Benjamin, Capitalism as Religion [Fragment 74], in: Eduardo Mendieta (ed.), *The Frankfurt School on Religion. The Key Writings by the Major Thinkers*, New York/London: Routledge, 2005, pp. 259–262.

gaze that makes something measurable and quantifiable as well as controllable and reproducible.

But isn't that too big a thesis too? Are all god's eye representations the gaze of capital? I guess that would indeed be too strong. But in certain practices these types of representations become operational for domination – and it's precisely their non-linear-perspectival character and their non-positioning of viewers that is their strength. And the least one could say is that the potentially subjectivist and anthropocentric linear perspective is not the characteristic expression and technology for modern power.

Figures

1a–c 20th Century Fox, James Cameron, *Avatar*, USA/UK 2009, TC: 00:47:49, 00:48:03, 00:48:10.

2 Cheryl Rodewig, US Army.

3 John Fischer, US Navy.

4-5 Jens Schröter.

6 E. Parker, P. R. Wallis, Three-Dimensional Cathode-Ray Tube Displays, in: *Journal of the IEEE*, 95 (1948), pp. 371-390, p. 373.

7 Rodney Don Williams, Felix Garcia, Volume Visualization Displays, in: *Information Display 5* (1989), pp. 8-10, p. 9.

8 Star Wars Episode III – Revenge of the Sith (USA 2005, George Lucas).

9 Reprinted from Parviz Soltan et al., Laser-Based 3D Volumetric Display System, in: Richard M. Satava et al. (eds.), *Interactive Technology and the New Paradigm for Healthcare*, Amsterdam et al.: IOS Press, 1995, pp. 349-358, p. 352, with permission from IOS Press.

10–11 Parviz Soltan et al., Laser-Based 3-D Volumetric Display System (The Improved Second Generation), http://handle.dtic.mil/100.2/ADA306215 (accessed August 30, 2007), pp. 16-20.

Carolin Höfler

Sense of Being Here

Feedback Spaces Between Vision and Haptics

Vertiginous Bodies

Facebook, Samsung, Google, and HTC are making headway on the mass consumer market with their latest virtual reality glasses, promising users a total immersion into pictorial worlds. By means and with the help of head-mounted displays – as the advertising slogans unanimously stress – virtual reality (VR) will eventually become authentic and real. It is in particular the combination of two technological methods that causes hope for increasing the *realism* of an observer's visual experience: Stereoscopic 3D and precise, low-latency head tracking are supposed to boost the impression of being not just *in front* of an image but *being part* of an image space and interacting with things and processes happening *there*. If the movements watched via head-mounted displays do genuinely match the perception of body movements, this will not only alleviate the user's feeling of motion sickness but also strengthen the user's impression of realness of the visually perceived environment. In this perceptual situation, the user might stress that the technological system can be controlled, thus providing the virtual system with credibility and enhancing its acceptance.

The promise of a nullification of the discrepancies between virtual and physical reality immediately evokes associations related to the prophetic description of the *Ultimate Display* by Ivan Sutherland in the year 1965, according to which it would be possible to generate virtual experiences able to persuade and convince our senses:

> *The ultimate display would, of course, be a room within which the computer can control the existence of matter. A chair displayed in such a room would be good enough to sit in. Handcuffs displayed in such a room would be confining, and a bullet displayed in such a room would be fatal.*[1]

In the realm of the design and development areas of human-computer interaction, virtual and augmented reality, and game and interaction design today, the discourse centers more than ever around a *fusion* of the virtual and physical space into a *mixed reality* that is prevalent in research and discussion. In that respect, mixed reality is understood to consist of environments and systems aiming at coupling a real-physical action space to a synthetic, computer-generated image space – with those two spaces hardly distinguishable from each other.[2] In such a mixed reality, it

1 Ivan E. Sutherland, The Ultimate Display, in: *Information Processing 1965. Proceedings of IFIP Congress* 65.2 (New York, May 24–29, 1965), pp. 506–508, p. 508.

2 Paul Milgram, Haruo Takemura, Akira Utsumi and Fumio Kishino, Augmented Reality. A class of displays on the reality-virtuality continuum, in: *Proceedings of SPIE. Telemanipulator and Telepresence Technologies* 2351 (Boston, October 31–November 1, 1994), pp. 282–292.

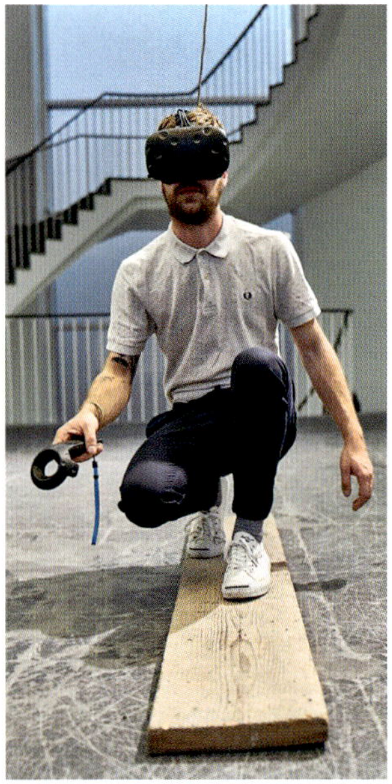

1 Toast, *Plank Experience*, 2016. Exhibition view, art association Frankfurt (Frankfurter Kunstverein), 2017.

is either the physical surroundings that are enriched with virtual information, like in augmented reality scenarios, or it is the virtual surrounding that is extended by physical information, like in virtual reality applications using tactile feedback. Both variations are considered to represent interim stages in the so-called “reality-virtuality-continuum”, as the researchers Paul Migram and Fumio Kishino stressed in their writings from the early 1990s.[3]

The discourse on a seamless fusion of the physical and virtual worlds grows in importance in the current discussion on “phygital” objects and surroundings which are touted as new forms for a computerized world access.[4] Having its origin in the USA, the marketing term *phygital* consists of the words *physical* and *digital* and refers to the linkage of physical objects and surroundings with their virtual representation, or vice versa the coupling of virtual things to a physical equivalent in a digital network. The network enables these things and surroundings to connect and exchange data. The development of phygital objects and spaces is in particular spurred by mobile display techniques, through which the interaction between display and user is enhanced. In contrast to augmented reality scenarios, in which digital elements can be blended into the physical environment, phygital applications instead use physical elements to be projected into a VR setting. In both cases, however, the key idea is a fusion of both worlds.

In the discourse on the melting of the boundaries between physical, virtual and imaginary worlds, much attention is paid to the idea of the holistic nature of the space as a whole being primarily *seen*. In current developments, this holistic view of space is to be supported by further sensory perceptions. What the new VR installations and experiments have in common is the relation to the recipient’s body: The recipient no longer serves as a pure watcher alone, instead their entire body should be capable of experiencing the virtual surroundings.[5] On the one hand, the impression of realness of the virtual world is supposed to be reinforced through real-world elements in virtual spaces. On the other hand, it is intended to let the virtual world appear even more intensely than its physical counterpart. In order to find stimuli to actually achieve the state of intensified perception, facets and spots of virtual 3D environments are brought into the game, physically materialized, and internalized into the VR setting.

One such work trying to intensify and enhance the user’s perception is delivered by the Australian game developing collective Toast with its project *Plank Experience*.[6] The recipient, with his VR glasses on his head, finds himself situated in a big city, heading up the elevator of a skyscraper. The door opens high above the skyline. A plank juts out over a deep abyss. According to the real physical presentation space, it is just a wooden board on which the VR user balances (fig. 1). But for the recipient, perception changes since the pictures of the VR glasses make him believe that he is stepping onto a loose and life-threatening balancing plank, while a ventilating fan blows into his face simulating wind (fig. 2). His glasses let him look into dizzying depths with the wind growing stronger and his body reacting more and more energetically.

3 Ibid., p. 283.

4 For example Alica Rosenthal, Phygital Marketing. Die analoge und digitale Welt verschmilzt, https://webmatch.de/blog/phygital-marketing-die-analoge-und-digitale-welt-verschmilzt (accessed January 1, 2018).

5 For example, see the exhibition *Perception is Reality. On the Construction of Reality and Virtual Worlds*, exhibition, curated by Franziska Nori (Frankfurt/M., Frankfurter Kunstverein, October 7, 2017–January 7, 2018), https://fkv.de/en/content/perception-reality-construction-reality-and-virtual-worlds (accessed January 1, 2018).

6 Toast, Richies’s Plank Experience, https://toast.gg (accessed January 1, 2018).

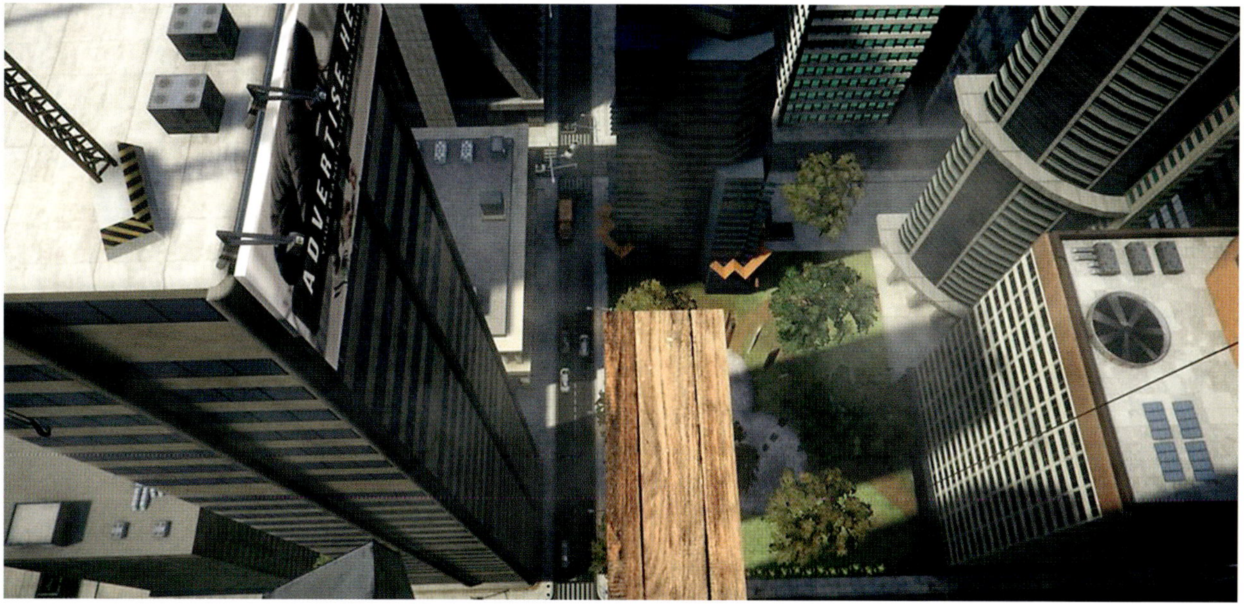

2 Toast, *Plank Experience*, 2016. Rendering.

With the display on his head, the user fumbles around with his feet. The physical ground is hard and stable, but the pictures give the impression that the user is about to fall.

There are many examples of similar projects and pieces of work that are willing not only to stimulate the eye but are trying to stimulate all senses, for instance the project *Swing* by Christin Marczinzik and Thi Binh Minh Nguyen. Here, wearing one's VR glasses, one can place oneself on a swing and gaze out at a wonderful landscape.[7] The stronger one swings, the higher one gets. Thus, the swing as an interface intensifies the physical perception in such a way so that not only the sense of sight and touch are stimulated, so is the sense of balance. Such a sensual state of involvement of the viewer leads to a special kind of strengthening and mobilizing one's physical senses and powers: the picture directly located in front of the viewer, diminishing the distance between subject and object, between the person watching and the things being watched, makes the (re)acting body become itself a medium for its individual imagination and self-deceit.

7 Christin Marczinzik, Thi Binh Minh Nguyen, *Swing VR. An Immersive VR Experience,* http://christin-marczinzik.de/portfolio/swing-vr (accessed January 1, 2018).

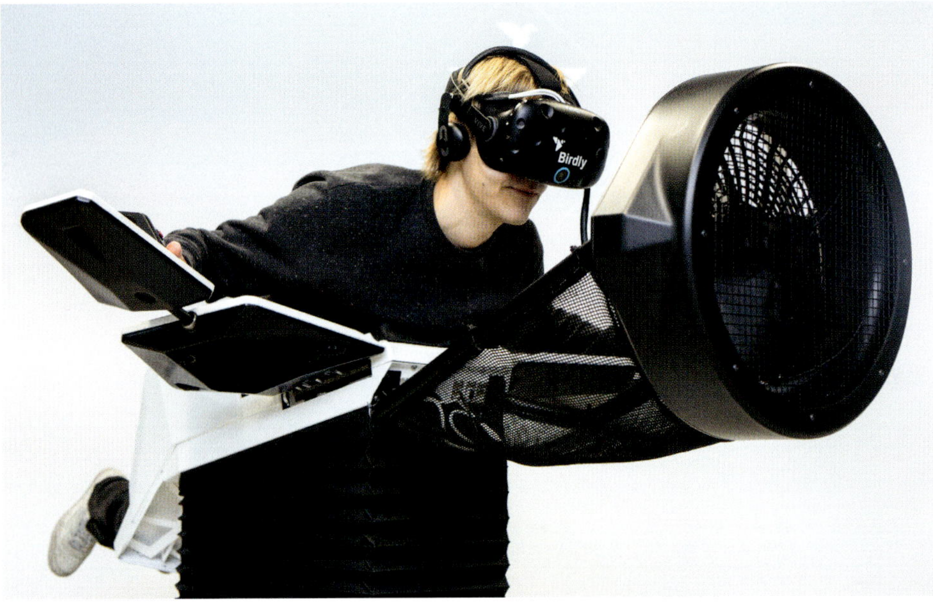

3 Somniacs, *Birdly*, 2015–2018. Photo.

Trompe-corps

A corporeal experience even further out there is the intention behind the installation *Birdly* by designer Max Rheiner, which provides the viewer with fictional pictorial spaces to explore along with a fictitious body as well.[8] *Birdly* is intended to make movements and situations instrumentally perceptible – movements that are not normally tangible for the human body. It is about modeling an artificial nature in which the user may have the intuitive experience of being a bird. In contrast to ordinary flight simulators, the user takes the flight position of a bird rather than simply piloting the run-of-the mill airplane or spaceship; and, in doing so, he interacts with the installation via whole body movements that mimic a bird in flight. To more intensely experience the scene presented in VR, the viewer is lying face down on a rack with two wings which he can grasp ahold of and move up and down through vigorous arm movements (fig. 3). With the palm of the hand spinning upward, one gains altitude, the palm of the hand spinning down, one loses height. During gliding flight, the viewer will have a soft wind blowing in one's face, generated, similar to *Plank Experience*, by a ventilating fan. The airflow increases with increased flapping. In order to achieve this state of acting, and being a genuine bird, *Birdly* uses photorealistic 3D graphics provided via head-mounted displays of the newest generation and sensorimotor couplings. The display shows pictures of a flight over Manhattan, Dubai, or Singapore (fig. 4). The bird's-eye perspective emphasizes panoramic views and monumentalizes the surrounding spaces. The recipient cannot see all of the scene before him at once but rather has to look back and forth, side to side. This gives the impression of a closer proximity and a stronger involvement for the recipient in flight.

But it isn't the immersive pictures alone that involve the user in this fictional space and special pictorial scene; rather it's a physical-material apparatus through which the special flight movements in these pictorial spaces can be steered, managed, and controlled that really solidifies the illusion. The body's physical presence experiences much greater feedback, reinforcing the user's experience of the physical world while – paradoxically – the actual physical space in which the apparatus and the display and the user

8 See Max Rheiner, *Birdly*, Zurich University of the Arts (ZHdK). Interaction Design, 2013–2014, http://iad.zhdk.ch/de/projekte/birdly (accessed January 1, 2018). In 2015, Rheiner founded out of ZHdK the company Somniacs which established *Birdly* as a commercial product.

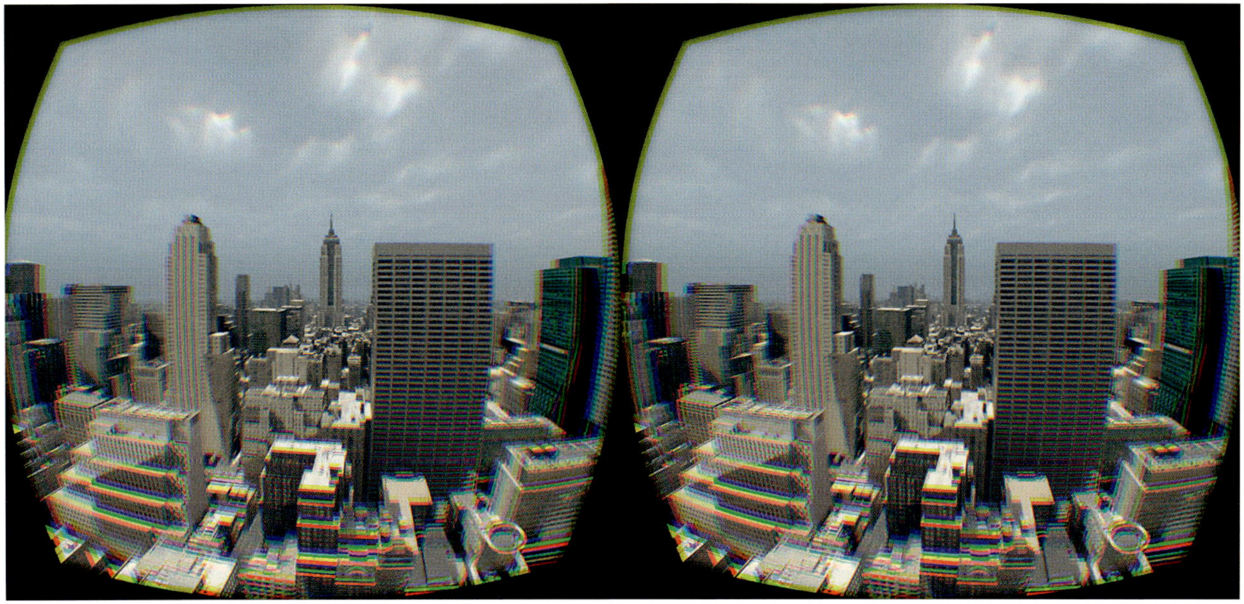

4 Somniacs, The *Birdly* aerial view of Manhattan, New York, 2015. VR Rendering (Oculus Rift). Generated by an earlier version of the simulator (current version *Birdly* Serial Edition, 2016).

are situated is left aside and replaced by totally different virtual surroundings, similar to typical flight simulators and VR games.[9] Thus, the rather small laboratory-like indoor space is turned into an urban, large-scale outdoor space being viewed from dizzying heights.

These and similar other media installations represent an attempt at intensifying bodily perception and playing down common knowledge about real-physical surroundings and situations as well, as German architectural critic Niklas Maak recently pointed out: "The classical 'Trompe-l'œil' is followed by the 'Trompe-corps'."[10] But do these VR scenarios amount to nothing more than deceiving the recipient's body, for example by provoking dizzying heights? Is not another main idea here to allow the body to (re-)identify with the physical world? *Plank Experience* not only aims at the body's deception through imagery, it also aims at the image's materialization through the body. The image

9 Principally Michael Friedman, Head-Mounted Display Screens. A (De) construction of Sense-Certainty, in: *MediaTropes* VI.1 (2016), pp. 114–136.

10 Niklas Maak, Kunst und virtuelle Realität. Der schwindelnde Körper, in: *Frankfurter Allgemeine Zeitung*, January 5, 2018, http://plus.faz.net/feuilleton/2018-01-05/der-schwindelnde-koerper/99295.html (accessed January 5, 2018). Translation from German by the author.

becomes real if the beholder is standing on physical ground, for instance, even if that ground is portrayed spatially and visually differently. The sensory and motor perceptions, the experiences of balance and imbalance, cause the body to reconnect to the physical space. However, the vision stays unbound by physical and practical limits. The visual perspective can be reinforced by other sensory impressions, for instance caused by a ventilating fan, serving as boosting elements for heavy bodily reactions to digital imagery.

The recollection of things, actions, and experiences in virtual space is often more intense than recollecting an experience in physical space. In the virtual space, it is possible to experience situations that can't be felt in the physical space, situations that are maybe too dangerous to expose oneself to in the real world. Involved in interactive moving images, and in line with the theory of enactivism, the body directly and actively frames and generates an experience that is stored not as a virtual imagination but as a real experience.[11] As a consequence, associations and meanings, which the recipient assigns to the physical and media-generated impressions, do blur and shift. What this shift can ultimately evoke and entail for our perception, consciousness, and memory is not yet clear and remains a little researched area.

Closely related to this is the question of how and how much our perception changes if the physical room where the user is located is not replaced by fictitious pictorial worlds but instead digitally constructed and can be operated and steered via VR glasses. By the same token, the question arises as to how our perception changes if facets of the virtual space are physically reconstructed and are thus perceptible for the body. VR glasses seem to appear transparent since they let the viewer look into a photorealistic, stereoscopic digital model of his direct environment. Thus, only a comparison between the contrasts and similarities of the virtual pictorial space and the physical model space can serve as a tool to gain reliable and informative insights on the construction of reality and virtual worlds, as the computer scientist and founder of the Media Research Lab at New York University, Ken Perlin, recently stressed.[12] It is rather through the analysis of these alternative experiences of space that the new conditions of human perception in relation to technically constructed realities can be made tangible. In order to illustrate the analytic potential of virtual-physical spaces, a case study from the game and leisure sector will serve as an example.

Phygital 3D Spaces

In 2016, a US start-up in the city of Pleasant Grove, Utah, opened an amusement park called The Void which features something quite different from the commonly known funfair rides and flying constructions. Visitors to this park are equipped, rather, with advanced VR hardware in order to interact with virtual and physical 3D settings and accomplish feats they could not otherwise experience due to spatial and physical limits of the real world and since the laws of physics cannot be repealed, like fighting dragons or even being hit by enemy bullets. Visitors wear data glasses called Rapture HMD with two curved, extremely high-resolution screens, integrated headphones, and a microphone.

11 Thiemo Breyer, Philosophie der Verkörperung. Grundlagen und Konzepte, in: Gregor Etzelmüller, Annette Weissenrieder (eds.), *Verkörperung als Paradigma der theologischen Anthropologie*, Berlin/Boston: De Gruyter, 2016, pp. 29–50, p. 43.

12 Munich ACM SIGGRAPH Chapter, Ken Perlin – Prototyping the Future, https://vimeo.com/145127565 (accessed January 1, 2018).

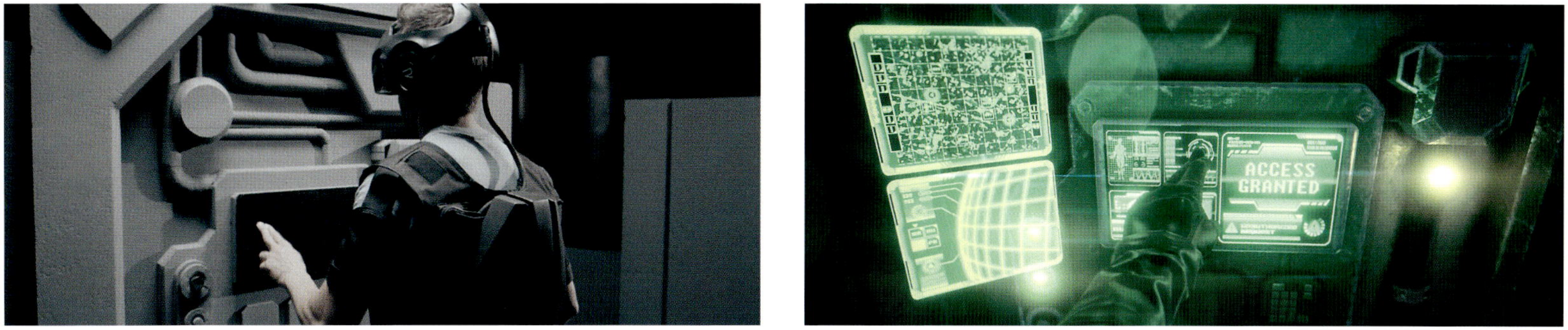

5 A player at *The Void*; Player's view, Pleasant Grove, Utah, 2015/16. Screen shots.

Additionally, they wear a waistcoat and gloves comprising numerous sensors for haptic feedback and body tracking.

With this virtual reality hardware, they pass through an ensemble of physically built rooms with numerous motion and interaction sensors. The Void's key component is a pitch called the *Gaming Pod*, an area of almost 330 square meters with a collection of moveable walls creating a maze of bending corridors. When players walk on virtual trails, through these corridors, they can move freely, without being pushed to an obstacle. The physical boundary and the objects of the pitch are integral components of the digitally constructed 3D image spaces. The physical game architecture is digitally remodeled and visualized and inserted into the VR glasses, where it overlaps with interactive moving images and 3D figures (fig. 5). These projections appear exclusively in the virtual space, whereas the player's actions are carried out simultaneously both in the physical and virtual space. Thus, head-mounted displays and tactile interfaces are ideally called upon to give consistent information in order to trigger parallel sensations. In this setting, the visitor, equipped with a display and wearable devices, physically and digitally intervenes in the game process, thereby exercising partial control and managing the moving images projected onto the digital 3D surfaces. The intended loss of control, primarily caused by the unfamiliar linkage between complex virtual-physical image space data, is thus an essential constituent of the game; it adds to the thrill.

"The goal is to attain total immersion", claims James Jensen, co-founder and Chief Visionary Officer of The Void.[13] In order to incorporate players more (and even fully) into the physical-digital settings of reality, special mechanisms, tools, and devices are provided to let players feel sensations like heat, cold, humidity, vibrations, height differences, and to allow them to touch objects or perceive bad or pleasant scents. *Gaming sickness*, as a common characteristic of many photorealistic PC or console games, is eliminated, the providers of The Void stress, citing the doubling of two different spaces as a main reason: "Any movement made in the

13 James Jensen, as cited in Angela Gruber, Virtual Reality Theme Park: The Void. Der erste virtuelle Freizeitpark, in: *Die Zeit*, June 8, 2015, http://zeit.de/digital/games/2015-06/virtual-reality-the-void-freizeitpark/komplettansicht (accessed January 1, 2018). Translation from German by the author.

virtual world is to the visitor also a movement made in the real world. The visitor doesn't feel any discrepancy, so he will not feel ill. The game is real."[14]

The Void's promise thus consists of a fully-immersive imaginary world, of an extension of images brought into the depths of space for viewers, with their *visual perception* being boosted through given and complementary, mutually supportive, sensory impressions in the physical space. But what kind of idea and conception of the senses and their peculiarities is this promise motivated and influenced by? What does it mean for our perception, orientation, and navigation if digital real-time images reproduce the surroundings but deprive us of a direct view of the physical space – if the visual space (Sehraum) and the tactile space (Tastraum) are disconnected, only to be reconnected through digital, moving images of the surrounding space generated in real time?[15] What dependencies and reciprocal relationships exist between the *physical here* and the *pictorial there*? What categories are being developed if the depicted and disconnected space complies with its physically real spatial dimensions, but differs in its qualities? What perception shifts and scaling effects come into play or originate if – as it is the case with regard to The Void – a stage set architecture of simplified form and materiality can be recognized by touch, while a space of high density, fluidity, light, and information is experienced visually? How does the visual perception influence our tactile perception if the VR player sees his gloved hands on the display, but does not wear gloves at all (fig. 5)? Is it possible to replace sensation with imagination?

Hierarchy of the Senses

Phygital experiences like The Void are based on a holistically oriented model of the structure of the senses according to which the senses, being part of different fields of perception, render synthetic performances in the process of the constitution of space. Until today, this idea of an entity of the senses, and also of its hierarchical structure, has characterized the interpretation and the application of imaging techniques. Although sensory perceptions are closely interconnected, physical-digital VR environments, in particular in the gaming and entertainment area, are based on a model of perception in which vision plays the key role and ranks first, followed by the senses of hearing, touch, taste and smell. For example, this is expressed by the fact that the geometric forms of physical objects and surfaces are radically reduced and simplified, whereas the interactive moving images gain in detail and complexity. The method of screen-based, stereoscopic vision, which aims at heightening and intensifying the visual perception, can thus be regarded as a possible starting point for the reconceptualization and radical expansion of the traditional hierarchy of the senses with vision at its top. The classificatory scheme, in which priority is given to the sense of sight, is now motivated and influenced by the use and interpretation of mobile display techniques.

This revived dispute on the interaction between physical and cognitive performances in the experience of space can be regarded as the continuation of a tradition stemming from the ideas of sensualistic aesthetics in the 18th century. Following that tradition, theorists and architects

14 Ibid.

15 The matter of the reciprocal relationships between the visual and tactile space in VR settings was a subject at the interdisciplinary conference *Mit weit geschlossenen Augen. Virtuelle Realitäten entwerfen* (*Eyes Wide Shut. Designing Virtual Realities*), May 31–June 1, 2017, at KISD – Köln International School of Design of TH Köln, organized by Carolin Höfler and Philipp Reinfeld in cooperation with the Institute of Media and Design of TU Braunschweig.

developed a perspective towards the idea of built space, stressing that it only exists in dependence upon the recipient and the entirety of his perceptions and emotions.[16] The decisive idea behind it was the approach of an emotional merger of subject and object in aesthetic perception, which later, in 1872, would be coined by the German philosopher Robert Vischer as the term of "Einfühlung" or "empathy", as it is often translated.[17] Instead of following traditional proceedings of representation, ornamentation, and iconography, aesthetic efforts at that time were rather aimed at reconceptualizing architecture in the context of a synaesthetic, optical-tactual perception. Insight into that shift of perspective provide, for example, the essays *Prolegomena zu einer Psychologie der Architektur* (Prolegomena to a Psychology of Architecture) by Heinrich Wölfflin, *Ueber den Werth der Dimensionen im menschlichen Raumgebilde* (On the Importance of Dimensions in Human Spatial Creation) by August Schmarsow, and *Das räumliche Sehen* (The Spatial Vision) by Paul Klopfer.[18] Instead of having an idea of space as an immutable entity, the named authors advocate for a more dynamic principle according to which space is generated in the very moment of perception. The idea of a moving, active recipient is thus a prerequisite for space and its creation. From the bodily movement on – in transition from the "tactile space" to the "face space", as Schmarsow points out – space comes into being as a concatenation of mental imagery.[19] It was philosopher Edmund Husserl who recapitulated – skipping the idea of psychologism – the approach of a sensomotoric linkage of all senses and the sensuous-bodily state of perception more systematically. Modern phenomenology of space, as Husserl has established it in his lectures at the beginning of the 20th century, assumed the impression of spatiality to be in connection with the awareness of one's own body movement and thus the result of a lasting sequence of perception.[20]

Referring to these phenomenological and psychological approaches of the first half of the 20th century, Hungarian philosopher Alexander Gosztonyi tried to define the peculiarities of the senses in his *Grundlagen der Erkenntnis* (Fundamentals of Knowledge) from 1972.[21] In his analysis, he not only took into account the classical senses like sight, hearing, taste, smell, and touch but also the "senses of bodily feeling" categorizing the senses of vibration, temperature, balance, gravity, and proprioception as the most important ones.[22] His interest was mainly aimed at analyzing how the different senses either compete or diffuse.

A constitutive element for the dominant physiological-rational understanding of human-computer interaction, virtual and augmented reality, as well as game and interaction design, is primarily Gosztonyi's emphasis on the "quality of reality" according to which every sense is supposed to have a "quality of realness" (Wirklichkeits-

16 An overview on the aesthetics of empathy is given by Jörg H. Gleiter, *Architekturtheorie heute*, Bielefeld: transcript, 2008, pp. 113–126.

17 Robert Vischer, *Über das optische Formgefühl. Ein Beitrag zur Ästhetik*, Leipzig: Hermann Credner, 1873.

18 Heinrich Wölfflin, Prolegomena zu einer Psychologie der Architektur (1886), in: Idem, *Kleine Schriften (1886–1933)*, Basel: Benno Schwabe & Co., 1946, pp. 13–47; August Schmarsow, Ueber den Werth der Dimensionen im menschlichen Raumgebilde, in: *Berichte über die Verhandlungen der Königlich Sächsischen Gesellschaft der Wissenschaften zu Leipzig. Philologisch-Historische Classe* 48 (1896), pp. 44–61; Paul Klopfer, Das räumliche Sehen, in: *Zeitschrift für Ästhetik und Allgemeine Kunstwissenschaft* XIII (1919), pp. 135–149.

19 Schmarsow 1896 (as fn. 18), p. 50, pp. 54–55.

20 Edmund Husserl, *Ding und Raum. Vorlesungen 1907*, Den Haag: Martinus Nijhoff, 1973.

21 Alexander Gosztonyi, *Grundlagen der Erkenntnis*, Munich: C. H. Beck, 1972, pp. 67–97.

22 Ibid., pp. 67–68.

wert), subdividing it into a "quality of reality" (Realitätswert) and a "quality of evidence" (Evidenzwert).[23] According to Gosztonyi, the sense of touch serving as sense of nearness (Nahsinn) therefore has a high quality of reality since it allows for a feeling of material resistance, whereas the sense of sight has a lower quality of reality. The sense of sight serving as sense of farness (Fernsinn), however, is supposed to have a high quality of evidence as it allows having oversight of and insight into complex formal contexts. But, as Gosztonyi posits, only the interplay of quality of reality and quality of evidence determines the degree of realness of the environment being perceived.

Even if Gosztonyi identified the reciprocal effects of the senses as prerequisites for the construction of reality, he assumed a hierarchical structure to exist: "The sense of touch is not dominant. [...] The one who sees subordinates the things touched, ranks qualities and forms of touch, and arranges it in order according to his field of view."[24] Such traditional approaches of the peculiarity and the hierarchy of visual and tactile-haptic sensory perceptions had a lasting effect on the debate on physical-virtual realities. It is the idea of the tactile sense as a simple pressure sense with a high quality of reality that currently dominates the design of mobile devices and interactive surroundings whose interfaces give haptic feedback. In contrast to the forms of sight, the forms of touch are rather poorly developed.

23 Ibid., p. 68.

24 Ibid., p. 81. Translation from German by the author. Original quote: "Der Tastsinn ist nicht dominant [...]. Der Sehende ordnet das Ertastete den Sehformen unter und ordnet die Tastqualitäten und die Tastformen in das Sehfeld ein."

Feedback Design

The new entanglements between the physical "form of being" (Daseinsform) of the tactile space and the digital "form of effect" (Wirkungsform) of the visual space in phygital VR environments fundamentally alter the idea, concept, and design of architectural spaces.[25] With regard to the realm of designing, the construction of specific spaces of action characterized by the interplay between human bodies, technical things, and physical surroundings is currently coming more and more to the focus of attention. On the one hand, the built spaces are being cross-linked through chips, tags, and sensors; on the other hand, they are designed to evoke specific sensory experiences. This means that the material surfaces and objects of the physical space are modeled in such a way that certain (expected or desirable) sensory perceptions, environmental experiences, and behavior patterns come into being in the virtual space. Relevant to the design and the construction of the physical space is the question of how to develop its parameters so that the VR glasses user will accept the virtual space as a *real* space. How can one design, create, and arrange a physical space and its form so that the impression of a sensory and emotional immersion, of control and intervention in virtual environments can be strengthened and best achieved?

This question is based on the assumption that the display user will accept the virtual surroundings as realistic and authentic as possible if he can move through them as naturally as possible. However, body perception can sometimes be deceptive, especially if the recipient wearing the head-mounted display has no visual access to the space being identified by touch. The intense visual perception

25 Schmarsow 1896 (as fn. 18), p. 50.

can alleviate other physical experiences: Visually perceived paths are therefore usually different from physical paths. It is this discrepancy between physical and virtual movement that the approach of *redirected walking* takes into account. Redirected walking assumes that the display user, while passing through virtual worlds, is relatively insensitive to rotations and curves and underestimates egocentric distances.[26] No matter how hard a test person tries to walk in a straight line while blindfolded or with VR glasses on his head, he often ends up going in circles without realizing it. These observations are utilized for the construction of physical-digital VR environments, in particular in those cases where the physical space is limited compared with the potentially infinite virtual space. As a result, the display user is physically guided around a curve while he thinks he is moving straight. According to current knowledge, it only requires a 22-meter radius in order to make the user think that he is walking a straight line, while he is actually walking in circles.[27]

A typical space configuration that permanently redirects the walking direction is the *Unlimited Corridor* developed in 2016 by engineers and computer scientists of the University of Tokyo in cooperation with the US company Unity Technologies (fig. 6).[28] In this spatial installation, the display user touches the corridor wall with one hand in order to enhance the virtual environment's credibility.

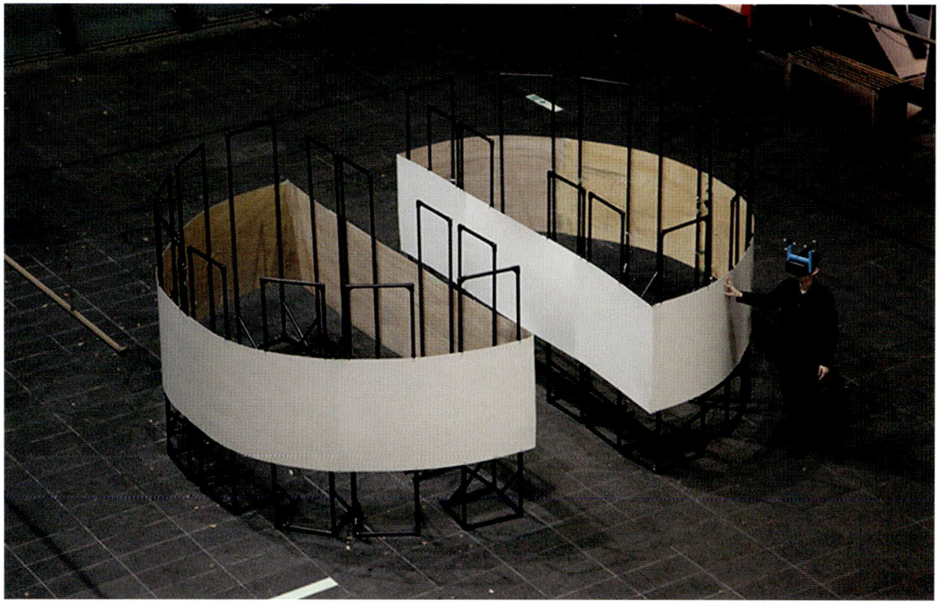

6 Keigo Matsumoto & Team, *Unlimited Corridor*, University of Tokyo, 2016. Photo.

Virtual crossroads and turnoffs are physically reproduced through an additional corridor in the center. The Void is an advocate of the redirected walking principle, too. Its Executive Illusionist Curtis Hickmann developed a similar endless corridor for the VR game hall.[29] At the very same time, the system can be used by several users simultaneously, movement sensors and images actively guide users not to bump and crash into each other, with virtual doors serving as barriers.

26 Frank Steinicke, *Being Really Virtual. Immersive Natives and the Future of Virtual Reality*, Cham: Springer, 2016, pp. 59–86.
27 Ibid., p. 77.
28 Keigo Matsumoto, Yuki Ban, Takuji Narumi et al., Unlimited Corridor. Redirected Walking Techniques Using Visuo-Haptic Interaction, in: *Proceedings of ACM SIGGRAPH 2016 Emerging Technologies*, Article No. 20 (Anaheim, CA, July 24–28, 2016), https://dl.acm.org/citation.cfm?-doid=2929464.2929482 (accessed January 1, 2018); see also the video of VR experiments: Keigo Matsumoto, Unlimited Corridor, https://youtube.com/watch?v=THk92revlVA (accessed January 1, 2018).
29 See the model of The Void's playing field https://theverge.com/2016/7/1/12058614/vr-theme-parks-disney-six-flags-the-void-ghostbusters-virtual-reality (accessed January 1, 2018).

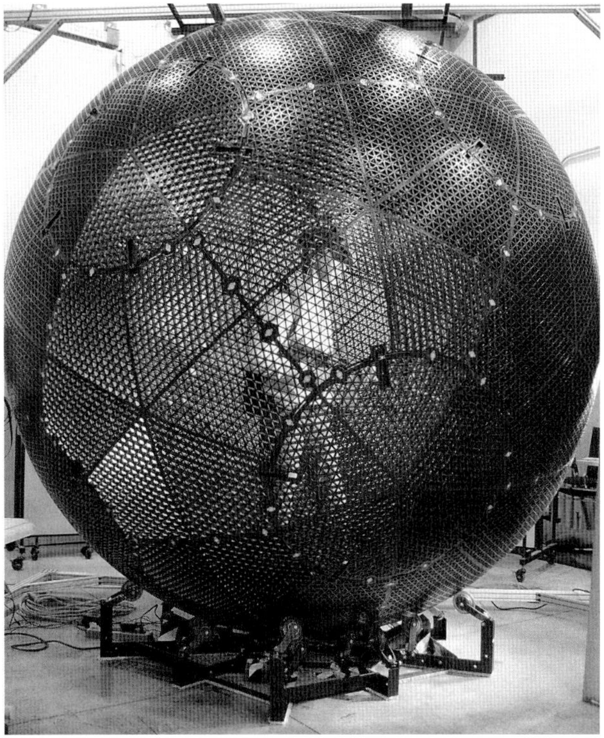

7 *VirtuSphere*, Mounted Warfare TestBed at Fort Knox, Kentucky, 2007.

A multidirectional version of the *Unlimited Corridor* is the so-called *VirtuSphere* originally used by the US military for training purposes (fig. 7). It is a ball to walk-in with a diameter of 3 meters supported on rollers. After entry, users can walk in any direction without changing their position in the physical space. With the help of a head-mounted display, test persons are transferred to virtual worlds in which they can move about freely. Sensors beneath the ball record any step and transmit the information to the display. In that respect, the space of the ball is rather a gigantic joystick managed, operated, and steered with the users' feet. Three of these balls are located in a Las Vegas casino, one ball can be found at the University of Bremen in the department for cognitive neuroinformatics, for the purpose of research on human orientation in virtual and physical spaces.[30] This experimental system also supports the idea that the VR user has the best orientation in unknown virtual worlds if he is provided with as many sensory impressions as possible. Besides the feeling of moving and watching 3D pictures of the virtual world, the setting can be complemented with sounds and odors. Such a maneuverable ball is characterized by an ideology insinuating that space perception and space movement follow sensory impressions. That idea has a strong impact on the development of today's phygital VR settings in the realm of game and interaction design. Yet, it is quite astonishing though that the material and atmospheric characteristics of real space receive so little attention in those settings.

Hallucinatory Interface

It even seems that the contradiction between physical spatiality and virtual imagery will sharpen: The visual-material, tactile-haptic, and olfactory features of space are being decoupled from their physical shape and materiality and transferred into visual information. The physical diffuses into the virtual for the benefit of a performance that permanently reconfigures the potential relationship between physical and virtual reality.

30 Kerstin Schill, *Räumliche Exploration (VirtuSphere)*, University of Bremen, working group Cognitive Neuroinformatics, http://cognitive-neuroinformatics.com/en/research/projects/raeumliche-exploration (accessed January 1, 2018).

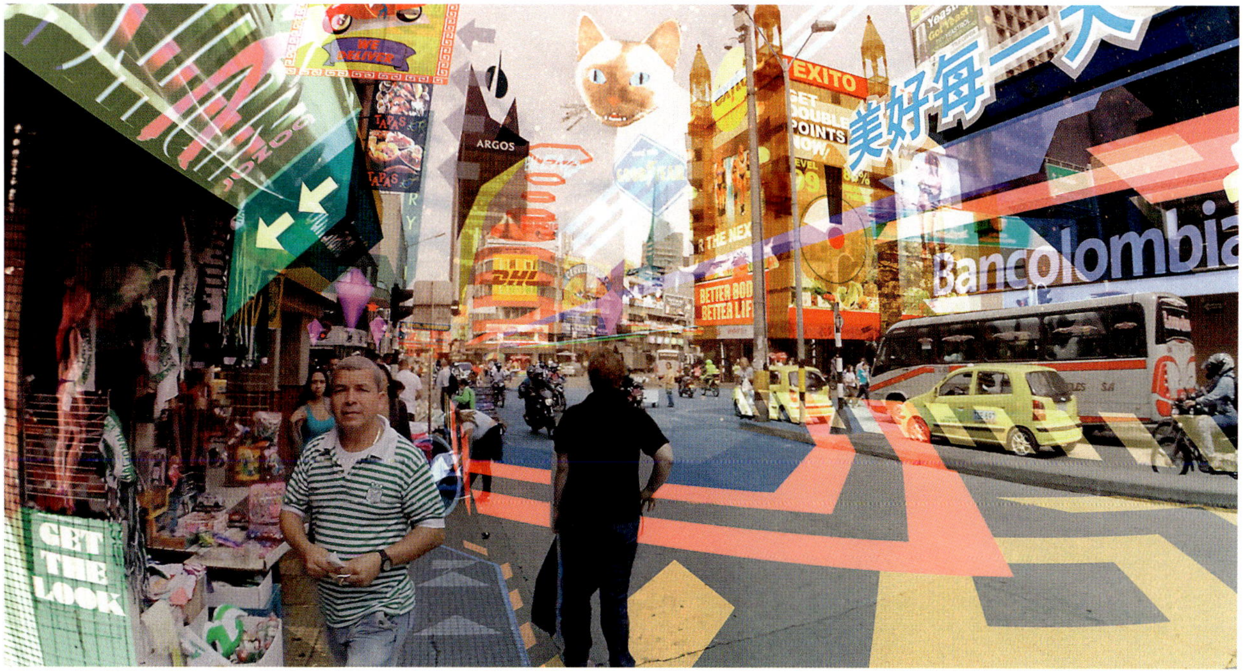

8 Keiichi Matsuda, *Hyper-Reality*, 2016. AR rendering.

Number and complexity of this projected imagery will in the near future certainly increase. Given the ubiquity of cameras and the availability of a billion photos online, recent years have witnessed new options for image-based 3D data capture and 3D reconstruction of physical spaces and objects.[31] With the help of photogrammetric techniques and procedures in computer graphics and computer vision, it is possible to generate, from a variety of images, photorealistic and editable digital models of those scenes, which can then later be integrated into virtual reality surroundings. In the ideal case, the display user can record his physical surrounding space via a digital video camera attached to his glasses, can process that information into 3D spaces and 3D objects through image-based geometry reconstruction software, and can eventually embed them into the VR scene – and all this in real time.

31 Simon Fuhrmann, Fabian Langguth, Michael Goesele, MVE – A Multi-View Reconstruction Environment, in: *Proceedings of the Eurographics Workshop on Graphics and Cultural Heritage* (Darmstadt, October 6–8, 2014), https://gcc.tu-darmstadt.de/media/gcc/papers/Fuhrmann-2014-MVE.pdf (accessed January 1, 2018).

The possible consequences of this continuous transformation of physical spaces into virtual imagery is outlined by Japanese architect Keiichi Matsuda in his short movie *Hyper-Reality* (fig. 8).[32] Just like in a computer game, the viewer is watching the action from first-person perspective of a protagonist moving through a consumer's day-to-day world, totally saturated by media and promotions. He takes the bus, gambles online, chats with a job manager about current vacancies, and enters the question "Who am I?" into a search engine. In the supermarket, there are new ads and apps perpetually popping up, and a small dog on the shopping cart serves as a shopping companion, reporting special offers via GIF animation. While in this visual supersaturation, innumerable background sounds rumble, roar, and ring. In *Hyper-Reality,* Matsuda coped with a specific kind of the interplay of virtual and physical realities – with augmented reality (AR). In contrast to VR displays, visual access to the physical space is still possible, but it is modified through interactive 3D projections. While doing so, the user is looking through transparent screens on which he is watching the projections. They constitute the front. Behind it, a real-physical space opens up. Thus, this procedure interlinks a flat with a deep space.

The technique of combining detailed virtual image information up front, with less detailed physical space configurations in the back, is closely associated with postmodern strategies of space-formation. The *Hyper-Reality* architectures, for instance, can be regarded as contemporary versions of the *decorated shed* which Robert Venturi and Denise Scott Brown considered to be a postmodern building type par excellence. In *Learning from Las Vegas* (1972), they deciphered the aesthetic functionalism of a commercial entertainment industry whose symbolism and imagery were particularly oriented towards the visual perception of motorists and pedestrians.[33] From the buildings along the Las Vegas Strip, they derived the building type of the *decorated shed* which is simple in configuration and form, but whose message is intricately designed and offensively put to the front of the façade – as ultra-large illuminated panels and signs. *Hyper-Reality* then appears to be an even more exaggerated version of the Strip. The digital symbols covering the city like a virtual skin are individually customized to those who pass through, can be dynamically modified, and are, in the truest sense of the term, transparent and easily comprehensible.

Hyper-Reality emphasizes the leveling of deep structures as a constitutive feature of physical-digital worlds, even though the pioneers of VR and AR systems originally came forward with a contrary promise. Given the revocation of deep structures, one is tended to consider the *Hyper-Reality* environments as an epitome of the late capitalist culture of spectacles and simulacra, as Fredric Jameson in his 1984 book *Postmodernism, or The Cultural Logic of Late Capitalism* had stressed.[34] With regard to a postmodern space that appeals to all senses, Jameson coined the term of "hyperspace".[35] He understood it as a spatiality characterized by a plenty of indissoluble and interwoven surfaces. He exemplified this idea of "depthlessness" in particular via the lobby of the Westin Bonaventura hotel in Los Angeles: "I am tempted

32 Keiichi Matsuda, *Hyper-Reality. A New Vision of the Future,* https://vimeo.com/166807261 (accessed January 1, 2018).

33 Robert Venturi, Denise Scott Brown, Steven Izenour, *Learning from Las Vegas. The Forgotten Symbolism of Architectural Form,* revised edition of the 1972 publication, Cambridge, MA/London: The MIT Press, 1977, pp. 87–103.

34 Fredric Jameson, *Postmodernism. Or, The Cultural Logic of Late Capitalism,* Durham: Duke University Press, 1991.

35 Ibid., p. 44.

to say that such space makes it impossible for us to use the language of volume or volumes any longer, since these are impossible to seize."[36] He stressed that such a space was filled with diaphanous materials, illuminating phenomena, and ongoing, continuous movements to such an extent that its material-constructional spatial boundary is forced back to vagueness: "Hanging streamers indeed suffuse this empty space in such a way as to distract systematically and deliberately from whatever form it might be supposed to have [...]".[37] It is as if the effectual spatial boundary seems to dissolve, dilute, and overlay all spatial regions. As if it merges into diversified layers, into limiting structures being at once open and enclosed floating through space as aerial objects. From these observations, Jameson concluded the existence of a modified spatiality – one that substitutes "depth" with "surface" and allows for an intensified form of what in German would be called *Sehenlassen* or letting itself be seen.[38] The numerous activities, moods, and ambiances in the hotel lobby make the observer feel as if he had totally immersed into the space: "[...] a constant busyness gives the feeling that emptiness is here absolutely packed, that it is an element within which you yourself are immersed, without any of that distance that formerly enabled the perception of perspective or volume. You are in this hyperspace up to your eyes and your body."[39] The feelings of disorientation and dizziness which Jameson identified to be the very effects of the hyperspace are, by implication, fundamental experiences also detected by various observers – from Rem Koolhaas to Jean-François Lyotard – when faced with the forces of capitalism unleashed and the aesthetic experiences of postmodernism.[40] In the age of VR and AR and the Internet of things, *Hyper-Reality* can be understood as an amplified sequel and fearsome intensification of those experiences and emotions.

36 Ibid., p. 43.
37 Ibid.
38 Ibid., p. 49, p. 12.
39 Ibid., p. 43.

Other Images, Other Spaces

With the growing penetration of new VR and AR glasses onto the mass market, relations between bodies and outer and inner environments become increasingly interconnected via screen-based, interactive moving images. The interactions with and through screen images lead to a specific alignment of the body in both the virtual and physical space. Being a player in physical-digital theme parks like The Void or supposedly flying via the full-body installation of *Birdly* are aimed at generating an intense perception and a powerful body mobilization. The visualizations of objects and spaces installed through the glasses actively affect the user's position and movement in the physical space and also have an impact on the user's spatial disposition towards the display, the apparatus, and the architecture. If the physical space with its haptic surfaces and things to be perceived and identified is digitally reproduced and visually brought into the displays, the physical space will transform into phygital surroundings, inducing the beholder to go through specific experiences and actions. It is thus not only important how the phygital 3D space is visually presented and perceived; critical is how the given data of both environment and body interact and communicate – also with regard to an interaction pattern between display imagery and the beholder's perceptions and actions.

40 Jean-Francois Lyotard, *Leçons sur l'Analytique du sublime*, Paris: Galilée, 1991.

Considering the increasing integration of VR and AR systems into economic and design-related processes and given the growing impact of display images on a computerized world access, many questions arise, in particular with regard to the issues of authorship, final authority, and decision-making autonomy of the actors involved: What industries and sectors will further develop and advance the potentials of these physical-virtual spaces of experience and action, and for what purposes? How will the observer be influenced? Who designs and is in charge of the content and type of physical-virtual experiences? Which interactive screen images will be provided? What emotional and mental experiences shall be triggered? Even though the action taking place in the huge spectacle of games and artistic installations has, in the first place, no far-reaching effect for the real-physical world, the perceptions and experiences made in those physical-virtual settings do create an intensity of experience which must be perceived to be real. As a result, virtual reality experiences will certainly be establishing themselves as a new category of spatial experience and will thus give a strong impetus to sensory perceptions, actions, and decisions in non-digital spheres of life.

Further, the scientific utilization of virtual reality tools will have far-reaching implications for knowledge production and evidence acquisition, for instance in investigative analysis and 3D crime scene reconstruction. The fact that architecture and the arts also harness forensic methods based on virtual reality is illustrated by the interdisciplinary research group Forensic Architecture and its project *77sqm_9:26min*, which through imaging techniques tried to resolve the case around German intelligence officer Andreas Temme with regard to the NSU-murder of Halit Yozgat in 2006.[41] This case mingled 3D image space analysis with criminology, political enlightenment, and legal proceedings. Knowing that virtual reality tools provoke perception shifts and scaling effects, it is quite astonishing that the actors involved have such a high confidence in imaging techniques being applied in those cases for the purpose of fact- and truth-finding.

Criticism and skepticism regarding how the recipient can be captured, deceived, and manipulated in VR environments is more effective through designing critical spaces. This raises the question of how alternative forms to physical immersion into 3D spaces can be devised and developed. Is it possible to maybe reduce the intensity of virtual 3D surroundings? In coping with VR, which medial strategies exist that circumvent the almost total immersion into virtual spaces and which provide transparency about the techniques that users are succumbing to? How can more heterogeneous, fragmented, and conflicting formal systems be applied in immersive environments?

Those strategies of undermining and disturbance are based on a conceptional approach which tries to overcome the traditional antagonisms between body and mind and which interprets sensory perceptions neither individualistic-hierarchically nor collectivist-holistically. From this perspective, seeing, hearing, and feeling are not being understood as naturally given skills, but as specific effects of socio-technical assemblages and as medial design practices.[42] This involves not only experiences of visual-spatial relations, but experiences of social relations as well. How can display users get in touch and interact with each other?

41 Forensic Architecture, *77sqm_9:26min*, investigation, 2016–2017, http://forensic-architecture.org/case/77sqm_926min (accessed January 1, 2018).

42 Beate Ochsner, Robert Stock (eds.), *senseAbility. Mediale Praktiken des Sehens und Hörens*, Bielefeld: transcript, 2016.

And how can they get in touch and interact with those not equipped with a display?

In such discontinuous settings, what is and will be the function of architecture? In hyper-realistic phygital VR environments, the materially bound spatial forms primarily serve as image carriers and step behind the digital interface. The question arises as to how to envision, design, and develop an architecture that does not constantly validate the perception of being in a virtual space (*sense of being there*), but rather challenges it through creating a sense of presence in the physical space (*sense of being here*). What other perceptions could be triggered by physical spatial forms if only they provided meanings and messages beyond the moving images of the displays – to be decoded and interpreted by the recipient in relation to the visually perceived 3D spaces? What if these perceptions, meanings, messages, codes, images did not rely on affirmation and affective adaption, but on disturbance and doubt instead?

Figures

1 Toast, *Plank Experience*, Frankfurter Kunstverein, Photo: N. Miguletz. https://fkv.de/sites/default/files/styles/680x540/public/FKV_Perception%20is%20Reality_Ausstellungsansicht_Toast_2.jpg?itok=YHOUdOBg (accessed January 1, 2018). Reprinted with permission.

2 Toast, *Plank Experience*, https://toast.gg/release/ (accessed January 1, 2018). Reprinted with permission.

3 Somniacs, http://somniacs.co/media.php (accessed January 1, 2018). Reprinted with permission.

4 Adam Savage, Tested, Flying the Birdly Virtual Reality Simulator, https://youtube.com/watch?v=gWLHIusLWOc (accessed January 1, 2018). Reprinted with permission.

5 THE VOID, First look at THE VOID, https://youtube.com/watch?v=cML814JD09g, 0:58, 0:59.

6 *Unlimited Corridor* Project Team (Keigo Matsumoto, Representative), Japan, http://cyber.t.u-tokyo.ac.jp/~matsumoto/image/uc/uc_web.jpg (accessed January 1, 2018). Reprinted with permission.

7 Paul Monday, https://en.wikipedia.org/wiki/VirtuSphere#/media/File:Virtusphere.jpg (accessed January 1, 2018).

8 Keiichi Matsuda, Still from *Hyper-Reality*, 2016, http://hyper-reality.co/assets/HQ_images/hyper-reality_03.jpg (accessed January 1, 2018). Reprinted with permission.

Timo Kaerlein and Christian Köhler

Around a Table, around the World

Facebook Spaces, Hybrid Image Space and Virtual Surrealism

On 6 October 2016, at the developer conference Oculus Connect 3 in San José, California, Facebook CEO Mark Zuckerberg presented the first working prototype of Facebook Spaces, the social media company's ambitious foray into the emerging virtual reality (VR) sector.[1] Unsurprisingly, Facebook's vision of VR is a *social* one, i.e. an attempt is made at translating the company's core business model of capitalizing social relations into a VR setting.

During the product demonstration on stage, which included Lucy Bradshaw and Michael Booth, two senior Facebook employees working in the VR team, Zuckerberg discussed all major features of Facebook's VR proposal that is available in an open beta version at the time of this writing. Only Zuckerberg himself was physically present wearing a head-mounted display, while Bradshaw and Booth were somewhere off-stage using a similar set-up to meet with Zuckerberg within Facebook Spaces, Facebook's social VR application. While the audience could only see Zuckerberg on stage talking into the air, they could turn to giant screens on which Zuckerberg's perspective of the virtual space was projected: It displayed what the headset let him see, Bradshaw and Booth virtually present via their avatars. After having showcased the comic-figure-like avatars, including their palette of facial expressions and hand gestures, a series of immersive photo and video environments – a deep-sea scenario, the surface of the planet Mars, and Zuckerberg's own office –, and interactive affordances like playing games at a virtual table and creating 3D objects in space, the audience experienced a dizzying moment of referential disorientation. Zuckerberg transported the group of three into a different setting again, this time (supposedly) a live video-feed of his actual living room. Then he suddenly received a video call from his wife via the Facebook Messenger application on his virtual wristwatch: While her moving image appeared in VR on an oversized virtual phone display for all participants to see, she in turn saw on the display of her phone her husband's VR avatar standing in their living room with two other comic characters, the unsuspecting family dog Beast reclining on the couch in the background. The climax of the product demonstration induces at the same time a media theoretical reeling: Zuckerberg turned around for a "modern family selfie"[2], using a virtual selfie stick handed to him by Michael Booth's avatar, aligning the gigantic phone display with his wife's image next to his virtual avatar and the moving image of the dog, in the same act suturing diverse layers of referentiality into a series of photo-like static representations that appeared on the virtu-

1 Road to VR, Facebook Social VR Demo – Oculus Connect 2016, https://youtube.com/watch?v=YuIgyKLPt3s (accessed November 3, 2017).

2 Ibid.

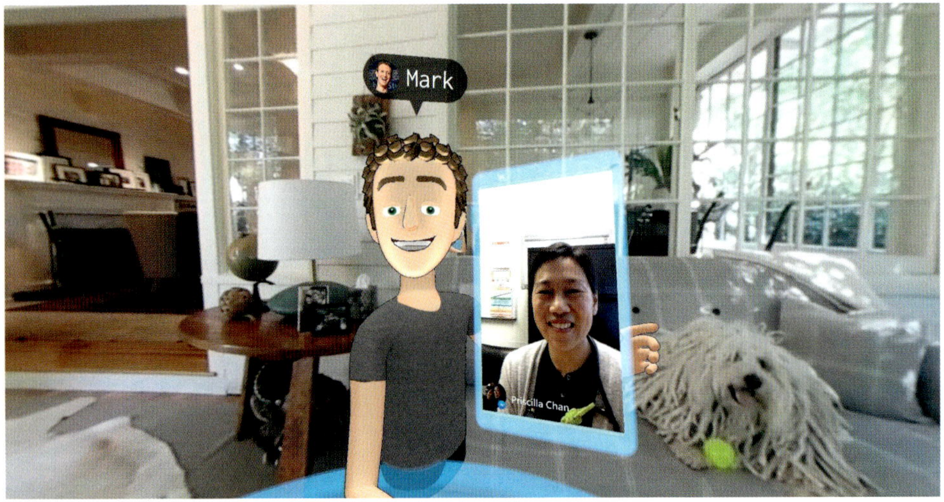

1 Mark Zuckerberg taking a "modern family selfie" in Facebook Spaces.

al table in front of him (fig. 1). Afterwards, the couple chose their favorite picture that was then effortlessly delivered to Zuckerberg's Facebook feed via the push of a virtual button on his other wrist. In Facebook's vision of social VR, the old dichotomies of actual and virtual, real and imagined, perception and action that structured major debates in the VR discourse of the 1990s seem to have irretrievably collapsed.

The above-described scene taking place in Facebook Spaces raises a series of questions concerning the character and shape of visual practices that are intended to constitute sociality in VR. Whereas screen-based practices in VR are often associated with anti-social behavior, Facebook Spaces, advertised with the slogan "VR is better with friends", promises to change this situation.[3] What people will eventually *do* in a social VR scenario, how they will interact with each other and with diverse media content, is first and foremost an as yet unsettled question of interface design. This is evident to the designers of Facebook Spaces,[4] but we will address this question not from a design perspective, but from a media-theoretical point of view that follows Johanna Drucker's definition of "interface" understood not as an object, but as "a set of conditions, structured relations, that allow certain behaviors, actions, readings, events to occur".[5] This allows for an analysis that considers the interface mise-en-scène of Facebook Spaces as constituting a proper media *dispositif* or apparatus.[6] This apparatus assigns subject positions, orders relations between participants and – crucially – sets the stage for a radically hybrid image space, in which

3 See Thilo Hagendorff, Virtual-Reality-Datenbrillen im Spannungsfeld zwischen Empathie- und Isolationsmaschinen, in: Institut für immersive Medien (ed.), *Jahrbuch immersive Medien 2016*, Marburg: Schüren Verlag, 2017, pp. 71–79.

4 Christophe Tauziet, leading designer in Facebook's social VR team, makes this point explicit in an extensive and insightful article on *medium.com*: "One of the biggest challenges for our design team was to design the user interface of Spaces. Unlike with traditional web, desktop or mobile design where we can rely on existing UI elements and interaction patterns that people have learned over the years, most of those patterns have *yet to be invented* for VR." Christophe Tauziet, Designing Facebook Spaces, https://medium.com/@christauziet/designing-facebook-spaces-part-4-creating-a-vr-interface-821861159495 (accessed November 3, 2017). We will refer to his development report repeatedly throughout our contribution.

5 Johanna Drucker, Performative Materiality and Theoretical Approaches to Interface, in: *digital humanities quarterly* 7.1 (2013), http://digitalhumanities.org/dhq/vol/7/1/000143/000143.html (accessed November 3, 2017).

6 See Jan Distelmeyer, *Machtzeichen. Anordnungen des Computers*, Berlin: Bertz + Fischer Verlag, 2017, pp. 81–82. Distelmeyer introduces the concept of interface mise-en-scène in explicit reference to film studies to signify the need to pay closer attention to the heterogeneous aesthetic arrangements organizing the use of computers. Analyzing the complex staging of interface processes can offer different approaches to digital cultures than just assuming the computer to be a functional tool.

the line between physical and virtual entities increasingly loses significance. Applying the designers' own language, this image space can best be characterized as giving rise to a *virtual surrealism* in which long contested dichotomies concerning the status of images, perception and action dissolve into a scenario of reality-agnosticism that is equally frightening and exhilarating.[7] Our main interest concerns the ways that the ensemble of interface techniques and processes of Facebook Spaces attempts – and ultimately fails – to keep together this heterogeneous action space, especially by rather surprisingly falling back to an almost-forgotten "old medium", namely the virtual table around which the users gather.[8]

The essay is divided into three parts. First, we give an introductory account of Facebook Spaces. The focus of our description lies in understanding the *brandscape* of Facebook Spaces as a *dispositif* as described in the works of Jean-Louis Baudry, i.e. as a spatial arrangement that regulates the behavior of participants and favors specific psychic dispositions.[9] The second part of our contribution then delves deeper into the theoretical ramifications of this setup: The hybrid image space constituted by Facebook Spaces is reminiscent of a heterotopia in that it involves a confrontation of widely disparate image spaces that have to be sutured together to constitute a reliable and secure action space. We will explore the strategies employed by the designers of Facebook Spaces to achieve this end, primarily the virtual table that acts as the central interface element and center of control. One major finding of our analysis amounts to the observation that the apparatus of Facebook Spaces, contrary to popular rhetorics of presence and immersion associated with VR, creates a strong impression of unreality by decontextualizing images and severing referential links. Finally, we discuss over-arching theoretical and normative concerns raised by our analysis: Facebook Spaces is understood as a step towards an emerging virtual surrealism – a scenario in which the affordances of digital media are taken very seriously to the extent that the referential status of images and actions in VR altogether ceases to be a relevant parameter for design and use. This also has implications for the subject positions assigned by the apparatus: In a scenario of mixed unreality, a tendency towards moral indifference can be observed and criticized.

Setting the Table – The Dispositif of Facebook Spaces

Due to the commercial availability of affordable VR hardware for the consumer electronics market, several companies have developed applications and platforms for social interaction in VR.[10] The discussions around these offerings are reminiscent of the ones that accompanied the early text-based MUDs (multi-user dungeons/dimensions) and MOOs (MUD, object-oriented) common in the 1980s and

7 See Gabriel Valdivia, Identity Transfer and the Rise of Virtual Surrealism, https://artplusmarketing.com/identity-transfer-and-the-rise-of-virtual-surrealism-bac751e6342c (accessed November 3, 2017).

8 See Walter Seitter, Möbel als Medien. Prothesen, Paßformen, Menschenbildner. Zur theoretischen Relevanz Alter Medien, in: Annette Keck, Nicolas Pethes (ed.), *Mediale Anatomien. Menschenbilder als Medienprojektionen*, Bielefeld: transcript, 2001, pp. 177–192, pp. 184–187.

9 See Jean-Louis Baudry, The Apparatus. Metapsychological Approaches to the Impression of Reality in Cinema, in: Philip Rosen (ed.), *Narrative, Apparatus, Ideology. A Film Theory Reader*, New York: Columbia University Press, 1986, pp. 299–318.

10 For an overview discussing different services like Bigscreen, vTime, AltspaceVR and Rec Room, see Adario Strange, Social Networking in VR is Here, and it Feels Like the Future, http://mashable.com/2017/01/12/virtual-reality-social-networks-vr/#iaf1.9tSSOqq (accessed November 3, 2017).

early 1990s, which was also the period when the cultural imaginary concerning virtual reality peaked in the all-encompassing term *cyberspace*.[11] Similar rhetorics concerning the general idea of online communities are applied in the Spaces context, including an understanding of virtual tribes defined "not by proximity but personal choice".[12] These notions are accompanied by a set of body and identity politics that were already common in the VR discourse of the 1990s, like the idea of experimenting with different types of embodiment, a fragmentation of the sense of self experienced in the relationship to one's virtual avatar, and a prevalent logic of mentally being somewhere else while the body is left behind in the physical world.

But there are also marked differences in how social VR is imagined in the present. In fact, one could go as far as to claim that VR companies apply metaphors and mental images concerning the affordances of the new medium partially dressed in the language of the 1990s, while something entirely different is happening.[13] We will focus our analysis on Facebook Spaces because the multinational social media enterprise already has access to a base of two million monthly active users, a fact that makes it especially well-positioned in the emerging social VR market. In contrast to older VR discourses that stressed possibilities of identity play and experimentation with different body types and shapes – including the wish to experience a virtual embodiment as an abstract geometrical shape like a triangle proposed by Jaron Lanier –, the premise and imperative of Facebook Spaces is simply: "Be Yourself in VR!"[14] Rachel Franklin, head of Social VR at Facebook and former general manager for the Sims series at Electronic Arts, further qualifies this statement: "It's easy to create an identity that represents the real you in Facebook Spaces. This helps people recognize you and makes VR feel more like hanging out in person. [...] You can change your eye color, hairstyle, facial features and more until your look fits your identity. It's all about being yourself."[15] On the one hand, this idea differs greatly from the (supposedly) wildly experimental character of virtual identities in the 1990s, while on the other, it hints at a notion of idealization and purification of the self that is tightly connected to the necessities of social media self-curation.[16] It is noteworthy in this context that avatars in Facebook Spaces can neither look unhappy nor have a body that deviates too far from the norm set by its designers.[17]

Topologically speaking, Facebook Spaces can further be understood as a curious kind of virtual brandscaping: Whereas Lev Manovich discusses examples of companies giving their brand a material shape via architecture – e.g. the design of OMA/Rem Koolhaas' Prada store in New York

11 See Sherry Turkle, *Life on the Screen. Identity in the Age of the Internet*, New York: Simon & Schuster, pp. 9–19, pp. 180–186.

12 Yaser Sheikh from the Oculus research team, as cited in Matt Weinberger, Facebook's Vision of the Year 2026 is Scary and Awesome, http://businessinsider.de/facebooks-world-of-virtual-reality-in-2026-2016-4?r=US&IR=T (accessed November 3, 2017).

13 Concerning the role of such legitimizing ideologies in the history of the internet, see Patrice Flichy, *The Internet Imaginaire*, Cambridge, MA: MIT Press, 2007, pp. 10–12.

14 Facebook, https://facebook.com/spaces (accessed November 3, 2017). See Jaron Lanier, Technology, http://globetrotter.berkeley.edu/people5/Lanier/lanier-con2.html (accessed November 3, 2017).

15 Rachel Franklin, Facebook Spaces: A New Way to Connect with Friends in VR, https://newsroom.fb.com/news/2017/04/facebook-spaces/ (accessed November 3, 2017).

16 See Mark Zuckerberg: "You have one identity. [...] Having two identities for yourself is an example of a lack of integrity." As cited in Karl Wolfgang Flender, #nofilter? Self-Narration, Identity Construction and Meta Storytelling in Snapchat, in: Florian Hadler, Joachim Haupt (ed.), *Interface Critique*, Berlin: Kulturverlag Kadmos, 2016, pp. 163–182, p. 172.

17 See Kyle Riesenbeck, Facebook Won't Let Me Be Fat in VR, http://revvrstudios.com/facebook-fat-in-vr (accessed November 3, 2017).

that makes heavy use of electronic displays to create an "augmented space" –, the challenge for Facebook lies in translating their product into a VR environment that serves as the condition and backdrop of user interactions.[18] Where the brandscaping described by Manovich worked by integrating screens into physical architecture, within the apparatus of VR screens create an image space which must be provided with virtual architecture. Although this environment is a thoroughly virtual one, the actions that are possible in it are fundamentally physical, i. e. they comprise a set of gestures and interface operations involving the whole body. In the following, we will first give an introductory description of the environment of Facebook Spaces by understanding it as a media *dispositif* or apparatus in the sense of Baudry. This serves to prepare a more detailed analysis of the hybrid image space constituted by Facebook Spaces in the following chapter. Baudry distinguished the

> *basic cinematographic apparatus [l'appareil de base], which concerns the ensemble of the equipment and operations necessary to the production of a film and its projection, from the apparatus [le dispositif] […], which solely concerns projection and which includes the subject to whom the projection is addressed.*[19]

It is the latter apparatus in the sense of a spatial arrangement of objects and bodies that interests us most; although it is impossible to separate this completely from the data infrastructures and economies constituting the position of Facebook in the contemporary social media business. In Baudry's account, the apparatus of cinema served first and foremost to create an "impression of reality […] dependent on a subject effect", i. e. the apparatus tends to make itself invisible in order to constitute a simulation of the real.[20] This only works because the cinematographic subject enters a "state of artificial regression" which leads to "a lack of differentiation between the subject and its environment" and thus a "partial elimination of the reality test".[21] Without reconstructing the intricacies of apparatus theory at this point, it suffices to say that the main thrust of the argument is to claim that the technical apparatus of cinema produces ideological effects independently of what is projected.

As others have shown, apparatus theory can deliver an adaptable conceptual framework to describe the ideological effects not of media content but of media themselves.[22] But, of course, cinema and VR are two fundamentally different media. Not only is the production process of VR applications, at best, only partly comparable to the production of movies, but the act and context of reception differs widely from cinema. The apparatus of cinema consisted of viewers who were physically restricted in a darkened room to watch unreachable images projected from behind their back.[23] In contrast, users of VR look onto light-emitting screens directly in front of their eyes locking out non-screen reality to see images which they can interact with. In fact, the positioning of screens is one of the core differences between the apparatus of cinema and the apparatus of VR/Facebook Spaces: Whereas spectators in the cinema are always in principal

18 Lev Manovich, The Poetics of Augmented Space, in: *visual communication* 5/2 (2006), pp. 219–240, pp. 234–235. The term *brandscaping* is here attributed to Otto Riewoldt.
19 Baudry 1986 (as fn. 8), p. 317.
20 Ibid., p. 312.
21 Ibid., p. 313.
22 See Knut Hickethier, Dispositiv Fernsehen. Skizze eines Modells, in: *montage a/v* 4/1 (1995), pp. 63–83.
23 See Jean-Louis Baudry, Ideological Effects of the Basic Cinematographic Apparatus, in: *Film Quarterly* 28.2 (1974), pp. 39–47, p. 44f.

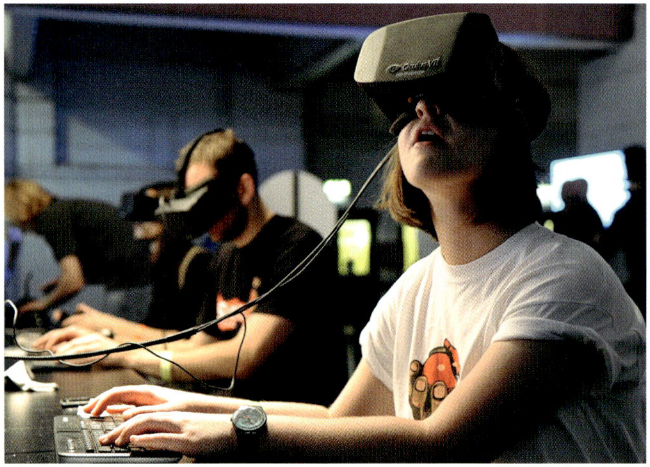

2 Apparatus_1: Users with VR headsets.

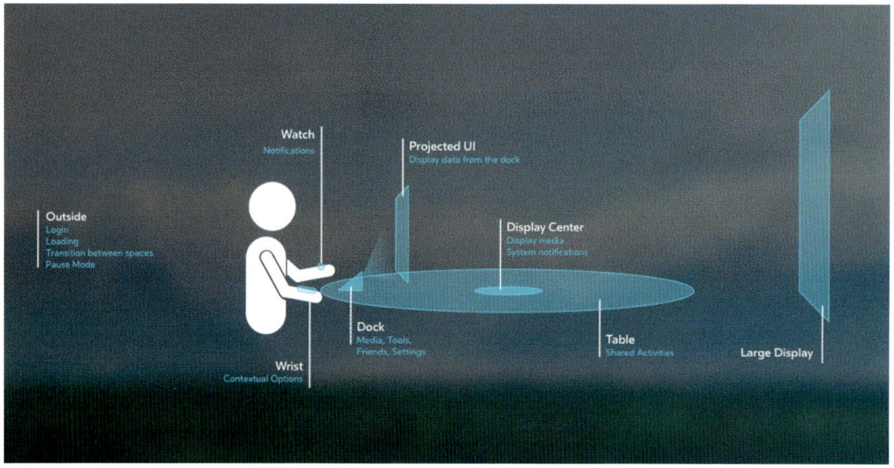

3 Apparatus_2: The *dispositif* of Facebook Spaces.

able to see the edges of the screen, screens in VR occupy the user's whole field of vision which even makes it necessary to simulate screens virtually to allow for specific operations. Therefore, one defining trait of VR one must consider if its ideological effects are to be analyzed is that, unlike cinema, it not only consists of the technological apparatus and content of the images but also of another mediating virtual layer.

In the case of Facebook Spaces, one would thus have to assume a double apparatus in the sense that the user first has to set up a space for the hardware, don a VR headset, and get proficient with a set of physical controllers. These taken together comprise a material interface arrangement or apparatus_1 that has become an iconic visual reference to VR technology in press reporting and advertising alike (fig. 2). The hardware in turn allows entry into a virtual action space that will be addressed as apparatus_2 in the following: It constitutes a visual setting with specific affordances and limitations that assigns subject positions and regulates participants' social behavior. We are interested in the set-up of this second apparatus and will further specify its various components as the interface mise-en-scène of Facebook Spaces (fig. 3).

In the most recent public beta version, the environment of Facebook Spaces is imagined not so much as the "infinite wonderland" of cyberspace so prevalent in the cyberpunk imaginary of the 1980s and 90s but rather more closely resembles the familiar surroundings of a "dinner party" with family and friends.[24] That is, if one is willing to ignore the highly technical composition of this virtual

24 The juxtaposition of wonderland and dinner party is taken from Rachel Rubin Franklin. See Peter Rubin, Facebook's Bizarre VR App Is Exactly Why Zuck Bought Oculus, https://wired.com/2017/04/facebook-spaces-vr-for-your-friends/ (accessed November 3, 2017). Both, of course, can be traced back to Carroll's literary blueprint.

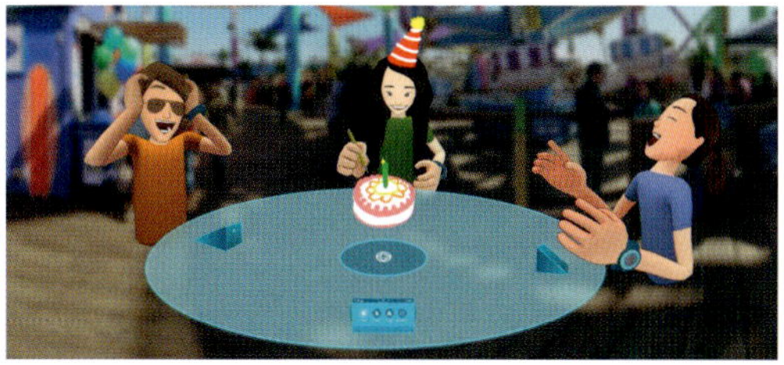

4 Birthday party at the virtual table of Facebook Spaces.

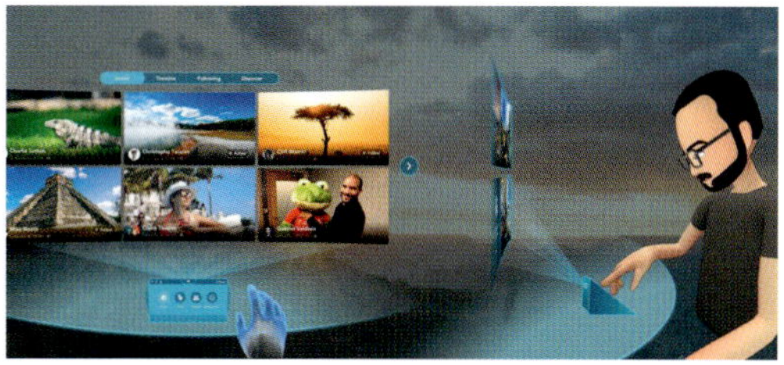

5 VR Dock 2.0, main navigation tool of Facebook Spaces.

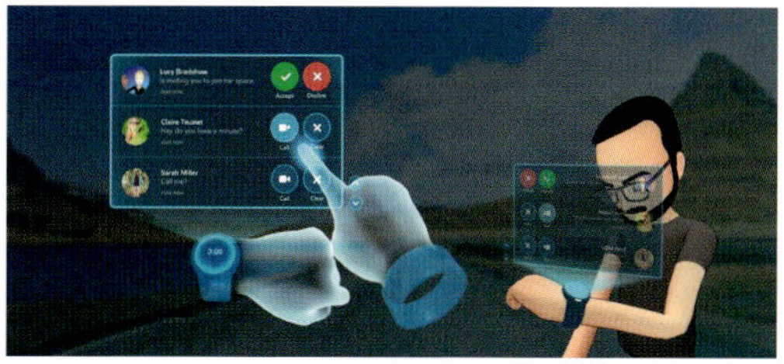

6 VR Watch 2.0, notification center in Facebook Spaces.

dinner party: The participants' cartoon-like avatars – that can be modelled after the users' likings using a photograph in the initial setup phase – gather around a virtual table that acts as the center for a diverse range of practices (fig. 4).[25] Each avatar is assigned a fixed position at the table of which four are available in total. Directly in front of each avatar is a small projector – dubbed VR Dock 2.0 by the developers – that acts as an individual access point to the users' personal Facebook content like images and videos, as well as third party content and a palette of interface tools like a pencil, a selfie-stick with a camera and a mirror (fig. 5). Apart from the dock, each user has at his or her disposal a virtual watch fitted to the avatar's left wrist that delivers notifications, e. g. about incoming calls via the Messenger app, and a user interface placed on the right wrist which gives contextual options equivalent to a right mouse-click in a desktop graphical user interface (fig. 6). The middle of the table houses the so-called Display Center: Media content can be placed in this circular area via hand gesture and is then either projected onto a large display visible to all participants in the case of traditional photographs or videos, or – in the case of 360° video content – onto the (imagined) walls of the virtual enclosure to constitute a shared environment. Sociality in Facebook Spaces is fundamentally and purely a screen-based practice, with the added twist that screens only ever appear as virtual constructs inside the user's perceivable action space. The scenario could also be described as an echo chamber of projections cohabited by up to four subjects that share traces of their memories in a consensually constructed dream world (fig. 4–6).

While a lot of the elements of the interface mise-en-scène of Facebook Spaces are quite innovative, the designers of Facebook Spaces draw on several interrelated HCI (human-computer interaction) conventions and established action patterns to define the operational modalities of the virtual environment. First among these is the idea of "direct manipulation" that has been popularized by the desktop metaphor of graphical user interfaces (GUIs) since the 1960s: Users can handle graphical representations of data like personal files and folders with intuitive gestures without the need to attain expert programming skills.[26]

25 The following description of the various interface elements of Facebook Spaces is mainly adopted from Tauziet's informative article mentioned above.

26 See Florian Hadler, Daniel Irrgang, Instant Sensemaking, Immersion and Invisibility. Notes on the Genealogy of Interface Paradigms, in: *Punctum* 1.1 (2015), pp. 7–25.

Many action routines situated in this paradigm have been further simplified and extended with the popularization of touchscreen interfaces, especially in mobile devices like smartphones.[27] Thus, many of the surfaces inside the apparatus_2 of Facebook Spaces are "touch-sensitive" and react to gestural inputs.[28] The designers also resort to general ideas from the tangible interaction paradigm first introduced by Hiroshi Ishii and Brygg Ullmer from the MIT Media Lab: Whereas the original vision of "tangible bits" aimed at augmenting physical objects to bridge "the gap between the worlds of bits and atoms", in Facebook Spaces users act in a completely virtual environment inside which abstract data processes are translated into physical activities with a spatial dimension.[29] For example, when using an in-built feature to live-broadcast from Facebook Spaces, a stream of friend's comments is visualized in the virtual environment and users can pull single comments out of this stream and interact with them spatially as if they were large sheets of paper.[30]

27 See Timo Kaerlein, Aporias of the Touchscreen. On the Promises and Perils of a Ubiquitous Computing, in: *NECSUS. European Journal of Media Studies* 1/2 (2012), https://necsus-ejms.org/aporias-of-the-touchscreen-on-the-promises-and-perils-of-a-ubiquitous-technology/ (accessed November 3, 2017).

28 The question of whether interactions inside a virtual environment can and should still be addressed as screen operations or whether it makes more sense to treat them as a new category in HCI has been debated as early as 1991 in Meredith Bricken, Virtual Worlds. No Interface to Design, in: Michael Benedikt (ed.), *Cyberspace. First Steps*, Cambridge, MA: MIT Press, 1992, pp. 363–382. Bricken wholeheartedly affirms a paradigm shift "between traditional interface design and designing virtual worlds" that is compared to the difference between watching the ocean from a boat and diving into it with a scuba gear set. Ibid., p. 364.

29 Hiroshi Ishii, Brygg Ullmer, Tangible Bits. Towards Seamless Interfaces Between People, Bits and Atoms, in: *Proceedings of the ACM SIGCHI Conference on Human Factors in Computing Systems (CHI '97)*, New York: ACM, 1997, pp. 234–241, p. 240.

30 See Mike Booth, Live from Facebook Spaces: A New Way to Share VR with Friends, https://newsroom.fb.com/news/2017/07/live-from-facebook-spaces/ (accessed November 3, 2017).

Other user-created objects like sketches, drawings and photographs made with the selfie-stick, or drawn from users' accounts constantly and increasingly litter the shared space of the virtual table or float freely around the avatars. All in all, this quickly leads to a dizzying array of visual elements that can get overwhelming and messy, which is documented by user experience videos uploaded to YouTube.[31] In the next section, we will generalize from these usability issues and understand them as indicating a representational crisis of the hybrid image space constituted by Facebook Spaces. The virtual table, employed as an element of the interface mise-en-scène to constitute a "space of affordances and possibilities structured into organization for use", inadvertently produces this crisis in the first place.[32]

Plights of the Round Table – How to Control a Hybrid Image Space

While Facebook has been discussed as a heterotopia before, we propose that this holds true even more for Facebook Spaces.[33] Its heterotopic character can be described on several levels: We would like to argue that VR in general constitutes a heterotopia on the level of apparatus_1, whereas the social VR scenario on the level of apparatus_2 intensifies this heterotopic character by drawing together and juxtaposing diverse types of images and screens. The hybrid image space thus constructed is the subject of interface design efforts to make it cohere and counter its diverging tendencies.

31 See TWit Netcast Network, Facebook Spaces VR Test Drive, https://youtube.com/watch?v=_kGRpSd4vnc (accessed November 3, 2017).

32 Drucker 2013 (as fn. 5), p. 31.

33 See Robin Rymarczuk, Maarten Derksen, Different Spaces. Exploring Facebook as Heterotopia, in: *First Monday* 19.6 (2014), http://firstmonday.org/ojs/index.php/fm/article/view/5006/4091 (accessed November 3, 2017).

First and foremost, Facebook Spaces is built on a system of openings and closures that is fundamental for its operations. As Foucault describes in his well-known lecture *Of Other Spaces*, entering a heterotopia is often regulated by rites or acts of purification, which play an import part in setting these places apart by isolating them from normal places.[34] In the case of Facebook Spaces, as with any other virtual environment, this division is implemented by the apparatus_1, i.e. the hardware and software necessary to enter VR. In particular, the head-mounted display serves to exclude the user's perception of the actual surrounding space while enclosing him or her in a virtual image space.[35] At the same time, the user's body moves in a space measured by several sensors, which capture head and hand movements to map them onto the avatar's body in virtual space, thereby inducing a strong reality effect.[36]

But once they enter Facebook Spaces, users find that they are not allowed to wander around in virtual space at their leisure, rather they are embodied as leg-less avatars gathered around a virtual table. These avatars are fixed in a position from which every user could at any time see any of the up to three other users in their instantiation of Facebook Spaces and with whom they could now engage. The possible interactions themselves are highly preformatted by the interface, making it difficult to talk of actions in an emphatic sense. Especially screen practices within the virtual environment mimic highly conventionalized actual practices with the help of virtual objects characterized by certain affordances, e.g. taking a selfie with a virtual selfie-stick.[37] Glitches aside, the interface arrangement of Facebook Spaces creates a "regime of control" which, at first glance, contradicts the rhetorics of limitless freedom generally applied to VR.[38] But, as became apparent during the design process, another system of opening and closing, not unlike the one granting access to the greater heterotopia of VR itself, had to be established to connect users to their friends via Facebook Spaces. The designers soon discovered that one particularity of VR is that many problems of actual space repeat themselves within the virtual space they created. The reproduction of more traditional social settings chosen in older virtual worlds and other contemporary social VR applications (like living rooms or bars) did not bring their users together effectively enough to let them engage in social interactions. On the contrary, "when able to freely move around, people tended to get lost and weren't really interacting with each other", according to Facebook Spaces lead designer Christophe Tauziet.[39] Therefore, the seemingly rigidly controlled virtual action space we addressed earlier as apparatus_2 was implemented to more closely define the range of possible social connections and interactions granted by the user interface. The key design element of this solution is the virtual table. Media theorist Walter Seitter acknowledged the mediality of tables early on, describing their ability to keep things – and people,

34 See Michel Foucault, Of Other Spaces. Utopias and Heterotopias, in: Neil Leach (Hg.), *Rethinking Architecture. A Reader in Cultural Architecture*, New York: Routledge, 1997, pp. 330–336.

35 See Michael Friedmann, Kathrin Friedrich, Moritz Queisner, Christian Stein, Conceptualizing Screen Practices. How Head-Mounted Displays Transform Action and Perception, in: *Media Tropes* VI.1 (2016), pp. i–v.

36 See Hartmut Winkler, Reality Engines. Filmischer Realismus und Virtuelle Realität, http://homepages.uni-paderborn.de/winkler/reality.html (accessed November 3, 2017).

37 The status of the resulting pictures remains unclear: Should one consider them as photographic images or rather as screenshots?

38 Sabine Wirth, Between Interactivity, Control, and 'Everydayness'. Towards a Theory of User Interfaces, in: Florian Hadler, Joachim Haupt (ed.) *Interface Critique*, Berlin: Kulturverlag Kadmos, 2016, pp. 17–35, p. 18.

39 Tauziet (as fn. 4).

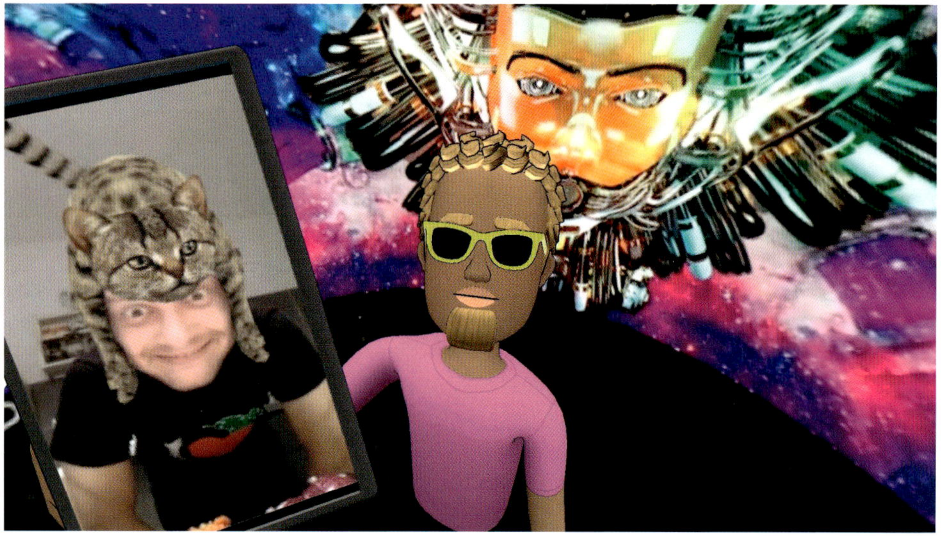

7 Hybrid image space – Video chatting within Facebook Spaces.

one might add – together and present them to make them available for communication ("*Verkehr*").[40] The designers of Facebook Spaces tried to employ exactly this ability to keep things together in VR.

However, the fundamental logistical capabilities of the table to draw things and people together in social VR are challenged by the same acts of communication it makes possible in the first place. One of the ways users can interact is by taking and sharing pictures and videos, thereby perforating the virtual space and linking it with other media and actual spaces. Foucault described the ability of heterotopias to juxtapose several spaces in one space that are in themselves incompatible – a heterotopia is not just a different space, it also brings together different spaces.[41] Tellingly, among the examples he chose to illustrate the concept of heterotopia are the theater, which brings several places onto the stage, and the cinema, "a very odd rectangular room, at the end of which, on a two-dimensional screen, one sees the projection of a three-dimensional space".[42] Foucault's examples seem rather tame compared to Facebook Spaces. In fact, if you have stood around a virtual table floating in a neon-colored shifting psychedelic space age scenario while video chatting with a friend wearing a digital cat on his head and then attempt to document the action using a virtual selfie stick, you might wish yourself back in the manageable space of a cinema (fig. 7).

All these heterogeneous spaces, like the users themselves, are centered around the virtual table to create a disturbing onslaught of many different screens and images with varying degrees of realism that seem to collapse onto the user. The designers foresaw this possibility which led to the decision to support users with the ability to pause their experience should it become too overwhelming.[43] What the designers did not anticipate, or at least not explicitly provide for, is the effect that the interface design has on the referential status of images churning through the apparatus.

We would like to argue that the effects of apparatus_1, the HMD and sensors in your living room, and apparatus_2, the virtual table and its plethora of gateways into

40 Seitter 2001 (as fn. 8), pp. 178–179.

41 See Foucault 1997 (as fn. 34), p. 334.

42 Ibid.

43 See Tauziet (as fn. 4): "Whenever people want to *take a break* from their experience, whether that's because the pizza delivery guy is knocking at the door, there's a destabilizing shaky 360 video around them, or they're feeling uncomfortable, they can pause their experience by pressing the *pause button* located on the inside of their wrist, or by taking their headset off. This teleports them out of the space momentarily and into a 'paused space', giving them a chance to catch their breath and take action if needed (resetting the space, muting/kicking people out, reporting content...)."

other image spaces, interfere in a specific way. Not unlike the apparatus described by Baudry, the apparatus_1 of VR evokes an *impression of reality* that is, if anything, stronger than in cinema, because the subject in VR actually occupies the perspective focal point from which space is constructed. As perception and action are tightly coupled in VR, the ensuing sensorimotor coupling between the image and the user's body can create highly convincing illusions of embodiment.[44] But at the same time apparatus_2 leaves a strong *impression of unreality* as it cuts any referential links images might have held in the past. This effect is increased by the CGI-based user avatars whose positioning and appearance is completely arbitrary with reference to the photographic image backgrounds. Images in Facebook Spaces may change places, be replaced or be subjected to post-production effects at a whim, whether or not they themselves were calculated or taken. Even though it is well-known that digital or digitized images can be manipulated in this way, *traditional* screen practices constituted a symbolic space explicitly distinct from actual space to enable these kinds of operations. The same does not hold true for VR which does not place its screen before a user as a manipulable object but wraps itself around the user's head.

Whereas cinema, as described by Baudry, depended on an interplay of psychological and architectural mechanisms to render its apparatus invisible, apparatus_1 of VR is physically invisible because it is situated outside of the user's perceivable space while simultaneously constituting this perceivable space.[45] HMDs make it specifically their point to place the user within a symbolic space which claims to be real while shutting out non-symbolic space – screens are no longer an object within the users' field of perception but their only means of visual perception.[46] The distinction between symbolic and non-symbolic space is further undermined by Facebook's advertising language, which reproduces well-known topoi of presence and immersion, by promising to bring people together in one room and to "transport you to new places" with the help of 360° videos.[47] In effect, one may say that Facebook Spaces, due to the interference of apparatus_1 and _2, generates a *real* symbolic, within which the distinction of real and symbolic collapses, thereby evoking an *impression of surrealism*, i. e. seemingly realistic representations of an unlikely and often bizarre character.

44 To some extent, this has already been the case for videogames. See Serjoscha Wiemer, Körpergrenzen. Zum Verhältnis von Spieler und Bild in Videospielen, in: Britta Neitzel, Rolf F. Nohr (eds.), *Das Spiel mit dem Medium. Partizipation – Immersion – Interaktion*, Marburg: Schüren, 2006, pp. 240–260.

45 In turn, the general invisibility of apparatus_1 often makes it necessary to simulate visual representations of physical controllers, keyboards and other input devices inside apparatus_2. This leads to the effect that users interact with images of devices they are actually holding in their hands because their field of vision is blocked by the head-mounted display.

46 This observation is supported by an article on wired.de reporting from Facebook's developer conference F8. Visitors who wanted to test Facebook Spaces were presented the headset by Facebook employees with the words: "Here are your eyes." Elisabeth Oberndorfer, F8. Die neue Social-VR-App von Facebook im Test, https://wired.de/collection/tech/facebook-spaces-vr-virtual-reality-oculus-rift-app (accessed November 3, 2017).

47 Facebook, https://oculus.com/experiences/rift/1036793313023466/ (accessed November 3, 2017).

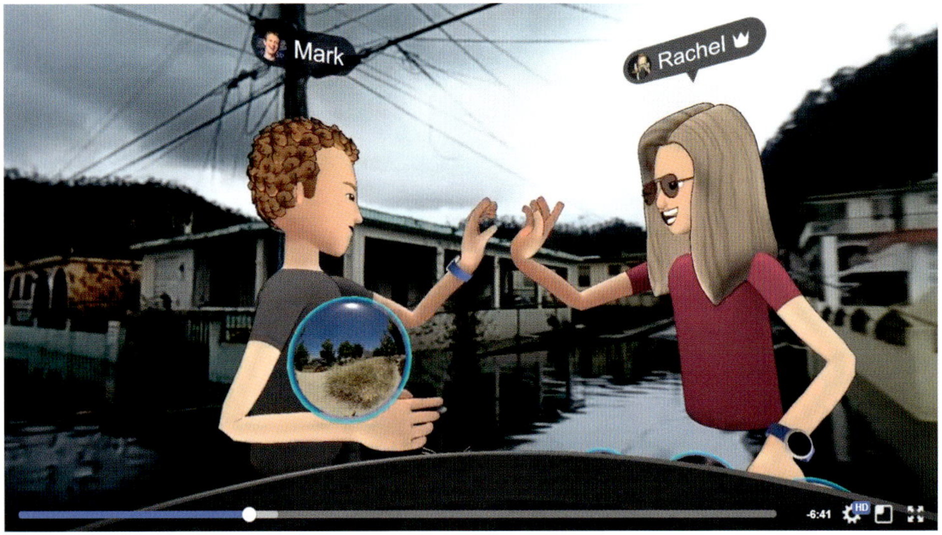

8 Mark Zuckerberg and Rachel Franklin visiting Puerto Rico in VR.

"A new era of reality has arrived."[48] Virtual Surrealism and the Loss of Referentiality

A different scene, uploaded to Facebook on 9 October 2017, shows Mark Zuckerberg and Rachel Franklin in a promotional video, advertising the features of Facebook Spaces while teleporting to different locations around the world, most remarkably a 360° video of Puerto Rico devastated by hurricane Maria. Their good-humored avatars frolicking around and high-fiving in front of images of the catastrophe, Zuckerberg and Rubin indulge in the experience of being there without leaving the comfort of their respective offices – "one of the things that's really magical about VR is you can get the feeling you're really in a place" (fig. 8).[49] It is here, in this "bizarre"[50] and "awkward"[51] video that was almost instantly perceived as a huge PR disaster, that the apparatus of Facebook Spaces instantiates what can most accurately be described as a disturbing kind of virtual surrealism. Despite all assurances of experiencing a sense of presence, the effect of the apparatus amounts not so much to an illusion of transparency, but an *illusion of homogeneity* of the images acting as the background for the virtual sociality of Facebook Spaces. (fig. 8)

This illusion – the sense that there are no substantial differences between the referents of the images processed by the apparatus – is ultimately grounded in the digital character of these images: Digital photography has finally lost all traces of indexicality in the world of Facebook Spaces. The smartphone camera has indeed advanced to "the first [mass-distributed, TK/CK] augmented reality platform" that includes possibilities of enhancing one's images with various special effects and of manipulating photos with the help of advanced object recognition capabilities, e.g. to remove elements from them that disturb the staging of a perfect vacation shot.[52] Accordingly, images on Facebook

48 HTC Vive, Vive Pre CES 2016, https://youtube.com/watch?v=CB9ecPgZlq0 (accessed November 3, 2017).

49 Mark Zuckerberg, as quoted in: Olivia Solon, Mark Zuckerberg "Tours" Flooded Puerto Rico in Bizarre Virtual Reality Promo, https://theguardian.com/technology/2017/oct/09/mark-zuckerberg-facebook-puerto-rico-virtual-reality (accessed November 3, 2017).

50 Ibid.

51 Alison Maine, Mark Zuckerberg Took His VR Avatar to Puerto Rico, and It Was Just So Awkward, http://mashable.com/2017/10/09/mark-zuckerberg-virtual-reality-fail-puerto-rico/#Zgoe6d4m_qqC (accessed November 3, 2017).

52 Mark Zuckerberg in Engadget, Facebook Spaces Announcement | F8 in Under 10 Minutes, https://youtube.com/watch?v=JXYmpqdhHzg (accessed November 3, 2017). In the same video, Zuckerberg explains another feature with the words: "You can add a second coffee mug so it looks like you're not having breakfast alone."

are becoming more and more self-referential in that they cease to point to any external referents whatsoever and instead exhibit the near limitless possibilities of digital image manipulation.

Currently, influential voices like VR filmmaker Chris Milk talk of VR as an "empathy machine", and some attention has been attracted by the application of VR technology to treat US soldiers' PTSD by confronting veterans with vivid simulations of traumatic events.[53] Our analysis of Facebook Spaces runs counter to these expectations and conceptualizations by grounding the unsettling moral indifference documented by the Zuckerberg/Rubin video media-theoretically in the fundamental disinterest towards the referential status of the images that are made part of the apparatus of Facebook Spaces. Virtual surrealism in the sense described here entails a form of reality-agnosticism: The question if and how something is grounded in any materiality outside the apparatus seems to have lost all relevance in Facebook Spaces. By providing the sort of interface criticism we have undertaken in our article, we can shed light on some of the ways that the operationality of VR interfaces itself contributes, once again, to an agony of the real.

53 Chris Milk, How Virtual Reality Can Create the Ultimate Empathy Machine, https://ted.com/talks/chris_milk_how_virtual_reality_can_create_the_ultimate_empathy_machine, (accessed November 3, 2017). See also Kathrin Friedrich, Therapeutic Media: Treating PTSD with Virtual Reality Exposure Therapy, in: *MediaTropes eJournal VI.1* (2016), pp. 86–113.

Figures

1 Facebook, Mark Zuckerberg.

2 BagoGames, https://flickr.com/photos/bagogames/26104037025, Creative Commons Attribution 2.0 Generic (CC BY 2.0).

3–6 Facebook, permission to use granted by Christophe Tauziet.

7 Jan Kaerlein, Timo Kaerlein, Christian Köhler.

8 Facebook.

Sarah Atkinson

Synchronic Simulacinematics

The Live Performance of Film Production

When I think back, the notion of a *simulacinematic* space was first invoked in my own mind by an unpleasant memory I had whilst visiting Universal Studios Hollywood in 2000 where I experienced the *Backdraft* theme park attraction. *Backdraft*, a 1991 film directed by Ron Howard, starring William Baldwin and Kurt Russell, focused on fire fighters and their dangerous encounters with extreme conflagrations. The theme park attraction simulated the most dramatic conditions of the film replete with flames, smoke and danger. The *Backdraft* attraction follows a lineage of theme park attractions that involve inferno-like conditions – for example "Fighting the Flames" was one such attraction at Dreamland in Coney Island in the early 20th century.[1] The main *Backdraft* experience took place upon a 500,000 cubic foot soundstage, dressed as the empty warehouse simulating the climactic moment of the film. As the director yells *Action* a fire is seen to start in an office at the other side of the warehouse. When searching for documentation relating to the *Backdraft* attraction, I found this quote from the director Ron Howard:

> *The screenwriter, Gregory Widen, was a former fireman and he made it very clear that in our movie the firefighters had to be right there in the middle of those blazing infernos. But I knew that getting these shots would be very hazardous for the actors and the entire film crew. Everyday, they found themselves in the middle of flame, smoke, flying ash and toppling scenery.*[2]

This quote is indicative of the frequency, during interviews, where film industry practitioners seek to call our attention to the drama of the film's making, and thus, by extension, how materials and attractions, such as *Backdraft* not only seek to position audience members in the fictional world of the film, but into the subjectivity of one of the film's makers – to vicariously experience the drama of the film's production.

I stood there in abject terror throughout, shuffling myself toward the exit door, as most of the other thrill-seekers shouted and cheered in appreciation, I experienced a visceral sense of panic, what if this is real? – no one will know or realize. Unbeknown to me at the time of that experience, but on recent research, I discovered that on 24 September 1992, a fire had broken out in the air conditioning ducts above

1 Andrea Stulman Dennett, Nina Warnke, Disaster Spectacles at the Turn of the Century, in: *Film History* 4.2 (1990), pp. 101–111.

2 Ron Howard, Scene by Scene Walkthrough. Scene 1, http://thestudiotour.com/wp/studios/universal-studios-hollywood/theme-park/past-attractions/backdraft/ (accessed January 23, 2018).

the attraction. Around 500 people were in the attraction at the time, and to quote a news article – "No members of the public panicked as they thought the black smoke was part of the presentation."[3] So my unease was entirely founded!

This particular experience is not just of interest to me, because of its conflation of the filmic text and the conditions of its making (the presence of a director's voice, and the acknowledgement by the director of the same sense felt on set), but because of the experiential affective space that is occupied by the audience. It is a space that is routinely inhabited by performing artists, actors, film production personnel (as the example of Backdraft illuminates) – it is the assimilation by the audience into a simulacinematic space which is the central concern of this essay.

Simulacinema is a portmanteau term – a combination of the words *simulation* and *cinema* – which I am using to account for a phenomenon in which an audience simultaneously experiences both the space of the filmic diegesis and/or the cinematic spectacle, and the attendant, but crucially, *simulated* space of its production. Simulacinematic spaces are characterized by the uncanny sense of inhabiting two conflicting ontological spaces (fiction and reality) whilst also embodying two diametrically opposed subjectivities (observer and participant). Simulacinematics refers to the aesthetic and affective qualities of these spaces that merge film style and visual cinematic codes with production aesthetics, as well as the live and the mediated elements of their experience. The making and reception of a film tend to be chronologically displaced moments – but within simulacinematic phenomena, by contrast, the two moments are folded together into simultaneous experiences in which the two temporalities converge in an experiential modality.[4]

Simulacinema is becoming an increasingly frequent phenomena as a result of digital technologies and their use and application in cinema spectatorship, as well as a result of the evolution of cinematic commodification – the expansion of the cinematic text across different forms and platforms – coupled with the commodification of cinematic *experience*. Where Thomas Elsaesser has previously made a distinction "between 'cinema' (event and experience) and 'film' (text and work)",[5] I would introduce a third axis – that of filmmaking (process). In simulacinema, I would contend that the dividing line has become increasingly blurred between these three – text, experience and process.

Simulacinemic phenomena, which are characterized by both aesthetic and affective qualities, have manifested in a number of different forms and contexts whereby the ontologies of cinematic production and reception are experienced by the audience. Instances of simulacinema have occurred on the set of film locations in major cities, within immersive theatrical cinema experiences (such as Secret Cinema) and on-stage theatrical performances that blend stage and screen techniques and aesthetics.

3 Backdraft On Fire, http://thestudiotour.com/wp/studios/universal-studios-hollywood/theme-park/past-attractions/backdraft/ (accessed January 23, 2018).

4 I here build on Guy Debord's concept of the spectacle, Jean Baudrillard's notion of simulacra and the work of Tom Gunning, in examining the history of fairground and cinematic trajectories of showcasing of technological apparatus from the birth of cinema. See Guy Debord, *The Society of the Spectacle*, Detroit: Black & Red, 1970; Jean Baudrillard, *Simulacra and simulation*, Ann Arbor: University of Michigan Press, 1994; Tom Gunning, The Cinema of Attraction. Early Film, Its Spectator and the Avant-Garde, in: Thomas Elsaesser, Adam Barker (eds.), *Early Cinema. Space, Frame, Narrative*, London: British Film Institute, 1990, pp. 56–62.

5 Thomas Elsaesser, Digital Cinema. Convergence or Contradiction?, in: C. Vernallis, A. Herzog, J. Richardson (eds.), *The Oxford Handbook of Sound and Image in Digital Media*, Oxford: Oxford University Press, 2013, pp. 13–44, p. 25.

Within simulacinematic phenomena, I have identified three different types of subjectivities or experiential modalities relating to temporalities – future, retrospective and present, I refer to these as: Prochronistic, Parachronistic and Synchronic. The first two types relate specifically to the marketization of cinematic experience, and are symptomatic of the confluence between audience and fan practices and their exploitation by film marketers and the film industry. The simulacinematic in these two cases emerges as a symptom and as effect of these two phenomena, as opposed to a deliberate intervention or strategy on the part of the filmmakers or distributors. I position this in the wider trend towards the exploitation of the economy of film production (what John Caldwell has referred to as the *Para-Industry*) where making-of content becomes the marketing material.[6] The *synchronic* manifestations, which are the central concern of this essay, are at the very creative edges and manifest as a deliberate experimentation in the mediation and manufacture of screen-based texts. In all three instances we are able to examine what happens in the conflation between the live and the mediated, what happens between the screen and the physical space – and in each instance there is a different relation. As I will go on to examine, in prochronistic moments, the *peripheral* screen practices and engagements of the audience characterize these moments, in parachronistic, the cinema screen is *embedded* in the experience, the screen is *centralized* in synchronic simulacinema.

I will firstly outline the key principles and characteristics of both prochronistic and parachronistic phenomena before examining synchronic simulacinema in more detail.

Prochronistic simulacinematic moments are created and experienced during the production of the film. I point to examples of the manifestation of prochronistic simulacinema, *Transformers 4: Age of Extinction* (2014) and *Suicide Squad* (2016).[7] These are films that were both filmed in various cities across the world. In the case of *Transformers 4*, multiple locations were used in the USA, Hong Kong and in mainland China. *Suicide Squad* was filmed at various locations across Canada (in particular in downtown Toronto) and Chicago. In both cases, spectacular car-chase sequences, crashes, explosions and destruction were filmed in the inner-city locations. Audience members were given a glimpse of the stunts and effects that were to come in the final films, and in close enough proximity to capture these moments on their portable devices and then to distribute them across various social media channels.[8] Given the context of the co-production between China and America, particularly in relation to *Transformers 4*, these highly public choices of location appear deliberate. In the case of *Suicide Squad*, the sets were left in-situ on the streets of Toronto, after filming had completed, as on-street installations which audience members could visit and be photographed against.[9] Here, there is a complete collapse between production, promotion and reception. Audience members are present at the time of the spectacle whilst are also witness to the point of its capture. They are immersed in the milieu of the fictional space at the same time at the point of its making. And this is the key principle of simulacinematic

6 John Thornton Caldwell, Para-Industry, Shadow Academy, in: *Cultural Studies* 28.4, 2014, pp. 720–740.

7 Michael Bay, *Transformers 4. Age of Extinction*, USA/China 2014; David Ayer, *Suicide Squad*, USA 2016.

8 Kevin B. Lee, *Transformers: The Premake*, 2014, https://vimeo.com/94101046 (accessed February 20, 2018).

9 Aynne Kokas, *Hollywood made in China*, Oakland: University of California Press, 2017.

phenomena – the simultaneous presence of the audience in both the manufacture and the experience of the cinematic spectacle. As Stephen Heath stated in 1980: "Resting on an industrialisable technological base, cinema, different to theatre, offers the *possibility of an industry of spectacle.*"[10] This notion of the industry being a spectacle in and of itself is bound up within the aesthetic and affective pleasures of the film theme park, which I alluded to in the introduction to this chapter. James Moran also alludes to this sensation at Universal Studios written in 1994:

> *Universal's blockbusters spill over the screens as interactive spectacles, which in turn sprawl onto the studio lots where they were originally spawned in a cycle that increasingly blurs production and exhibition, 'reality' and representation, 'art' and entertainment.*[11]

It is through the presence of digital technologies that these lines are not only being blurred, but they collapse entirely. In prochronistic moments this occurs through the unification of the screen and the space of production through the presence of audience members' screen capture equipment. The use of mobile phones and recording devices pre-mediate the onscreen action in these moments. In parachronistic moments the cinema screen acts as the linking screen interface between the action taking place around it.

Parachronistic moments are created long *after* production has taken place – during the formal reception phase of a film. Such instances emerge under a different commercial (pre-promotional) imperative to the former category. They have predominantly emerged in contexts of the generation of (3rd party) retrospective revenue, through the screening of old, cult films. These instances can be aligned to the film theme park modality, where films, and their making, are repackaged and re-experienced for audience's years after their release – i. e. the *Jaws* exhibit at Universal Studios.

UK-based organization Secret Cinema deliver immersive experiences around a film screening through a recreation and reinterpretation of the fictional world of the film. On the surface, these productions encourage and engender a variety of fan practices such as singing, dancing, cos-playing and quoting-along to the film being screened. Furthermore, and as a by-product, these productions, in their elaborate restaging of a cinematic fictional universe, invariably mobilizes the mechanics of the film production industry machine, through the hiring of film production personnel (set builders, sound designers, actors and stunt performers), and through working with film distributors to secure licenses for screenings and in some cases new releases. In an article on Secret Cinema's instantiation of *Back to the Future*, we drew out the significance of the emulation of the filmic world and how an aesthetics of production was embedded within the experience:

> *As with the* Back to the Future *event, it became apparent very early on in the experience that as an audience member you are not actually immersed in the world of Hill Valley, you are immersed in the world of its making – such was the presence of the physical*

10 Stephen Heath, The Cinematic Apparatus. Technology as Historical and Cultural Form, in: Teresa De Lauretis, Stephen Heath (eds.), *The Cinematic Apparatus*, New York: St. Martin's Press, 1980, pp. 1–13, p. 7 [emphasis added by the author].

11 James Moran, Reading and Riding the Cinema of Attractions at Universal Studios, in: *Spectator* 14.1 (1994), pp. 78–91, p. 79.

> *evidence of its construction (scaffolding, light rigs and scenery), populated by stunt vehicles, production and security personnel.*[12]

Parachronistic simualcinematic moments therefore become characterized by these unintentional, accidental instances of staging filmmaking aesthetics. The presence of the screen calls to attention the audience's awareness of the construction and the artifice of film – as it arguably happens in all simulacinematic moments. In these cases, the screen is embedded into the experience itself (the screen is literally framed within the experiential space – in the screening of *Back to the Future*, the screen is centralized in the Hill Valley town hall façade; in *Moulin Rouge*, the screen is framed by the stage of the famous Parisian night club).

The reception of a filmic text is bound up in the appreciation of its making, so much so, that the two conflate. As with the former category – this is not necessarily a conscious decision undertaken on the part of the creators but rather a symptom of the always-intertwined nature of film production and film reception, as well as dual audience and fan pleasures of meta-filmic awareness.

The most sophisticated and complex form of simulacinema and the most conscious form through artistic intentionality (I have argued that the other forms are unconscious/ unintentional on the part of the creators) is synchronic simulacinema. These are moments created during production – designed to be appreciated in synchronicity with the output of the finished text. Synchronic simulacinematics are highly reflexive and afford a critical edge to understanding this phenomena of dual audience pleasures. In this manifestation of simulacinema we witness a celebration of the cinematic apparatus and the visual spectacle that this creates. It is a complex space where theatre and cinema coalesce and the cinematic production process is itself conceptualized as a form of live theatre. In synchronic simulacinematic case studies, the screen is absolutely central to the concurrent creation of an image for the screen. The creative actions of film production and practice come into focus and transform themselves into theatre-show.

I am putting forward two case studies where the act of production becomes the act of performance and there is a simultaneous collapse of production, performance, capture, transmission and reception. The first is a strand of work by theatre director Katie Mitchell – and its evolution over three productions – *Waves* (2006), *...some trace of her* (2008), and *Forbidden Zone* (2014).[13] The second is Kid Koala's *Nufonia Must Fall Live* (2014).[14] The analysis of these two different examples has been undertaken through the study of their associated documentary videos in order to examine both the form and techniques of the pieces, as well as the discourse of their description. The documentation and framing of these two examples are key to conveying their simulacinematic qualities, and themselves become part of the economy of film and theatre production. It is the simultaneity of screen

12 Sarah Atkinson, H. W. Kennedy, From Conflict to Revolution. The Secret Aesthetic and Narrative Npatialisation in Immersive Cinema Experience Design, in: *Participations. Journal of Audience & Reception Studies* 13.1 (2016), pp. 252–279, p. 274. See Sarah Atkinson, H. W. Kennedy, 'Tell no one': Cinema as Game-space. Audience Participation, Performance and Play, in: *G|A|M|E. The Italian Journal of Game Studies* 4 (2015), https://gamejournal.it/atkinson_kennedy/ (accessed August 20, 2017).

13 Katie Mitchell, *Waves*, London: National Theatre 2006; Katie Mitchell, *... some trace of her*, London: National Theatre 2008; Katie Mitchell, *Forbidden Zone*, London: 59 Productions 2014.

14 K. K. Barrett, Kid Koala (Erik San), *Nufonia Must Fall*, Montreal: Envision Management & Production 2016, Live Performances, Ann Arbor Center, March 11–12, 2016; K. K. Barrett, *KID KOALA, NUFONIA MUST FALL LIVE!* [Official Trailer], 2016, https://youtube.com/watch?v=s_DhuuHt76M (accessed August 20, 2017).

space and physical space showing the same events that is the principle difference to the previous instances of simulacinema that I have described.

The two case studies are drawn from a number of notable examples of projects, which have sometimes been referred to as "Live Cinema".[15] These include works by Film Live in Italy, a group of artists whose practice is to make and broadcast films live – "a movie that is filmed at the same time that it is screened".[16] Francis Ford Coppola has also worked in a similar mode with his project *Distant Vision*, which he also refers to as "live cinema" where a film was shot and broadcast live to screenings rooms on 22 July 2016 after 26 days of rehearsal.[17] Coppola positions live cinema in contradistinction to live multi-camera broadcast, associated with the televisual:

> *I felt the need to experiment in order to learn the actual methodology of live cinema, which is a hybrid of theater, film and television. The shot is the basic element, as in film; the live performance is from theater; and the advanced television technology to enable it is borrowed from TV sports.*[18]

59 Productions and Katie Mitchell refer to the third and final piece *Forbidden Zone* production as live cinema – a theatre production which is simultaneously being performed, filmed, projected and observed live on a screen above the stage, underneath which audience members are able to see the inner workings of the film set in which the on-screen action is being shot. Production crew, i.e. camera operators and sound recordists, negotiate the film set in full view of the audience as they frame the action and capture the performance. This viewing mode, in which the audience can constantly switch between the registers of fictionality and its construction invokes a metafictional experience and awareness, which can on the one hand create as Patricia Waugh states a "fiction that both creates an illusion and lays bare that illusion"[19] or as Thomas Elsaesser has noted: "the production process can take on a textual form".[20] There is a key distinction to be made here between *live cinema* and *simulacinema*. Synchronic simulacinema occurs when audiences have access and are witness to both the on and off screen spaces, and not just the on screen-output as is the case in the Coppola example. In synchronic simulacinema, the audience can take an active role in what they choose to focus their attention upon; they take the vantage point of a director who watches both the monitor of the camera output on set and the production itself. The distinction is that the director has the active power to make choices about where the camera directs its lens, whereas the audience are passive in this regard.

The format of *Forbidden Zone* is based on Mitchell's earlier productions *Waves* and *…some trace of her*, which at that point were referred to as a *multi media* productions.

15 Sarah Atkinson, H.W. Kennedy (eds.), *Live Cinema. Cultures, Economies, Aesthetics*, New York: Bloomsbury, 2017.
16 http://film-live.org (accessed January 18, 2018).
17 Bill Desowitz, Francis Ford Coppola Completes 'Distant Vision' Live Cinema Workshop at UCLA, http://indiewire.com/2016/07/francis-ford-coppola-completes-distant-vision-live-cinema-workshop-at-ucla-1201709229/ (accessed August 15, 2017).
18 Dave McNary, Variety Francis Ford Coppola Starts Experimental 'Live Cinema' Project at UCLA, http://variety.com/2016/film/news/francis-ford-coppola-experimental-live-cinema-ucla-1201820998/ (accessed August 15, 2017).
19 Patricia Waugh, *Metafiction. The Theory and Practice of Self-Conscious Fiction*, London: Routledge, 1984, p. 6.
20 Thomas Elsaesser, Fantasy Island. Dream Logic as Production Logic, in: Thomas Elsaesser, Kay Hoffman (eds.), *Cinema Futures. Cain, Abel or Cable? The Screen Arts in the Digital Age*, Amsterdam: Amsterdam University Press, 1998, pp. 143–157, p. 143.

Waves was a work devised from the fragmented text of Virginia Woolf's novel, *The Waves* (1931). *...some trace of her* is inspired by Fyodor Dostoevsky's *The Idiot* (1868–1869). Both later productions include the visual production of live sound effects and real-time video captured and projected on-stage.[21] For the purposes of this essay I am focusing upon the principles, techniques and aesthetics that are deployed in Mitchell's productions in relation to the staging of production aesthetics and mechanics. I posit that these are not transparent filmic productions-techniques but artistic techniques.

Within all three Mitchell productions, the performative aspects of film production are staged through the use of real-time production aesthetics. These provide a simulational tendency themselves as on the set of *actual* film productions action is fragmented across scenes and takes, action is cut for cameras, lights and production personnel to re-set. Classical narrative film production is a mode that has persisted since its establishment in the 1890s – very often dictated by the economics of production and the availability of locations and performers. Mitchell describes this approach as "fragmenting the stage picture, combining video output with the live construction of it", whilst Ben Whishaw who plays the character of Myschkin in *...some trace of her* explains it as "the juxtaposition of image and the artificialness of the way that image is made".[22] Whishaw goes onto describe how another performer plays his hands, how another speaks his thoughts and how he "just provides his face".[23] This proves consistent with traditional filmmaking conventions in what could be referred to as the "cinefication" of the theatre.[24] In these examples, we see how theatre explores and reveals, whilst cinema and film continue to conceal. These techniques raise questions around whether what is being produced is for screen or stage *consumption*, or in this case, the in-between – the simulacinematic space. In the most advanced and technically sophisticated iteration of the simulacinemtic aesthetic in *Forbidden Zone*, it is the digital which is foregrounded in both form and content. Digital technologies make possible the live and simultaneous capture and broadcast of image and sound, whilst digital aesthetics are laid bare in its presentation. In Mitchell's earlier works, it is the analogue, the craft-based, the *handmade* (foley production techniques) that are experimented with. Mitchell describes the approach to the live creation of sound effects through foley as the theatricalization of sound "where the image you see on screen is matched with something entirely different, the realization in the performance and then the audiences as they watched this creation live".[25] There is an example where a performer taps a piece of chalk on a blackboard in order to create the sound of fingers tapping keys on a typewriter. Here we witness the sound being created for the film soundtrack, the film audience. Performing it in this way creates an audio/visual disjuncture, whilst also invoking the loss of the real, the absence of the authentic.

21 For in-depth analyses of *Waves* see Louise LePage, Posthuman Perspectives and Postdramatic Theatre. The Theory and Practice of Hybrid Ontology in Katie Mitchell's The Waves, in: *Cultura, lenguaje y representación. revista de estudios culturales de la Universitat Jaume I* 6 (2008), pp. 137–149; Janis Jefferies, '... some trace of her'. Katie Mitchell's Waves in Multimedia Performance, in: *Women. A Cultural Review* 22.4 (2011), pp. 400–410.

22 National Theatre Discover, *Katie Mitchell on directing multimedia productions*, 2011, https://youtube.com/watch?v=rAij9r9RvF0&t=2s (accessed August 20, 2017).

23 National Theatre Discover, *Ben Whishaw on Acting in a multimedia production*, 2011, https://youtube.com/watch?v=5hK0y8tN29w (accessed August 20, 2017).

24 Vsevolod Meyerhold, The reconstruction of the theatre, in: Edward Braun (ed.), *Meyerhold on Theatre*, New York: Bloomsbury, 2014, pp. 253–273.

25 National Theatre Discover, *Sound design for '...some trace of her'*, 2011, https://youtube.com/watch?v=THpcmuKNumY (accessed August 20, 2017).

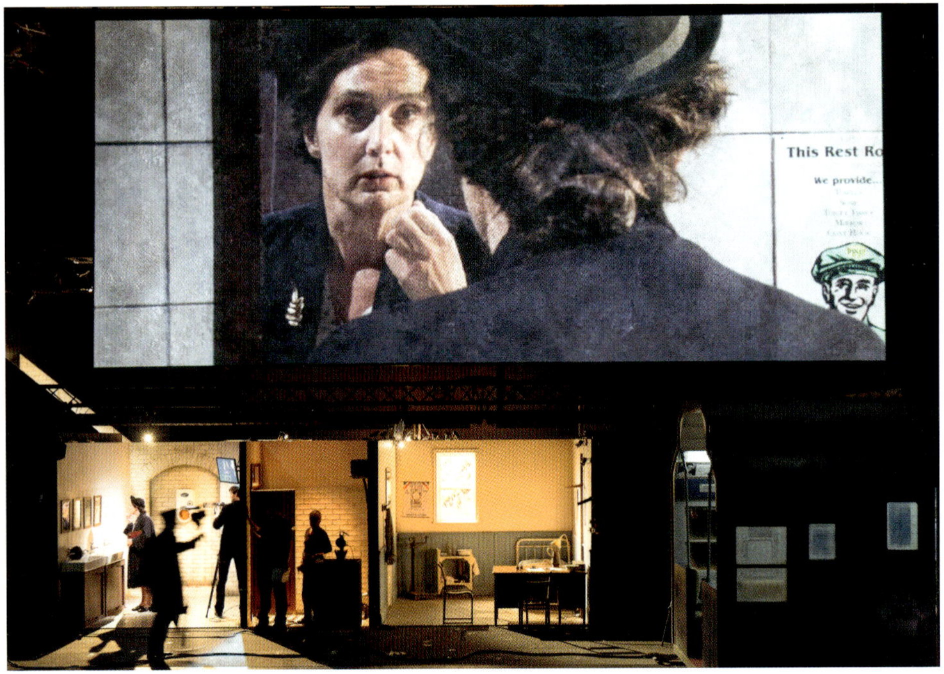

1 *Forbidden Zone.* The camera operator captures the performer looking in the mirror in the set in the bottom left.

As Figure 1 illustrates, the overall formal quality of Mitchell's stage production creates a visual multi-screen *picture-in-picture* effect, invoking a digital aesthetic (a trope in various multi-screened films that proliferated in the early 2000s with the advent of digital editing, and an aesthetic of digital postproduction edit interfaces). The viewer is positioned as voyeur and is witness to the various screened vignettes, reminiscent of the aesthetic of Alfred Hitchcock's *Rear Window* (1954).

A principle characteristic of synchronic simulacinema is that the productions are experienced in real-time – the seduction of this aesthetic prevails in one-take cinema.[26] Time on stage unfolds at the same time as time on screen, and it is the complexities and complications of this audiovisual achievement that are laid bare in the production. The fluid on-screen action and the flowing movement of the characters is juxtaposed and contradicted by the on-stage presence of the fragmented set and the urgent choreography of the production personnel as they negotiate the presence of the apparatus – the cabling, lights, set etc. This invokes the labour in the viewer who also has to visually negotiate and cohere the audio/visual complexity – this affective labour is a key trait of simulacinema: while time is parallel in the theatre-space and screen-space, dimensionality or space is not.

The accompanying documentary by 59 Productions presents the drama of the production and the high stakes that at any moment – anything can go wrong, all at once deliberately inhabiting, celebrating and performing "the commercial drama of a movie's source".[27] This sense of liveness is intrinsic to the synchronic simulacinematic. The presence of the spectator at the moment of capture is central to the experience, whilst the live is also a marketing and promotional tool, serving to historicize these instances as *on-off*, *unique* and *ground-breaking*. Similarly, in the next case study under consideration, *Nufonia Must Fall Live*, the director K. K. Barrett emphasizes the importance of liveness:

26 See Sarah Atkinson, "You sure that's a film, man?" Audience anticipation, expectation and engagement in Lost in London LIVE, in: *Participations. Journal of Audience and Reception Studies* 14.2 (2017), pp. 697–713.

27 Timothy Corrigan, *A Cinema without Walls. Movies and Culture after Vietnam*, New Brunswick: Rutgers University Press, 1991, p. 118.

> *[I]n this world of being able to get anything on-demand, you have to come to the theatre currently to see this show, that's what makes it special [...] it's going to evaporate after this until its next performance.*[28]

Nufonia Must Fall Live was a simultaneous stage and screen performance described as *a live silent film* conceived by Eric San, better known as Kid Koala adapted from his silent comic book of the same name. In the 60-minute performance piece, the on-screen animation was performed live through puppetry which the audience could watch on the screen above the stage and below the screen (fig. 2). The stage consisted of 12 different model sets lit with LED, which were animated by puppets and puppeteers, all visible to the audience, along with a camera operator, sound engineer and video editor. The performance was accompanied by a string quartet and Koala's own scratch DJ music. San describes it as

> *the most complicated show that I've ever been a part of, it's like this eight-ring circus, amazing energy and chaos on stage; but what we hope about this eight-ring circus is this fluid, unified feeling that we create on screen and for the people in the audience just to drift away with that story; but then they can have that picture-in-picture vibe where they can look down and see how it's all happening in real time.*[29]

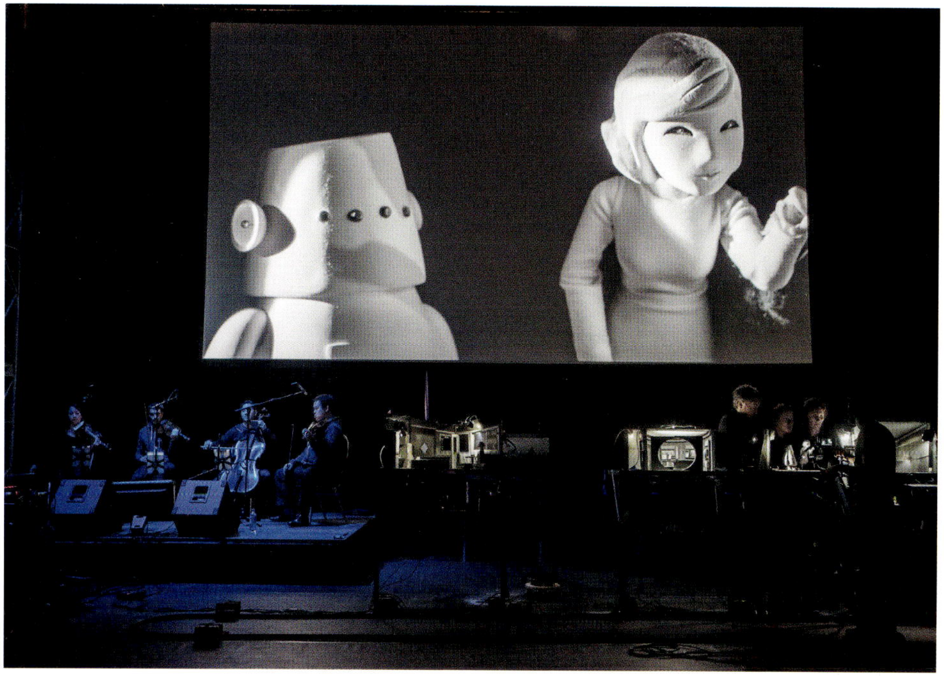

2 *Nufonia Must Fall Live.* The three puppeteers to the right of the stage are visibly manipulating the puppets within the set that we see on the screen above the stage. The four musicians on the left play the accompanying soundtrack.

Here San alludes to a similar effect created in Mitchell's work using the language of new digital media – *real-time* and *picture-in-picture*. The unification which San seeks to achieve between the performance and its making on the surface appears to be highly challenging since animation production is the antithesis of live viewing – it requires extensive and timely production work. The meticulousness and effort of stop-frame animation processes is very often celebrated and foregrounded in industry discourses (i.e. Aardman animation feature films reportedly take eighteen

28 Banff Centre for Arts and Creativity, *Nufonia Must Fall: A Making of the Stage Production*, 2014, https://youtube.com/watch?v=K01BWCWk6ek (accessed February 4, 2018).

29 The Creators Project, *Puppets, Turntables, And A String Quartet | K.K. Barrett & Kid Koala's "Nufonia Must Fall Live"*, 2014, https://youtube.com/watch?v=HFOImWFUL7k (accessed August 20, 2017).

months to shoot with 25 to 30 animators working across 25 sets at once).[30]

In *Nufonia Must Fall Live*, it is a theatrical form of animation – live puppetry – that is produced and staged. As such, the performance follows two different cultural trajectories, the first being that of the fairground and the showcasing of film apparatus and the second that of the making-of and behind the scenes, the fascination with the process and *magic* of filmmaking which is as old as the history of cinema itself.

Firstly, San alludes to the spectacle of the moving image through his reference to the circus. The simultaneous viewing of the illusion and its source has its trajectory in cine-fairground attractions as Gunning states "the earliest years of exhibition the cinema itself was an attraction",[31] using the term "attraction" to emphasize "that of exhibitionist confrontation rather than diegetic absorption".[32] Film has always showcased its techniques and artistry since its early history and has always delighted through this. Today's viewer knows how film is made and thus takes pleasure in film-production being made into its own form of spectacle, thus becoming an image for a screen without being presented on a screen. With the zoetrope for example, the audience first see the mechanics and the machinery kick into action before looking deeper and immersing themselves in the illusion within. It is here that we see a return to fairground practices of cinema exhibition, where the technology is revealed and showcased before the illusion itself.

Secondly, the revealing of the secrets of animation production has its historical antecedents at the turn of the 20th century where craft-based techniques and the human labour behind the production were revealed in early proto-making-of documentaries.[33]

Through this study, the emergent functions, leitmotifs and aesthetics of simulacinema phenomena have emerged through a fusion of forms that merge the live with the mediated, the screen with the stage, and – through the mobilization of film production machinery – its production mechanics, processes and techniques. The creation of the screen image is central: in parachronistic simulacinema, the screen image is cohered by the spectators, in prochronsitic the screen image is recreated by performers and scenography, and in simulacinema the screen image is subject to live creation.

In Mitchell's work, attention is deliberately placed on the artifice of film-based production; the inauthentic artifice of the construction of the screen image is highlighted. In *Nufonia Must Fall Live*, it is the spectacle of screen-image creation that is foregrounded – the production is a performance and the artform and craft of animation production is celebrated.

Simulacinema is a concept which most usefully helps to extend understandings of the complexities of audience viewing pleasures; the evolution of creative practice in performance and the complexities of the commodified ecosystem of film and cinema including its intrinsic and endemic

30 Kate Abbott, Nick Park, Peter Lord, How we made Wallace and Gromit, in: *The Guardian*, March 3, 2014, https://theguardian.com/tv-and-radio/2014/mar/03/how-we-made-wallace-and-gromit (accessed January 18, 2018).

31 Gunning 1990 (as. fn. 3), p. 58.

32 Ibid., p. 59.

33 Examples include Wallace Carlson, *How Animated Cartoons Are Made*, USA 1919; *How Walt Disney Cartoons Are Made*, USA 1939; Dave Fleischer, *How the Fleischer Studios, Miami, Florida, made 'Aladdin and His Wonderful Lamp'*, USA 1939; Alfred L. Werker, *The Reluctant Dragon*, USA 1941. See a detailed study in Sarah Atkinson, *From Film Practice to Data Process. Production Aesthetics and Representational Practices of a Film Industry in Transition*, Edinburgh: Edinburgh University Press, 2018.

politics. The synchronic simulacinematic examples are suggestive of a politicization of production through the performance of labour and the simultaneous "pseudo visible, hyper invisible" economy of film production.[34]

There is significant labour required of the audience to comprehend and to process these complex experiences. Simulacinema is affectively taxing, and laborious, the viewer has to always shift in focus between two realities to grasp the overall production.

As cinema and theatre continue to evolve, merge and converge, the continued dissimilation of simulations becomes an increasingly complex, yet important task, particularly in relation to the decoding of mediations of film production so that we may evolve critical understandings of the emergent and seductive economy of *live* within the contemporary film and cinema ecology.

Figures

1 Katie Mitchell, *The Forbidden Zone*, Photo: Stephen Cummiskey.

2 Kid Koala, *Nufonia Must Fall* Live, Festival Noorderzon, Photo: Pierre Borasci.

34 Sarah Atkinson, *From Film Practise to Data Process: Production Aesthetics and Representational Practices of a Film Industry in Transition*, Edinburgh: Edinburgh University Press, 2018, pp. 161–168.

Ina Neddermeyer

I Want to See How You See

Curatorial Practices of Exhibiting Virtual Reality

Displaying time-based media art poses major challenges for the presentation and reception in exhibitions. Sound overlays, distraction effects or the missing reset button, which would make it possible to start an artwork as soon as visitors enter the room are everyday problems in curatorial practice. Virtual reality (VR) installations can be understood in this tradition of time-based media art, while at the same time a fundamental change is taking place with the increasing availability of head-mounted displays (HMD) in exhibitions. While VR installations can be displayed on traditional screens (on VR monitors or as cave installations), most artists aim to let their virtual worlds be accessible through HMDs (fig. 1–2). So even if *the* VR artwork cannot be clearly defined, its usual combination with such a screen-set up has to be taken into account discussing the changes in the use and function of VR artworks.

Approaching the topic through curatorial practice gained in the exhibition *Beautiful new worlds. Virtual realities in contemporary art* in the Zeppelin Museum Friedrichshafen, this essay aims to take stock of the possibilities and challenges of VR installations exhibited with HMDs, reflecting the presuppositions and specific modalities of individual artworks. The spectrum ranges from interactive, experience-orientated works to linear-narrative works, from VR displayed on headsets to room-filling installation settings, which also critically deal with the technology and the process of seeing. Despite their diversity all works in the following essay can be characterized as VR-HMD thus virtual realities consisting of CG Imagery generated in real time and displayed on HMDs.

VR Headsets as Unbounded Displays

In VR installations with HMDs, the displays are no longer presented in a statically installed form, e.g. a monitor or a video projection, with which the visitors are usually confronted in time-based media exhibitions. Instead, the mobile display is flexibly available. The application starts whenever the visitors put the headset on and stops as soon as the HMD is taken off.

In this moment the screen as a physical object fades into the background. While video projections and monitors as displays are clearly delineated and always visible, the head mounted displays seem to disappear. In the immersive installations the visitors perceive the virtual images primarily as unframed image spaces. The boundary between image and non-image blurs as the virtual image offers a 360° view in which the screen is not visible anymore. As soon as the VR-HMD is put on, there is no possibility to look away; viewers are completely in the virtual world and with the only possibility to leave being to take the HMD off. Instead of merely viewing an art work from an outside view, virtual

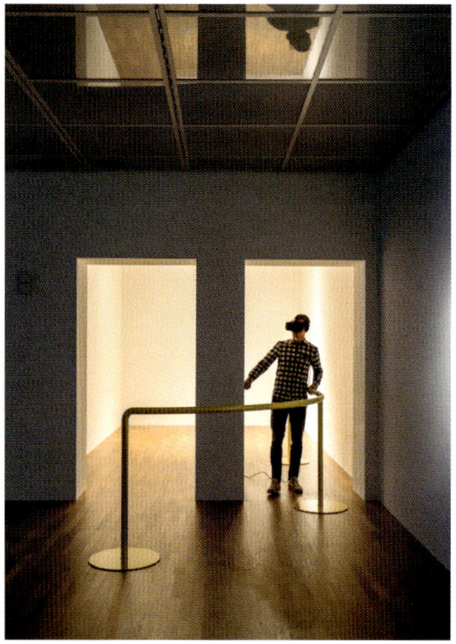

1 *Beautiful new worlds. Virtual realities in contemporary art*, Zeppelin Museum Friedrichshafen, November 11, 2017–April 8, 2018, exhibition view.

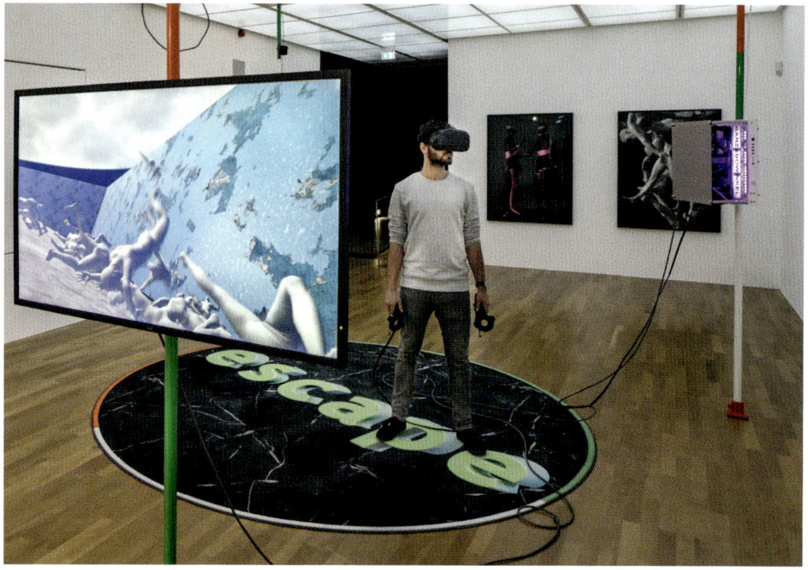

2 Banz & Bowinkel: *Palo Alto*, 2017, anodized and powder coated aluminium tubes, 55" LCD monitor, HTC Vive VR, VR PC, custom printed carpet, installation view.

reality means experiencing works of art from an inside view and thus in a supposedly immersive way. Visitors become a part of the virtual world.

Further, viewers have an active part. They are directly addressed and involved by wearing VR headsets or experiencing interactive elements within the VR installation. Theoretically, it is even possible to move with the HMD relatively freely through the exhibition space. This mobility relates not only to the visitors' external freedom of movement but also to a large area of actions within the VR installation – when, for example, there is no longer an auctorial narrative, but the viewers become protagonists who interact with the virtual environment that is created by the technology. In the installations of the artist duo Banz & Bowinkel the viewer navigates through virtual worlds by means of controllers operated by touching a button that leads to action sequences, such as movements through the space (fig. 2). The narrative is therefore developed non-linearly and always different depending on the choices that were made. These three elements, the inner and outer freedom of action as well as the disappearance of the technical equipment, lead to the impression that VR-HMDs are unbounded displays. But this is more a felt impression, as there are a lot of technical limitations that such installations, as almost all VR installations, impose.

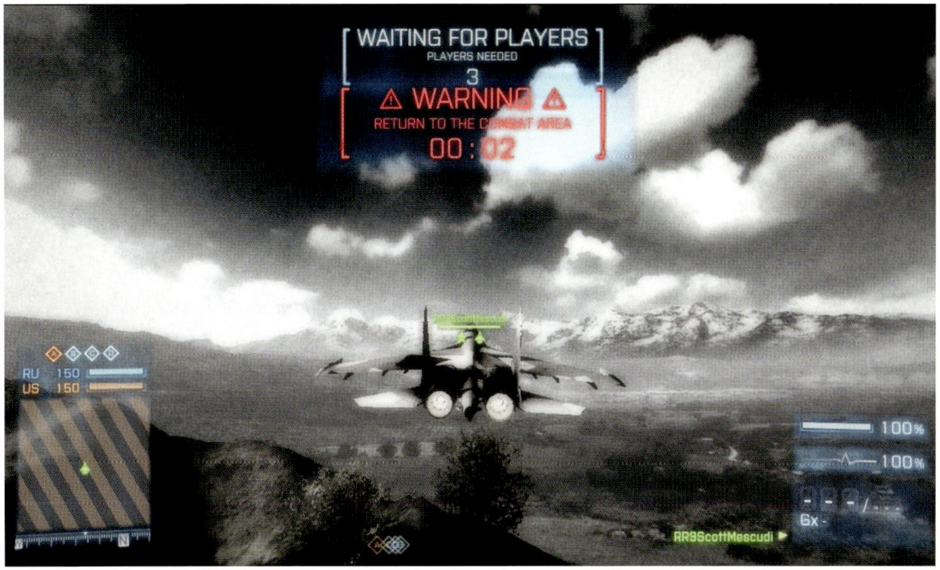

3 Harun Farocki: *Parallel II*, 2014, single channel video installation, color, 9 min., still.

First of all, the so-called invisible technology is revealed by cabling which curbs the movement space of the viewers. Secondly, there are the headsets themselves, which remain visible with their edges and noticeable in their weight. To create a perfect illusion, heavy displays are necessary that counteract with the immersive experience. Thirdly, the size of the exhibition space that can be tracked by infrared cameras is limited due to the bandwidth, and so the scope of action is limited as well. Standalone VR headsets might be launched within near future but quite possibly to the disadvantage of quality. Furthermore, elaborated VR installations that include, besides the VR environment, live simulation, tracking systems and interactive experiences are not as reliable and stable enough for everyday use. For instance, synchronization problems occur (moving one's head does not necessarily result in a simultaneous virtual movement), PC and HMD loose connection, the tracking system might not work or it is necessary to reset the PC due to technical problems.

With his series *Parallel* (2012–2014) Harun Farocki already showed very early that the seemingly unbounded virtual image spaces have perceivable boundaries (fig. 3). In these 2D videos the artist deals with computer generated images, especially with video games. The boundaries are, on the one hand, spatial ones that can be understood as consciously or unconsciously programmed endpoints. They limit the scope of action. Users are asked to return or come back if they are leaving the official part of the game, or insurmountable objects, barriers or parts of the landscapes appear. On the other hand, there are limits on social norms and interactions (i.e. which behaviors cause a reaction and how other players, authorities and leaders in the game impose sanctions). These experiences from video games can be transferred to virtual environments in which the boundaries of virtual grid cages are very quickly perceivable, since viewers might step out of the programmed environments and cause the VR installation to fail.

Immersion and Corporeality

Most of the VR installations tend to aim to sensorily overwhelm the user with immersive effects. Although, probably in the not too distant future, VR-HMDs will become an integral part of everyday life; presently, they represent a new technological setup for a large part of the population, providing an entirely different visual experience. These lead to a strong physical involvement, which is a challenge for the

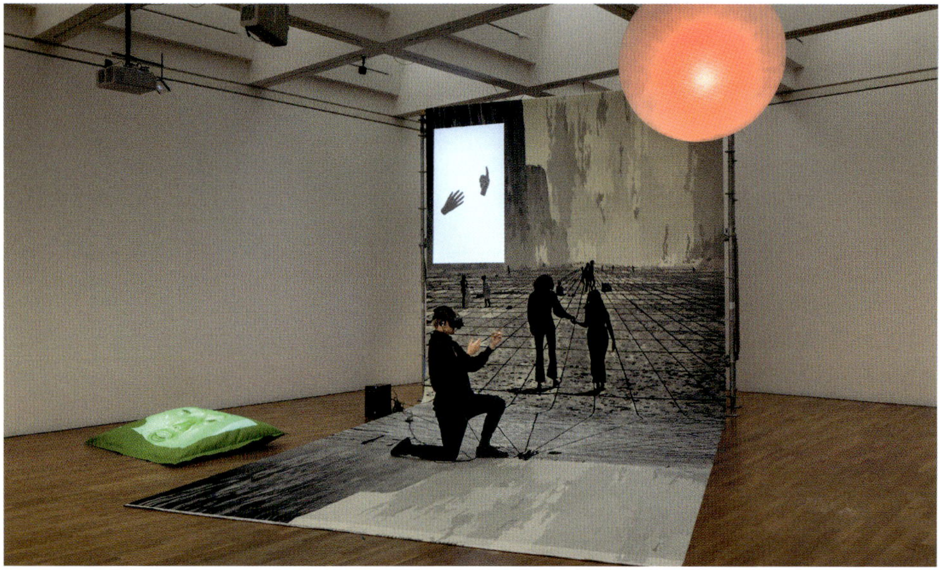

4 Florian Meisenberg: *Pre-Alpha Courtyard Games (raindrops on my cheek)*, 2017, VR-Installation, custom made VR Software build with Unity, VIVE VR Station, Leap Monitor Controller, custom made carpet, scaffold, plexiglass, projectors, installation view.

sense of balance, and reveals a different spatial experience of images as users move and change their perspective in the virtual worlds. At the same time this corporeal involvement is juxtaposed with a distance to the virtual images, because it is not possible to touch and feel objects as physical objects (The Swayze Effect).[1]

VR-HMDs change the relationship between body, space and display. It is not only the exhibition space that is omitted by the virtual reality simulations, but the body seems to disappear too. For the visitors this might offer a very unique experience: When they look down at their bodies, often there is no virtual representation; their bodies are invisible. Some artists therefore use avatars to overcome this difference of the real and the virtual world, and thus giving the invisible body a virtual stand in-representation. It is interesting to note that processes of identification take place very quickly, and the virtual body of an avatar is perceived as one's own physical body.

At the same time, the physical body is strongly emphasized by interactive elements that enable new physical experiences in a lot of VR installation. Head movements lead to a direct change in perspective and angle of view. With the aid of controllers, viewers can move through space. Or the other way around, in Florian Meisenberg's VR installation hand movements in the exhibition room are transferred directly into the virtual space by means of infrared cameras (fig. 4).

Meisenberg created a VR installation on site for the Zeppelin Museum in which he works with the Leap Motion technology as a tracking tool. Instead of a "traditional" VR installation in which visitors have to use a controller to move around, the Leap Motion device makes it possible to track the movements of the hands without using controllers or gloves. The hand movements are directly tracked and translated to actions in the virtual space, where visitors can shape the appearance of wireframe objects intuitively. Here the hands are represented as avatars, as it would be too confusing for the visitors to have no corporal representation in the virtual world. Furthermore, it would be difficult to interact with the technology if the user is unable to see what their hands are doing. In a second step the visitor can place a texture on the wireframe to create a sculpture that can be uploaded on sketchfab.com, an online platform for publishing 3D, AR or VR content. Both the wireframe objects and textures are part of pre-programmed data-

1 https://oculus.com/story-studio/blog/the-swayze-effect/ (accessed June 12, 2018).

base. Consequently, the viewers who are engaged actively are co-producers of these artworks. It has to be mentioned as well that Florian Meisenberg works closely together with a programmer to create the application, thus invigorating the interdisciplinary interface between art and technology.

The VR simulation is embedded in a space-filling installation of a giant carpet hanging from the ceiling that gives the impression of a photographic studio. So Meisenberg is weaving together the virtual and physical layers of the works. A live VR simulation is projected on this carpet. But viewers from outside can only see the isolated gestures of the hand avatars; they are not able to see the artworks.

Through movement and without further tools, one's own physicality becomes even more strongly represented in the virtual environment. This simultaneity of disembodiment on the one hand and at the enhancement of corporeality on the other is to be seen as a defining trait of VR installations, which is reflected on by numerous artists.

Challenges for Mediation

VR installations are a major challenge to master in the exhibition context. Just the simple fact that, the VR-HMDs used in exhibitions are regularly worn by many different people naturally raises questions about hygiene. Due to the close eye contact there is the danger of transmitting viruses and bacteria, e.g. a conjunctivitis. For proper care the use of special VR cleaning sets is necessary, also to maintain the function of the lenses. Despite the fact of producing an incredible amount of waste, we decided that every visitor should receive a personal hygiene cover that can be used at every VR station in the exhibition. Before putting the HMD on, the disposable hygiene cover is placed on the face.

VR works place new demands on art education especially. To be confronted with VR installations for the first time raises simple questions such as: How do I put on a VR-HMD? What about my belongings, if I cannot see (and hear) what is happening right next to me? Do I have to sit down while watching VR videos? The supervisory staff is now no longer solely responsible for guarding the exhibits but must also provide support as a *VR sitter*. They have to pay attention that the visitors do not get tangled up or stumble over a cable while they are *travelling* through virtual worlds.

This support is also necessary as the new spatial experiences make viewers feel their bodies in a very strong and direct sensation. In some cases they suffer from motion sickness. Nausea is caused by the lack of proprioceptive signals corresponding to the impression that you should be moving physically as your point of view changes. For this reason, the first-time or extensive use of VR-HMD is associated with a great deal of effort, especially for untrained visitors, which is revealed in quite physical reactions.

With only a few exceptions, it has generally been possible to watch time-based media art together with others. Viewing a VR installation is a thoroughly solitary experience. It is only possible that one person wears the VR-HMD at a time, so that another person is not able to see the video installation. The other visitors are thus excluded to a certain extend, and therefore the process of being observed while observing is reinforced. Furthermore, talking about the artworks thus becomes a challenge. The process of simultaneous art viewing and verbal exchange becomes a downstream process, and a very practical question for guided tour arises: How is it possible to speak about VR works that not everyone has seen or cannot even get a glimpse at together?

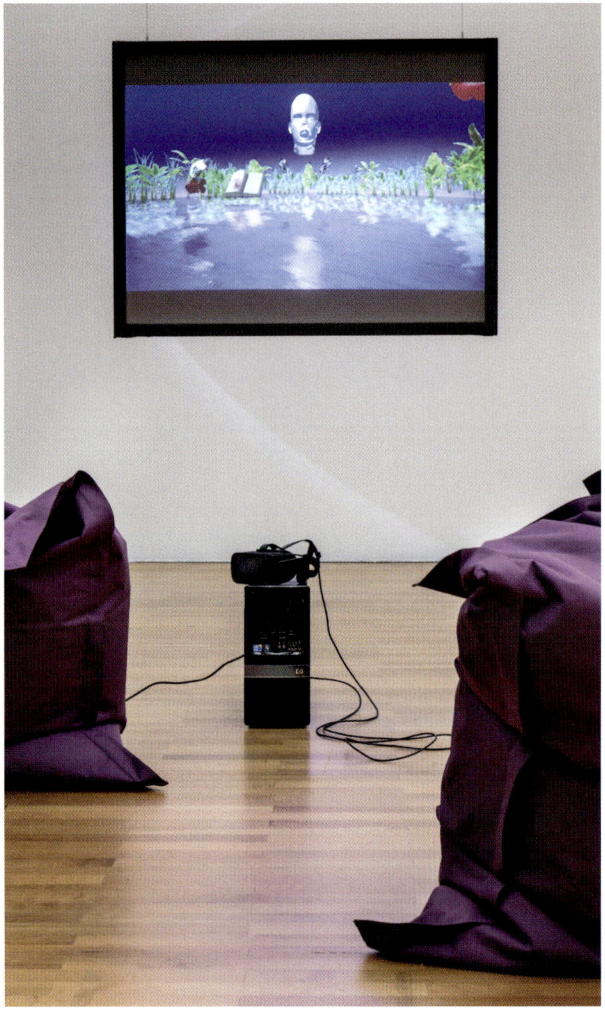

5 Salome Asega & Reese Donohue & Tongkwai Lulin: *ASM(V)R*, 2017, Oculus Rift VR, PC, headphones, projector, bean bag, 6.22 min., installation view.

This evokes the question of how many VR headsets are installed for each artwork in the exhibition space. If only one headset is available, the mediation is difficult at well-attended exhibitions or when groups are present. Projections or monitors have generally been available for (almost) all visitors; here a curatorial decision must be made in the case of works that are not bound to a certain setting. Installing more than one VR-HMD means also financial challenges. A perfect solution would be to enable as many individual art experiences as possible.

A provisional solution, which is already practiced in some exhibitions, is to show, in addition to the VR-HMD, a projection of what the visitor sees in the VR installation. The artists Salome Asega, Reese Donohue and Tongkwai Lulin decided to show their VR installation *ASM(V)R* combined with a welcome screen/image, before the visitor jumps into the Oculus Rift VR-HMD (fig. 5). This way the visitors can get an impression of the work by looking at a 2D video snippet of the VR simulation. It is, of course, important to note that a transfer from VR-HMD set-up to a conventional 2D screen is a profound intervention in the work, whereby the core characteristics of VR, the individual experience of the viewer and the specific reaction of the program to the viewer, are completely eliminated.

Discussing these challenges, it should also be mentioned that some common problems do not occur: Sometimes time-based media art raises questions about the protection of minors when watching videos with pornographic or violent content. This problem is not as virulent when using VR-HMDs, since actively donning the headset is necessary to watch the simulations, and additional warning signs will make sure that no one steps into the artwork by accident.

Artistic Potentials of VR Technologies

VR technologies open up many new possibilities for artists to reflect on their aesthetic potential while at the same time critically examining their technical conditions. Most VR artists give precise instructions how the artworks should be installed, which VR system they prefer, how many headsets have to be installed in an exhibition, if headphones are necessary, the sitting (or standing) accommodations (e.g. color and material of the bean bags) as these aspects are a fundamental part of the artworks. Of course these instructions depend on the specific spatial (and financial) situation of the museum and are adapted for every exhibition.

Sidsel Meineche Hansen's VR installation *Dickgirl 3D(X)* deals with VR pornography as the porn industry is one of the biggest engines of innovation for VR technologies. It is a critical pornographic VR production that examines not only post-human sexuality but also the voyeuristic view. By using VR-HMD and because of the isolation from the outside world, the installation of immersive pornography gives the impression of a private setting. The setting is specifically defined by the artist: The VR-HMD has to be positioned on a black vegan leather beanbag that allows viewers to sit comfortably. This setting of the installation creates the illusion of privacy, of watching porn on your own sofa or bed. At the same time, however, you find yourself in the public setting of an exhibition space, which allows other visitors to watch you watch porn. Thereby the act of waiting for a set of VR-HMD to be freed up while observing other visitors becomes an important part. But only by watching the videos themselves can the viewers understand what they just saw the person before them do. A special situation arises as the knowledge about the VR video generates an information advantage, which makes the visual process appear in a new light that underlines the performative aspect of the installation.

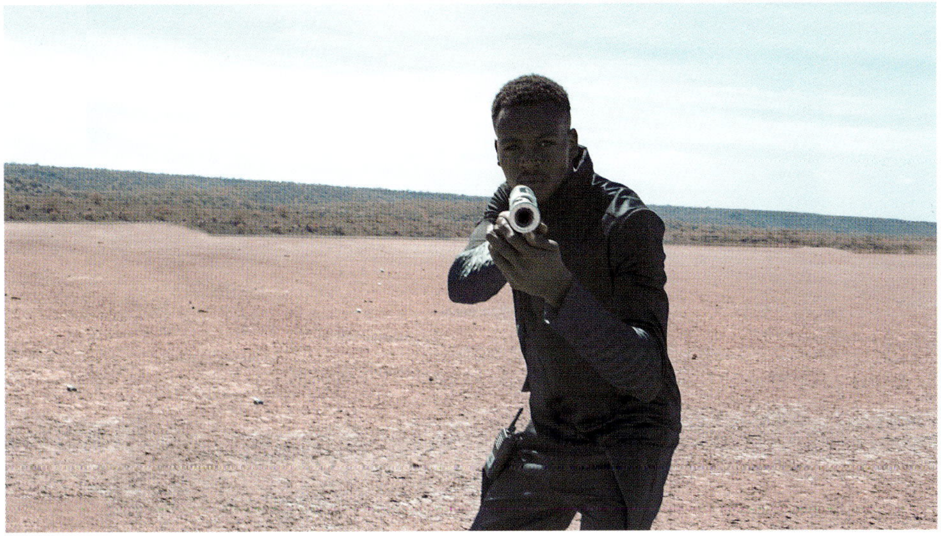

6 The Nest Collective: *Let This Be A Warning – A VR Short Film*, 2017, VR video, 10.43 min., Oculus Rift VR, still.

The artist duo Banz & Bowinkel deals with the question of how virtual image worlds and physical spaces can be connected to one another and how virtual reality can be made visible to outsiders. Their VR headsets are embedded in installations with sculptural elements and printed images. The virtual environment is transmitted to a monitor, so the actions are also visible to outsiders. Banz & Bowinkel thus discuss the visual process as a central element of VR works, which raises questions of participation: Who can see which video and who is excluded? Within their VR installations, there is no linear timeline. There is no particular starting or end point, so the visitors will become protagonists who act

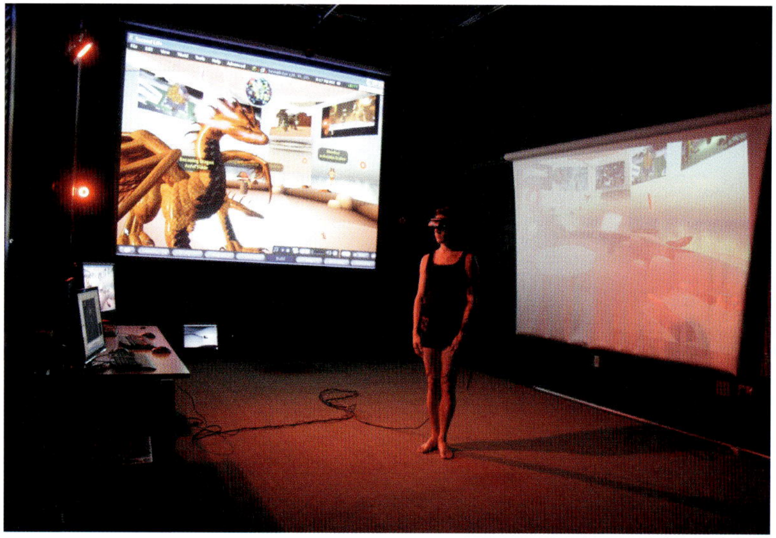

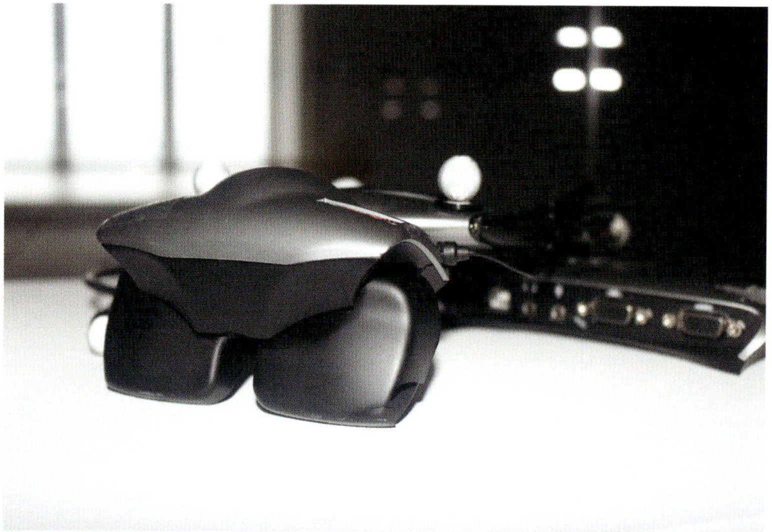

7–8 micha cárdenas: *Becoming Dragon*, 2008, VR-HMD, photographs, video, 3 min., still and exhibition view.

in the virtual and physical world simultaneously. Thus the boundaries of physical and simulated spaces blur.

VR technologies open up new forms of narration as well. As viewers are addressed and involved directly, they can become part of the narration. The Nest Collective, a Kenyan art collective, created a science fiction film as a VR experience. The storyline follows a group of African people who have left earth to colonize a distant planet. Putting on the VR headsets, visitors are instantly immersed in the installation as they arrive on the planet and have to ask themselves if they would be welcomed in a black world (fig. 6).

Curatorial Potentials (and Challenges) of VR Technologies

From a curatorial perspective, the aforementioned challenges underlie the pressing question of how VR works should be installed in an exhibition. Should the technical infrastructure be hidden according to the claims of invisibility or is it necessary to counteract at this point, to show the physical presence of technology? The latter choice enables viewers to adopt a distance to the artwork they are confronted with and hopefully to open up a critical point of view. Therefore, we decided to make the whole technical infrastructure, with all the cables, computers and displays, visible in the exhibition. Next to VR-HMD that can be used as displays, we show a headset that micha cardenás wore in her perfor-

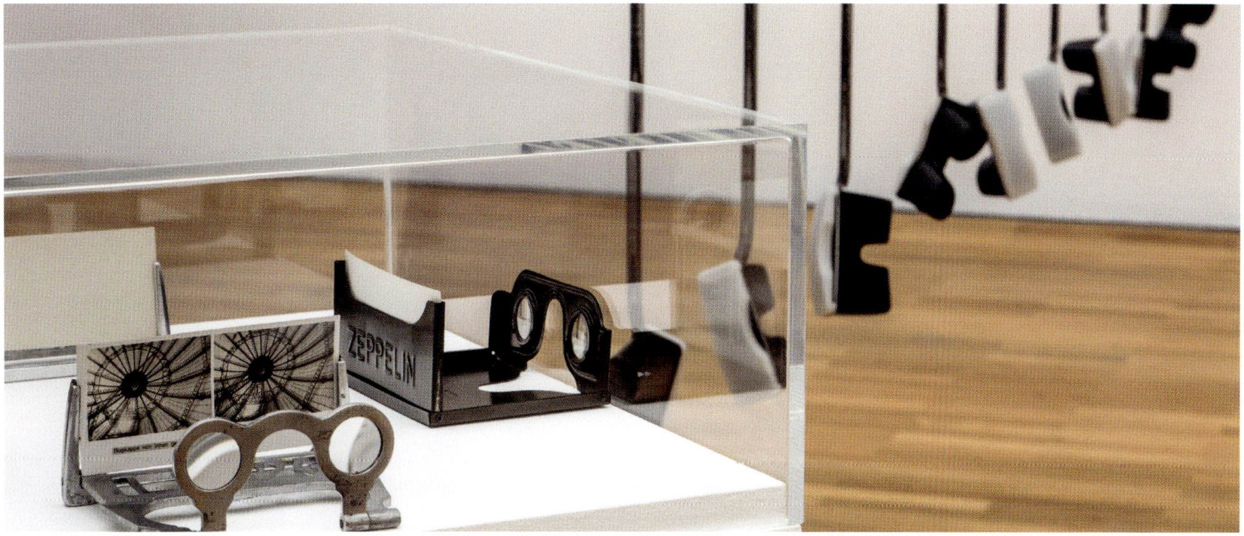

9 *Beautiful new worlds*, exhibition view.

mance (fig. 7–8). It is the only exhibited object that cannot be used. It is placed in a showcase, turning into a museum piece.

Another focal aspect of the exhibition is showing the long tradition of immersion in illusionary worlds that is by no means without its history. Panoramas, dioramas or stereoscopy illustrate the centuries-old history of mankind's interest in immersive media. We have therefore decided that the starting point of the exhibition should be stereoscopic images, which accompany the history of the Zeppelin from 1900 to the 1930s (fig. 9).

VR technologies open up new exhibition formats. In a very radical understanding, the Kunstsammlung NRW established a completely new exhibition format for VR technology with the exhibition *Unreal* in 2017. In contrast to the exhibition *Beautiful new worlds* at the Zeppelin Museum, a purely virtual exhibition space was developed (programmed), which can be entered through a virtual reality lounge in the physical museum. Once the visitors have put on the VR-HMD, they can move through the simulated exhibition rooms with the help of a controller, and zoom into the artworks in the different virtual rooms. Such display formats are, of course, only possible in the case of VR works, in which the installation in the surrounding physical exhibition room is irrelevant to the work. Artistic positions in which the physical exhibition space is part of a specific installation, e.g. with Halil Altindere, who installed his VR work *Journey to Mars* in an oval room and with an airspace wallpaper, are excluded. In *Journey to Mars* the HMD as a display becomes a fundamental part of the artwork as visitors look like astronauts who are flying to Mars. Therefore,

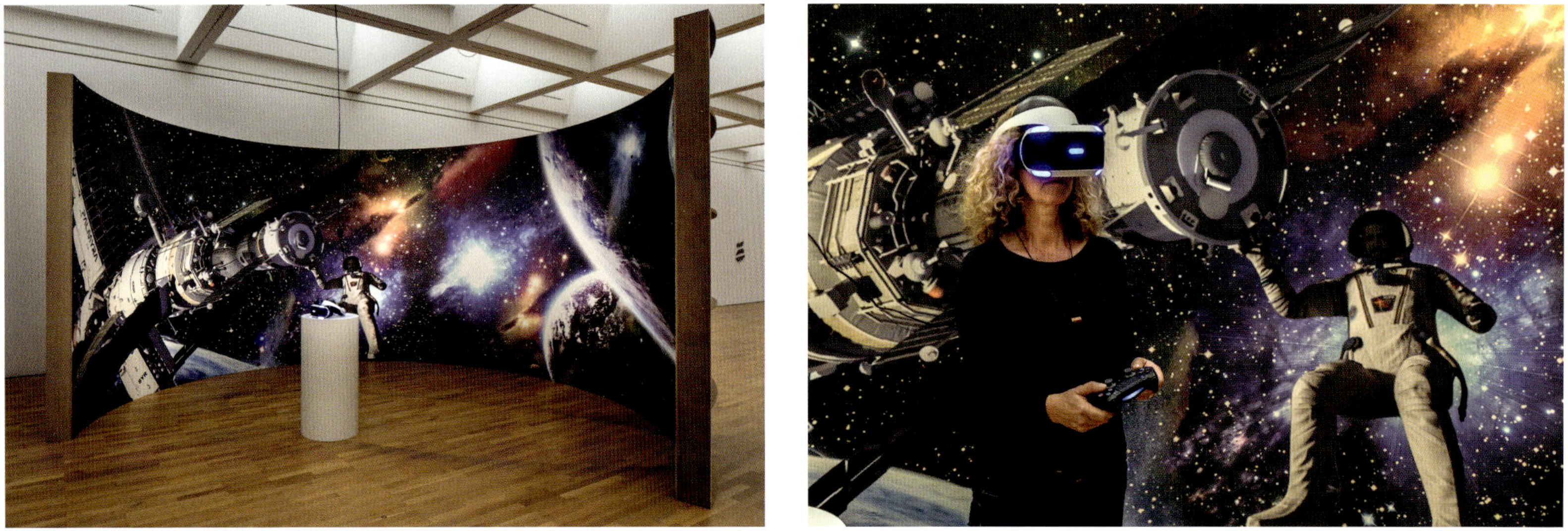

10–11 Halil Altindere: *Journey to Mars*, 2016, 360 VR Video, 4K, 5.10 min., wallpaper, installation views.

the display is transferring the storyline of the VR work (the artist ironically proposes that the refugees be relocated to mars, if no country will receive them: outer space as refuge for the refugees) in the physical exhibition room (fig. 10–11). Thus works which operate at the interface between analogue and digital to query traditional perceptions and definitions of virtual and real worlds cannot be integrated into these solely virtual exhibition formats.

Nevertheless, the exhibition *Unreal* has made it very clear that curators and artists have just begun to explore the potentials the new technology creates for the conception of exhibitions with VR works. With its Arts & Culture App, Google is offering virtual tours through traditional exhibitions of the MoMA in New York, the Naturkundemuseum in Berlin and the Louvre in Paris to make museums, their exhibitions and collection accessible for people far away. That the visit to the physical museum itself will be replaced completely seems rather doubtful nonetheless.

At the moment, it seems more sensible to open an exhibition for the visitor that focuses not only on the experience effect, overwhelming and immersion. Rather, a parcours is necessary, one that places the critical reflection in the foreground. To show the visitors the displays and technologies themselves, as the VR-HMD must be consciously raised and lowered, is a central component of the exhibition. For the exhibition *Beautiful new worlds. Virtual realities in contemporary art* in the Zeppelin Museum a circuit was developed together with the exhibition architects Kooperative für Darstellungspolitik that makes the entanglement of virtual and physical spaces perceptible for the audience

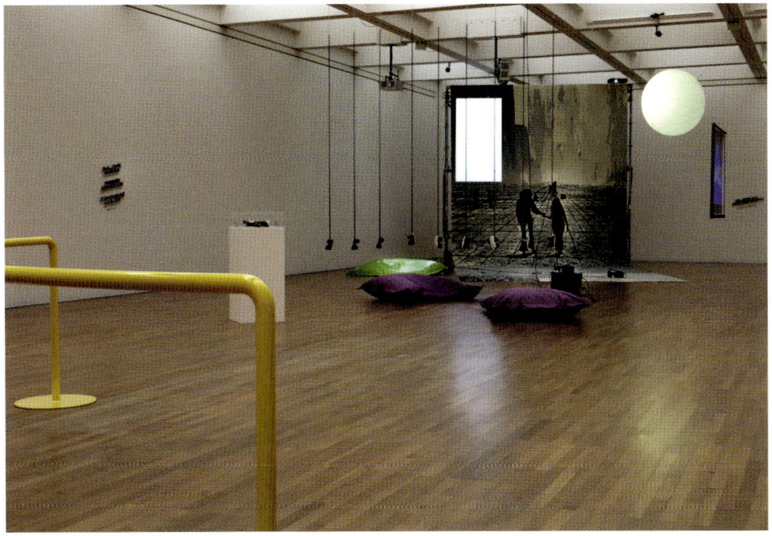

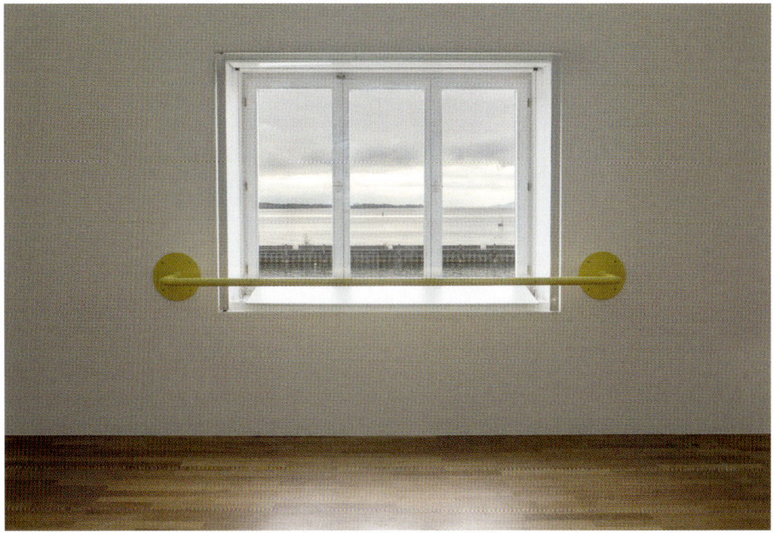

12–13 *Beautiful new worlds*, exhibition views.

through bodily experience. One focal point is to open the window shutters of the exhibition space, which are normally closed to create black boxes for the presentation of time-based media works. Thus it is possible to make a connection between the exhibition space and the world outside and to question what is real, what is virtual and what is just an exhibition space. In the exhibition a system of yellow handrails known from public transport is installed – a metaphor of orientation that counteracts the feeling of getting lost in virtual reality (fig. 12–13).

It is very likely that within the next years new exhibition formats and new ways of art mediation will be developed, even to the extent of purely virtual exhibition formats that allow for meeting other visitors in virtual environments in order to exchange ideas or to interact. At the moment many of these available applications do not offer multiuser experiences, i. e. Tilt Brush gives the chance to paint in 3D space with virtual reality but not as a multiplayer activity.

Critical voices already complain that the physical museum abolishes itself, but they are only partly right. It might be true that in the future it might become unnecessary to physically visit exhibitions to view virtual art, but these challenges are not uniquely linked to the development of VR technology. They have accompanied museums for many years, when one looks at the discussions about technical reproducibility and digitization. Nevertheless, physical museums have not lost their fascination. The museums, however, are now more than ever challenged and encouraged to rethink themselves as social places that offer and create discourse about art.

Figures

1, 9, 12–13 *Beautiful new worlds. Virtual realities in contemporary art*, Zeppelin Museum Friedrichshafen, November 11, 2017 - April 8, 2018, exhibition views. © Zeppelin Museum Friedrichshafen, photos: Markus Tretter.

2 Banz & Bowinkel: *Palo Alto*, 2017. © Zeppelin Museum Friedrichshafen, photo: Markus Tretter.

3 Harun Farocki: *Parallel II*, 2014. © Harun Farocki GbR.

4 Florian Meisenberg: *Pre-Alpha Courtyard Games (raindrops on my cheek)*, 2017. © Zeppelin Museum Friedrichshafen, photo: Markus Tretter.

5 Salome Asega & Reese Donohue & Tongkwai Lulin: *ASM(V)R*, 2017. © Zeppelin Museum Friedrichshafen, photo: Markus Tretter.

6 The Nest Collective: *Let This Be A Warning – A VR Short Film*, 2017. © Electric South.

7 micha cárdenas: *Becoming Dragon*, 2008. © the artist, photo: Elle Mehrmand.

8 micha cárdenas: *Becoming Dragon*, 2008. © Zeppelin Museum Friedrichshafen, photo: Markus Tretter.

10–11 Halil Altindere: *Journey to Mars*, 2016. © Zeppelin Museum Friedrichshafen, photo: Markus Tretter.